3rd International Conference on Communication Systems (ICCS-2017)

IOP Conference Series: Materials Science and Engineering
Volume 331

Rajasthan, India
14 - 16 October 2017

Editors:

Santosh Jangid
S.K. Rai
Manoj Pandey
Shridhar B. Dandin
L. Solanki
P.S. Bhatnagar

ISBN: 978-1-5108-6065-0
ISSN: 1757-8981

Printed from e-media with permission by:

Curran Associates, Inc.
57 Morehouse Lane
Red Hook, NY 12571

proceedings.com

Some format issues inherent in the e-media version may also appear in this print version.

Printed by Curran Associates, Inc. (2018)

Institute of Physics
Dirac House, Temple Back
Bristol BS1 6BE UK

Phone: 44 1 17 929 7481
Fax: 44 1 17 920 0979

techtracking@iop.org

Additional copies of this publication are available from:

Curran Associates, Inc.
57 Morehouse Lane
Red Hook, NY 12571 USA
Phone: 845-758-0400
Fax: 845-758-2633
Email: curran@proceedings.com
Web: www.proceedings.com

Table of contents

Volume 331

3rd International Conference on Communication Systems (ICCS-2017)

14–16 October 2017, Rajasthan, India

Accepted papers received: 5 March 2018
Published online: 29 March 2018

Preface

Papers

IOP Conf. Series: Materials Science and Engineering **331** (2018) 011001 doi:10.1088/1757-899X/331/1/011001

3rd International Conference on Communication Systems "ICCS-2017" October 14-16, 2017

B K Birla Institute of Engineering and Technology, Pilani (India)
https://bkbiet.ac.in/

In the persistence of successful completion of previous two International Conferences on Communication Systems ICCS-2013 and ICCS-2015 organized by BK Birla Institute of Engineering and Technology, Pilani in the years of 2013 and 2015 respectively, the 3rd International Conference on Communication Systems "ICCS-2017" aims to ignite our students & faculties towards research and innovations going on in the area of Electronics and Communication Systems across the world. This conference starts with opening new avenues for the students, participants and delegates that covers a broad range of Communication Systems, Microwave Engineering, Electronic Systems, Signal Processing, Semiconductors & allied areas like Automation, Robotics, IOT and Mathematical computation and optimization techniques. This time, in response to call for papers, as about 105 research papers have been received from every corner of country and abroad. Out of these, 86 papers have been scheduled for presentation. To accommodate all the shortlisted and registered papers, 25 sessions have been scheduled where three sessions were executing in parallel in three different halls. For the foreigners as well as far off state Indian scholars, two sessions were scheduled through video conferencing and 9 papers have been be presented through this mode. Each hall was controlled by three faculty members and two students from student chapters as in college. For the smooth execution of all programs 14 different committees were formed.

"ICCS-2017" happened at BKBIET Pilani during Oct. 14-16, 2017, has better accomplishments not only because of well organized, better co-operations, synchronized execution but also due to huge participation of students, excellent invited talks, papers presented by the students, participation of faculties and outside research scholars. As per the enclosed procedural program schedule, it has been executed as well as it was expected. It is summarized as below:

- International invited Speakers/Chair persons (France & Thailand) 09
- Indian Invited Speakers/Chairpersons/Co-Chairpersons/Guests: 41

IOP Conf. Series: Materials Science and Engineering **331** (2018) 011001 doi:10.1088/1757-899X/331/1/011001

- Research Papers Received from scholars: 105
- Research Papers Accepted for Presentation: 86
- Research Papers presented (Oral): 70
- Research Papers presented (Video conferencing): 09
- Research Papers Accepted for IOP Proceeding Publication: 32
- Participants from host institute Faculties & Students: 286
- Total Gathering (includes Delegates, external & host institutes participants): 386

Organizing committee is very thankful to all the guests and participants to have made "ICCS-2017" a success.

GLIMPSE OF INAUGURAL SESSION

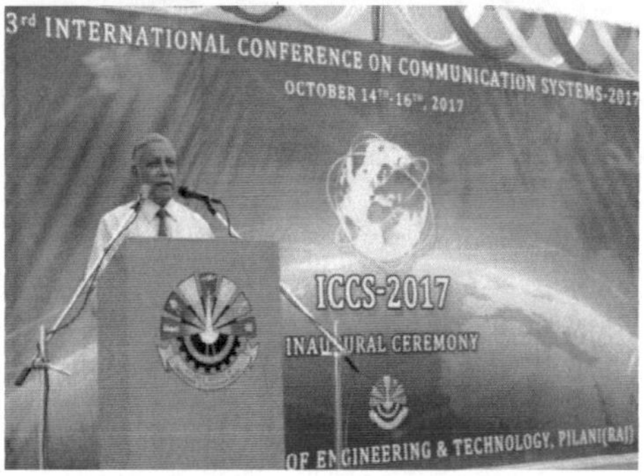

IOP Conf. Series: Materials Science and Engineering **331** (2018) 011001 doi:10.1088/1757-899X/331/1/011001

GLIMPSE OF INVITED TALK/PAPER PRESENTATION

Hall 1

Hall 2

Hall 3

Video Conferencing

IOP Conf. Series: Materials Science and Engineering **331** (2018) 011001 doi:10.1088/1757-899X/331/1/011001

EDITORS

Mr. Santosh Jangid
Assistant Professor, Dept. of ECE, BKBIET, Pilani, India

Mr. S K Rai
Assistant Professor, Dept. of ECE, BKBIET, Pilani, India

Dr. Manoj Pandey
Associate Professor, Dept. of ECE, BKBIET, Pilani, India

Prof. Shridhar B Dandin
Principal (Administration), BKBIET, Pilani, India

Prof. (Dr.) L Solanki
Principal (Academic), BKBIET, Pilani, India

Prof. (Dr.) P S Bhatnagar
Director, BKBIET, Pilani, India

IOP Conf. Series: Materials Science and Engineering **331** (2018) 011001 doi:10.1088/1757-899X/331/1/011001

Patron

Syt. B K Birla, Chairman, BKBIET, Pilani

Smt. Manjushree Khaitan, Trustee, SKCT, Kolkata

General Chair

Dr. P S Bhatnagar, Director, BKBIET, Pilani

Program Chair

Prof. (Dr.) L Solanki, Principal (Acad.), BKBIET, Pilani

Prof. S B Dandin, Principal (Admin.), BKBIET, Pilani

National Advisory Committee

Sh. R P Pansari, Sr. President, Aditya Birla Group, India

Dr. V D Tibrewal, Chancellor, JJTU, Jhunjhunu

Dr. S Pal, Vice Chancellor, DIAT, Pune

Dr. V N Dhaulakhandi, Director, Birla Museum, Pilani

Dr. S S Aggarwal, Director General, KIIT, Gurgaon

Dr. S P Kosta, Group Director, SRIT, Jabalpur

Dr. V N Waliwadekar, Ex-Scientist, CEERI, Pilani

Dr. S N Gupta, Director, JECRC, Jaipur

Dr. N M Devashrayee, Professor, Nirma Univ., Ahmedabad

Mr. Parag Bhatnagar, Director, Mindlance Technologies

Mr. Rakesh Adhikari, Director, Incise Infotech, Noida

Dr. S K Bhatnagar, Professor, SKIT, Jaipur

Dr. C B Gupta, Professor, BITS Pilani

Dr. L M Joshi, Chief Scientist, CSIR-CEERI, Pilani

Dr. N K Agrawal, Professor, KIIT, Gurgaon

Dr. S N Joshi, Ex-Scientist, CSIR-CEERI, Pilani

Dr. R C Dubey, Ex-Scientist, CSIR-CEERI, Pilani

International Advisory Committee

Prof. Banmali S Rawat, University of Nevada, Reno, USA

Prof. Tayeb Md. Brahim, Université de Rennes 1, France

Prof. Pramod Rastogi, EPFL, Switzerland

Prof. Chawin Chantharasenawong, KMUTT, Thailand

Prof. Natthaphob Nimpitiwan, Bangkok Univ, Thailand

IOP Conf. Series: Materials Science and Engineering **331** (2018) 011001 doi:10.1088/1757-899X/331/1/011001

Prof. Prapat Wangskarn, Thammasat Univ, Thailand

Prof. Supan Tungjitkusolmun, KMITL, Thailand

Prof. Thami Zeghloul, IUT, France

Prof. Jean Philippe Lauffenburger, UHA, France

Prof. Tchanguiz Razban, Polytech Nantes, France

Prof. Patrick Le Callet, Polytech Nantes, France

Prof. Mireille Ducasse, INSA Rennes, France

Prof. Karel L Sterckx, Bangkok University, Thailand

Prof. Michel Doisy, ENSEEIHT, Toulouse, France

Prof. Marc Renner, Director, INSA Strasbourg, France

Prof. Habib Abdulrab, INSA Rouen, France

Prof. Nesim Fintz, Director, EISTI, France

Prof. Yves Millou, INSA, Rouen, France

Prof. Raphaël GILLARD, INSA Rennes, France

Prof. Majdi Khouder, Director, IUT Poitiers, France

Prof. Montalee Sasananan, Thammasat Univ, Thailand

Prof. Kouroch Mahdjoubi, Université de Rennes, France

Prof. Ala Sharaiha, IETR, Université de Rennes 1, France

Prof. Julien Sarrazin, UPMC, France

Convener

Dr. Manoj Pandey, Associate Professor, BKBIET, Pilani

Mr. Satish Kumar Rai, Assistant Professor, BKBIET, Pilani

e-Mail: iccs2017@bkbiet.ac.in

Coordinator, Publication

Mr. Santosh Jangid, Assistant Professor, BKBIET, Pilani

e-Mail: publication_iccs2017@bkbiet.ac.in

Coordinator, Sponsorship

Dr. Smita Pareek, Associate Professor, BKBIET, Pilani

IOP Conf. Series: Materials Science and Engineering **331** (2018) 011001 doi:10.1088/1757-899X/331/1/011001

PROGRAM COMMITTEES

Mr Rakesh Saini
Mr Rajesh Singh Shekhawat
Dr R M Jha
Mr Reetesh Gautam
Mr Ravi Chaudhary
Dr (Mrs) Krishna Meel
Ms Chanchal Jakhodia
Ms Jyoti Purohit
Mrs Manusmriti Sharma
Ms Garima Rungta
Mrs Monika Jangir
Mr Ramakant Soni
Mr Abhishek Kumar
Dr Vipin Kumar
Mr Vipin Saraogi
Dr Hasibur Rahaman
Mr Nachiket Sainis
Mrs Reena Saini
Mr Sanjay Kr Saini
Mr Ramesh Jangir
Mr Sunil Kumar
Mrs Ruchi Sharma
Mr Gauttam Jangir
Mr Narayan Kumar
Dr (Mrs) Disha Meel
Mr Gaurav Sahu
Dr (Mrs) Sonam Mittal
Mrs Anupama Sharma
Mr Pradeep Singh Shekhawat
Mr Dinesh Kumar
Mr Keshav Khandelwal
Mr Suresh Kumar
Mr Anil Kumar
Mr Umesh Ratawa
Mr Surendra Kumar Sharma

IOP Conf. Series: Materials Science and Engineering **331** (2018) 011001 doi:10.1088/1757-899X/331/1/011001

3rd International Conference on Communication Systems

ICCS–2017

October 14-16, 2017

Organized by

B K Birla Institute of Engineering & Technology
Pilani, INDIA

Program Schedule

WELCOME

IOP Conf. Series: Materials Science and Engineering **331** (2018) 011001 doi:10.1088/1757-899X/331/1/011001

DAY 1: SATURDAY, OCTOBER 14, 2017

Registration-09:00 AM - 10:30 AM, **Venue**- Registration Desk, BKBIET, Pilani

INAUGURAL SESSION

TIMING-10:30 AM-12:30 PM

- **Lamp Lighting & Saraswati Vandana**
- **Welcome Address:** by **Prof. (Dr.) P S Bhatnagar,** Director, BKBIET, Pilani, India
- **About Conference :** by **Dr. Manoj Pandey,** Conference Convener, ICCS-2017
- **Welcome & Felicitation of Chief Guest & Guests of Honor:**
  by **Prof. (Dr.) P S Bhatnagar,** Director, BKBIET, Pilani, India
  - ⇒ ADDRESS (10min): by Chief Guest: **Dr. S P Kosta**, Director General, SRIT, Jabalpur, India
  - ⇒ ADDRESS (10min): by Guest of Honor: **Dr. Surendra Pal**, Vice Chancellor, DIAT, Pune, India
  - ⇒ ADDRESS (10min): by Guest of Honor: **Dr. Santanu Chaudhury**, Director, CSIR-CEERI, Pilani, India
  - ⇒ ADDRESS (10min): by Guest of Honor: **Dr. A K Sarkar**, Director, BITS, Pilani, India
  - ⇒ ADDRESS (10min): by Guest of Honor: **Maj.-Gen. S S Nair,** Director, BET, Pilani, India
  - ⇒ ADDRESS (10min): by Guest of Honor: **Dr. V N Dhaulakhandi**, Director, Birla Museum, Pilani, India
- **Welcome & Felicitation of Invited Speakers & Chairpersons/Co-chairpersons:**
  by **Prof. (Dr.) L Solanki,** Principal (Academic), BKBIET, Pilani, India
  by **Prof. S B Dandin,** Principal (Administration), BKBIET, Pilani, India
- **About Publication:** by **Mr. Santosh Jangid,** Publication Coordinator, ICCS-2017
- **Vote of Thanks:** by **Mr. Satish Kumar Rai**, Conference Convener, ICCS-2017

LUNCH-12:30 PM - 02:00 PM

SESSION 1

TIMING-2:00 PM - 04:00 PM

HALL-1	HALL-2	HALL-3
SESSION 1A **Microwave Devices & Antenna**	SESSION 1B **Industrial Engineering & Automation**	SESSION 1C **Space Communication & MEMS Devices**
Chairperson {Invited Talk-01 (30min)}: **Prof.(Dr.) Tchanguiz Razban** Polytech Nantes, France Co-Chair {Invited Talk-02 (30min)}: **Dr. L M Joshi** Emeritus Scientist, CEERI, Pilani, India Paper Presentation: **3 Papers:** 1A.1-1A.3	Chairperson {Invited Talk-03 (30min)}: **Prof. (Dr.) Chanikaran Wongviriyawong,** KMUTT, Thailand Paper Presentation: **4 Papers:** 1B.1-1B.4	Chairperson {Invited Talk-04 (30min)}: **Dr. S Pal** VC, DIAT, Pune, India Co-Chair {Invited Talk-05 (30min)}: **Dr. Kamaljeet Singh** Scientist, ISAC, Bangalore, India Paper Presentation: **4 Papers:** 1C.1-1C.4

TEA / COFFEE- 04:00 PM - 04:15 PM

SESSION 2

TIMING-04:15 PM - 06:15 PM

SESSION 2A **Microwave Devices & Antenna**	SESSION 2B **Semiconductor Devices**	SESSION 2C **Robotics & Automation**
Chairperson {Invited Talk-06 (30min)}: **Dr. S N Joshi** Emeritus Scientist, CEERI, Pilani, India **4 Papers:** 2A.1-2A.4	Chairperson {Invited Talk-07 (30min)}: **Prof. (Dr.) R C Dubey,** IES-IPS Academy, Indore, India Paper Presentation: **4 Papers:** 2B.1-2B.4	Chairperson: **Prof. (Dr.) Montalee Sasananan** Thammasat University, Thailand Co-Chair {Invited Talk-08 (30min)}: **Prof. (Dr.) L Solanki** Principal (Acad.), BKBIET, Pilani, India Paper Presentation: **4 Papers:** 2C.1-2C.4

Cultural Program-07:00 PM – 8.30 PM

DINNER-08:30 PM - 10:00 PM

IOP Conf. Series: Materials Science and Engineering **331** (2018) 011001 doi:10.1088/1757-899X/331/1/011001

DAY 2: SUNDAY, OCTOBER 15, 2017		
Registration-08:00 AM - 09:00 AM, **Venue**- Registration Desk, BKBIET, Pilani		
SESSION 3 **TIMING-09:00 AM - 11:00 AM**		
HALL -1	**HALL-2**	**HALL-3**
SESSION 3A **Image & Signal Processing** Chairperson {Invited Talk-09 (30min)}: **Prof. (Dr.) Majdi Khoudeir** Université de Poitiers, Poitiers, France Paper Presentation: **4 Papers:** 3A.1-3A.4	**Industrial Academia Meet**	SESSION 3C **Biomedical Engineering** Chairperson {Invited Talk-10 (30min)}: **Prof. (Dr.) Thami Zeghloul,** Director, IUT Angouleme, France Paper Presentation: **3 Papers:** 3C.1-3C.3
TEA / COFFEE- 11:00 AM- 11:30 AM		
SESSION 4 **TIMING-11:30 AM - 01:00 PM**		
SESSION 4A **Image & Signal Processing** Chairperson {Invited Talk-11 (30min)}: **Prof. (Dr.) Chaker Larabi** Université de Poitiers, Poitiers, France Paper Presentation: **4 Papers:** 4A.1-4A.4	**Industrial Academia Meet**	SESSION 4C **Microwave Devices** Chairperson {Invited Talk-12 (30min)}: **Prof. (Dr.) Kouroch Mahdjoubi** University de Rennes, France Paper Presentation: **3 Papers:** 4C.1-4C.3
LUNCH-01:00 PM - 02:30 PM		
SESSION 5 **TIMING-2:30 PM - 04:30 PM**		
SESSION 5A **Communication Systems** Chairperson {Invited Talk-13 (30min)}: **Prof. (Dr.) H M Gupta** Formerly, Professor, IIT Delhi, India Co-Chair: **Dr. Sneh Mohan** Executive Director, CDT Foundation Paper Presentation: **4 Papers:** 5A.1-5A.4	SESSION 5B **EMFT & Antenna** Chairperson {Invited Talk-14 (30min)}: **Prof. (Dr.) S K Bhatnagar** Director, SKIT, Jaipur, India Paper Presentation: **4 Papers:** 5B.1-5B.4	**International Coordinator's Meet-2017** **(Venue: Meeting Hall)**
TEA / COFFEE- 04:30 PM - 04:45 PM		
SESSION 6 **TIMING-04:45 PM - 06:45 PM**		
SESSION 6A **VLSI & Embedded Systems** Chairperson: **Dr. O. P. Wadhawan** Ex-Scientist, CEERI, Pilani, India Co-Chair {Invited Talk-15 (30min)}: **Dr. S C Bose** Chief Scientist, CEERI, Pilani, India Paper Presentation: **3 Papers:** 6A.1-6A.3	SESSION 6B **Renewable Energy** Chairperson {Invited Talk-16 (30min)}: **Prof. (Dr.) Ratna Dahiya** NIT, Kurukshetra, India Co-Chair {Invited Talk-17 (30min)}: **Prof. (Dr.) M S Soni** BITS, Pilani, India Paper Presentation: **4 Papers:** 6B.1-6B.4	SESSION 6C **Electronic Systems & Communication** Chairperson {Invited Talk-18 (30min)}: **Prof. (Dr.) S S Agarwal** Director, KIIT, Gurgaon, India Co-Chair {Invited Talk-19 (30min)}: **Dr. Kota Solomon Raju** Principal Scientist, CEERI, Pilani, India Paper Presentation: **3 Papers:** 6C.1-6C.3
Visit to Museums followed by Tea-7:00 PM- 8:00 PM		
DINNER-08:30 PM - 10:00 PM		

IOP Conf. Series: Materials Science and Engineering **331** (2018) 011001 doi:10.1088/1757-899X/331/1/011001

DAY 3: MONDAY, OCTOBER 16, 2017

Registration-08:00 AM - 09:00 AM, **Venue**- Registration Desk, BKBIET, Pilani

SESSION 7
TIMING-09:00 AM - 11:00 AM

HALL-1	HALL-2	MEETING HALL
SESSION 7A **Mathematical Computation & Optimization Techniques**	SESSION 7B **Automation, Robotics & IoT**	SESSION 7C **Papers Presentation via Video Conferencing**
Chairperson {Invited Talk-20 (30min)}: **Prof. (Dr.) C B Gupta** BITS, Pilani, India Paper Presentation: **7 Papers:** 7A.1-7A.7	Chairperson {Invited Talk-21 (30min)}: **Er. R L Agarwal** Fellow IEI, India Paper Presentation: **4 Papers:** 7B.1-7B.4	Chairperson: **Prof. (Dr.) Yues Millou** INSA Rouen, France Co-Chair: **Prof. S B Dandin** Principal (Admin.), BKBIET, Pilani, India Paper Presentation: **5 Papers:** 7C.1-7C.5

TEA / COFFEE- 11:00 AM- 11:30 AM

SESSION 8
TIMING-11:30 AM - 01:30 PM

SESSION 8A	SESSION 8B	SESSION 8C
Mathematical Computation & Optimization Techniques	**Automation, Robotics & IoT**	**Papers Presentation via Video Conferencing**
Chairperson {Invited Talk-22 (30min)}: **Prof. (Dr.) Pankaj Gupta** University of Delhi, India Paper Presentation: **8 Papers:** 8A.1-8A.8	Chairperson {Invited Talk-23 (30min)}: **Prof. (Dr.) K. S. Sangwan** BITS, Pilani, India Co-Chair {Invited Talk-24 (30min)}: **Prof. (Dr.) Rajesh P Mishra** BITS, Pilani, India Paper Presentation: **3 Papers:** 8B.1-8B.3	Chairperson: **Prof. (Dr.) Yues Millou** INSA Rouen, France Co-Chair: **Prof. S B Dandin** Principal (Admin.), BKBIET, Pilani, India Paper Presentation: **4 Papers:** 8C.1-8C.4

LUNCH-01:30 PM - 02:30 PM

Formally Closing of ICCS-2017 || Wish to see you again in

4th International Conference on Communication Systems
ICCS-2019
October 18-20, 2019

SUMMARY

International invited Speakers/Chairpersons/Co-Chair			08
National Invited Speakers/Chairpersons/Co-Chair			32
Research Papers Received			105
Research Papers Accepted for Presentation	Oral		77
	Video conferencing		09
Research Papers Accepted for IOP Proceeding Publication			40
Participants	Speakers + Authors		40+95
	Attendee	External	13
		Faculties & Students of Host Institute	285

IMPORTANT CONTACTS

Conference Convener	+91-7737435144 & 9785551210
Publication Coordinator	+91-9414367879
Hospitality	+91-9461813717, 9828765129

IOP Conf. Series: Materials Science and Engineering **331** (2018) 011001 doi:10.1088/1757-899X/331/1/011001

THANKS TO SPONSORS

Aditya Birla Group, India

Council of Scientific and Industrial Research, New Delhi, India

Defense Research & Development Organization, New Delhi, India

Institution of Engineers (IE), India

Indian National Science Academy, India

Defence Institute of Advanced Technology, Pune, India

IOP Conf. Series: Materials Science and Engineering **331** (2018) 011002 doi:10.1088/1757-899X/331/1/011002

Peer review statement

All papers published in this volume of *IOP Conference Series: Materials Science and Engineering* have been peer reviewed through processes administered by the proceedings Editors. Reviews were conducted by expert referees to the professional and scientific standards expected of a proceedings journal published by IOP Publishing.

A 2x1 Coplanar Monopole Antenna Structure for Wireless RF Energy Harvesting

Monika Mathur[1], Ankit Agarwal[2], Ghanshyam Singh[3], S.K. Bhatnagar[4]
[1]Associate Professor, Electronics and Communication Department, SKIT, Jaipur
[2]Assistant Professor, Electronics and Communication Department, SKIT, Jaipur
[3]Associate Professor, Electronics and Communication Department, MNIT, Jaipur
[4]Professor, Electronics and Communication Department, SKIT, Jaipur
E-mail: monikamathur16@gmail.com

Abstract. The objective of this paper is to design an efcient wireless energy harvesting (WEH) system to eliminate the problem to continuous charging of a battery operated electronics devices. Most of the devices are battery operated so charging of a battery time to time is serious issue. This WEH system receives the Radio Frequency (RF) and Microwave frequency signals present in the atmosphere and converting it into DC signal so that it can stores in Capacitor or charges a battery for utilize the power. For this RFenergy harvesting purposes 21 antenna array structure of the coplanar monopole antenna is presented. This structure shows the gain of 10.2 dBi and 83% efciency. This structure is designed for resonating on multiple bands (Radio, GSM, ISM, and UWB). It is useful for this application because it covers almost all useful bands in the maximum capturing area. The antenna can be connected directly to an RF-DC converter module and it uses about 60% of the total PCB area. And RF to DC convertor circuit can be implemented in that remaining 40% area on the same substrate so it eliminates the need of port connectors and impedance matching circuit between them. This design is worked in receiving mode only when used for energy harvesting purposes. This may be useful for the biomedical and satellite applications.

Keywords- Energy Harvesting System, Monopole Antenna, RF-DCConverter, Coplaner Antenna, Wilkinson Power Divider

1. INTRODUCTION

In the Field of energy harvesting the RF energy harvesting is the new and interesting era of the researchers because RF spectrum is free to use and easily available everywhere. The basic building blocks for the harvester for RF ranges are composed of three segments 1) RF receiv-ing antenna 2) Rectifying & amp; Booster circuits 3) Impedance matching circuits. Antenna designing is the important part of the harvesting circuit. The antenna may be designed for any band of RF For UWB band designing Coplanar antenna is suitable option. M. Yazdi et al. Presented an UWB planar antenna with band rejection characteristics [2]. For particular band rejection di erent variations in patch dimensions were investigated [3]-[5]. Parasitic patch is also one solution for band rejection [5]-[7].

The authors have already designed a compact coplanar monopole antenna [8] which is capable to resonant on multi bands e.g. Radio , GSM,. ISM and UWB band (900MHz-3.1GHz and 5.6GHz- 9.6GHz) with band rejection for WLAN(3.1GHz - 5.6GHz) bands.

This paper presents the extension of that work by array arrangement of that coplanar monopole

IOP Conf. Series: Materials Science and Engineering **331** (2018) 012001 doi:10.1088/1757-899X/331/1/012001

antenna structure to improves the gain .

2. ANTENNA DESIGN

Figure 1 illustrates the structural geometry of the proposed 2 1 array of coplanar monopole antenna. The power of these array structures has been combined by using Wilkinson power combiner. The benet of implementing this combiner comparison to others is that impedance matching with the antennas is easy. The matching of impedances at the antenna ports and the power combiner has been optimized.

The impedances at the antenna ports have measured and the two output ports impedances of the combiner have calculated from the formulas.
Antenna and combiner ports a matching by parametric analysis has been implemented at the end of the line feed. The substrate is taken FR4 epoxy here with a relative permittivity of 4.4 and a thickness of 1.6mm. The antenna is designed to cover some usable bands e.g. radio, GSM, ISM and UWB bands (900MHz - 3.1GHZ & amp; 5.6GHz - 9.9GHz).

In between these bands 3.1GHz - 5.6GHz has been rejected because it provides only inter-ference for energy harvesting application. The optimized values of the design parameters are as follows: L_S= 126mm, W_S =40mm. patch length L_p = 20mm and width W_p= 20mm. Designed frequency is 9GHz.

The Array arrangement of the structure by using Wilkinson power combiner is as shown in gure1.

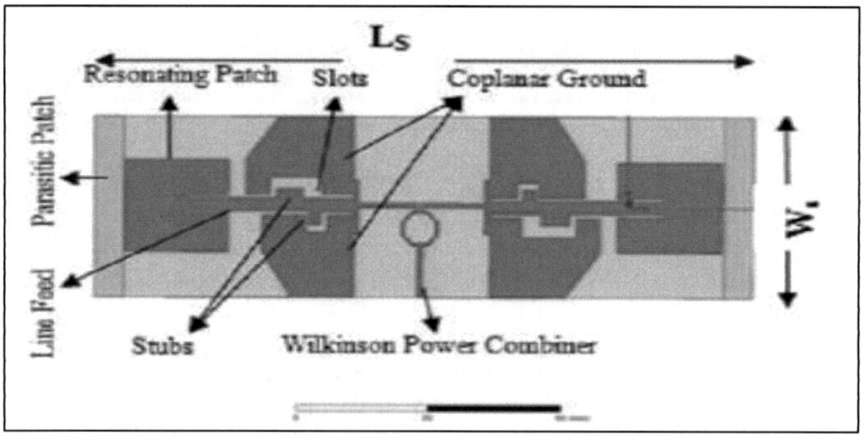

Figure 1. Array arrangement of Monopole antenna.

In the next section by using licensed software HFSS$_{TM}$ 16,the antenna performance has been simulated and analyzed.

3. RESULTS AND DISCUSSION

3.1. Parametric Study

Two patches of 20x20 mm^2 dimensions have been simulated on the substrate 126 x40 x1.6 mm^3 for the designed frequency 9 GHz. The application of this antenna is basically for RF energy harvesting. So for designing multiband (Radio, GSM, ISM, and UWB) antenna optimization of

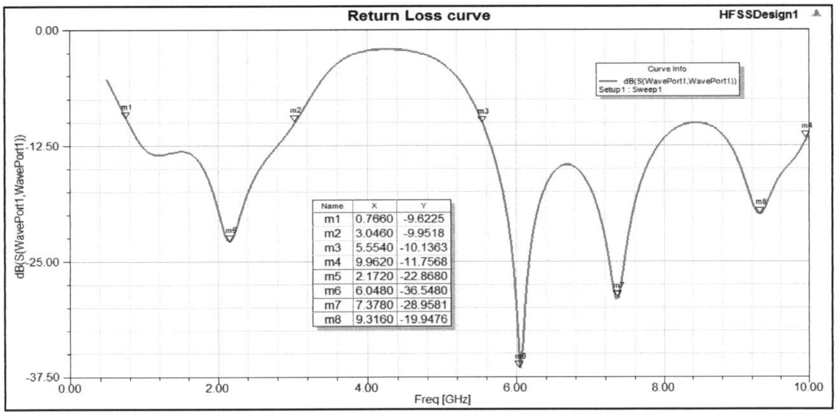

Figure 2. Return Loss vs. Frequency curve of UWB Antenna.

different parameters by parametric has been applied. In this structure parametric analysis has been applied on the following structures:

(a) The gap between the coplanar ground and resonating patch has been optimized for reso-nant the antenna on multi band e.g. Radio, GSM and ISM (900MHz - 2.4GHz).

(b) The coplanar structure has optimized for UWB range. The Corner edges cutting dimen-sions of the coplanar ground have been optimized for bandwidth enhancement of this UWB range.

(c) The location and dimensions of the slots in coplanar grounds and the stub of feed line have been optimized for impedance matching.

(d) The dimensions of the parasite patch above the resonating patch has been optimized for WLAN band (3.1-5.6 GHz) rejection for avoiding the interference of this band particularly for energy harvesting applications.

In Figure 2 the structure has shown resonance from 900 MHz - 3GHz and From 5.5GHz - 10GHz and also the band rejection from 3GHz - 5.5GHz.

3.2. Current Distribution

The current distribution in figure 3 shows that the current is maximum at the corner edges.

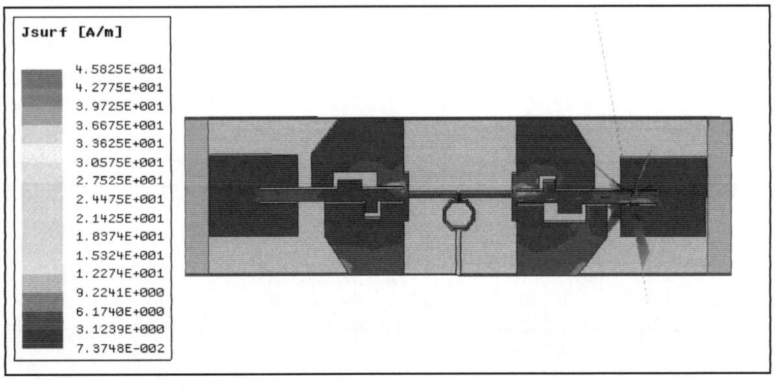

Figure 3. Current Distribution of Array of Monopole antenna

3.3. Radiation Pattern

The H- field patterns for the frequency 2.4 GHz and the values of θ= 90°, 60° and 180° has shown in figure 3. The E-field pattern for the same frequency 2.4 GHz and θ = 0°, 90° and 30° has been shown in figure 4. The H- field and E- field patterns shows the omni directional behavior of the structure.

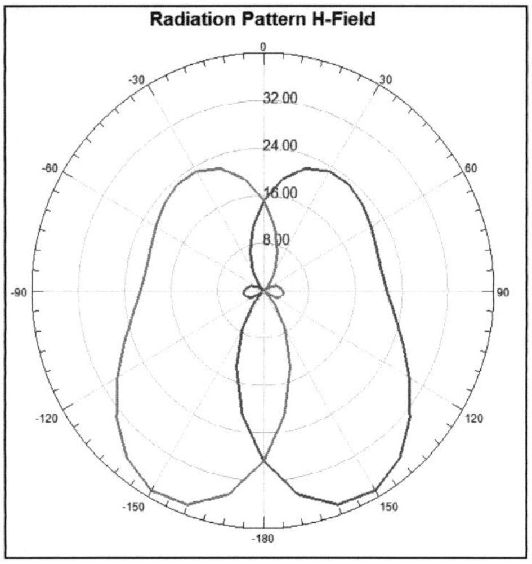

Figure 4. H- Field Pattern.

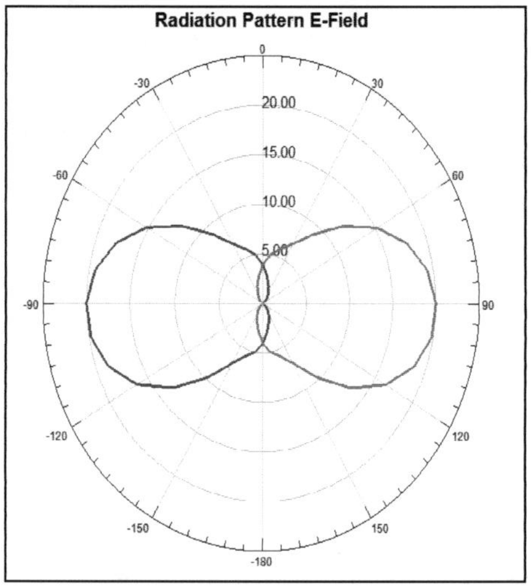

Figure 5. E- Field Pattern.

Gain and Efficiency

The main motive of this array arrangement is to increase the gain. The authors noticed the 7 dBi peak gain of single structure of the coplanar monopole antenna. This 2x1 array arrangement shows the peak gain of 9.2 dBi gain by Figure 6. The effciency of presented array arrangement is 83% as shown in Figure 7.

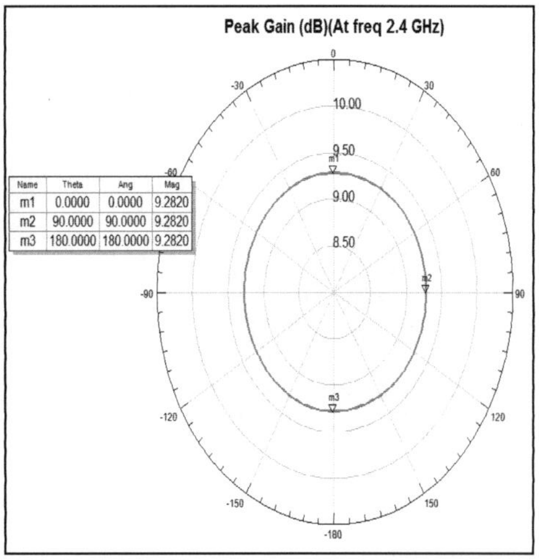

Figure 6. Peak Gain of Designed Structure.

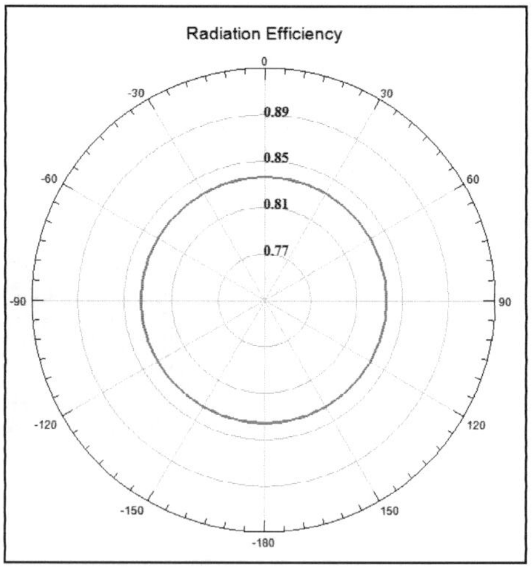

Figure 7. Radiation Efficiency Curve.

4. CONCLUSION

The paper presents an arrangement of array of coplanar monopole antenna for the RF energy harvesting applications. The uniqueness of this arrangement is that a Wilkinson power combiner is added to combine the power of two antennas. So it eliminates the need of impedance matching circuit between the antenna and rectiers circuits. The rectier circuit part of any RF harvester circuit may be directly connected to this antenna. The proposed design of array provides the gain of 9.2 dBi and the efficiency is 83%. The design is ready to fabricate and the measurement may be done and results will be soon compared and presented.

References

[1] Park Ji Yong, Caloz C., Qian Yongxi and Itoh T.2002, A Compact Circularly Polarized Subdivided Micro Strip Patch Antenna, IEEE Microwave and Wireless Components Letters,12(1),18-19

[2] Yazdi M. and Komjani N. 2011 A compact band-notched UWBplanar monopole antenna with parasitic elements Progress In Electromagnetics, Research Letters, 24, 129-138.

[3] Mehdipour A., Aghdam K. M., Dana R. F., and Khatib M. R. K. 2008 A novel coplanar waveguide-fed slot antenna for ultrawideband applications, IEEE Trans. Antennas Propag., 56(12) , 3857-62.

[4] Heshmat N. Farrokh, Nourinia J., and Ghobadi C. 2009 Band-notched ultra-wideband printed open-slot antenna using variable on-ground slits, Electron. Lett., 45(21), 1060-61.

[5] Subbarao A. and Raghavan S. 2012 A compact UWB Slot antenna with signal rejection in 5-6 GHz band, Microwave Opt. Technol. Lett.,54(5) pp. 129296.

[6] Kim, K. H., Cho Y. J., Hwang S. H., and Park S. O. 2005 Band-notched UWB planar monopole antenna with two parasitic patches, Electronics Letters, 41(14), 783- 85.

[7] Abdollahvand M., Dadashzadeh G. and Mostafa D. 2010 Compact dual band-notched printed monopole antenna for UWB application, IEEE Antennas Wireless Propagation Letters, 9, 1148-51.

IOP Conf. Series: Materials Science and Engineering **331** (2018) 012002 doi:10.1088/1757-899X/331/1/012002

A Comparative Analysis of Digital and Passive Filters for IFOC based Induction Motor Drive (EV) fed through ZSI

Tarun Bedi[1], **Dave Heema**[2], **Dheerendra Singh**[3]

[1] B.E. (Hons.) Electronics & Instrumentation, BITS Pilani, Rajasthan – India
[2] Research Scholar, Department of Electrical & Electronics, BITS Pilani, Rajasthan – India.
[3] Associate Professor, Department of Electrical & Electronics, BITS Pilani, Rajasthan - India.

Email: dhs@pilani.bits-pilani.ac.in

Abstract. It is known that harmonics are generated in any power electronics based application. Since presence of harmonics is not desirable, it is necessary to remove the harmonics. The IFOC is based on stator current regulation, and the stator currents are sensed and used in the speed control algorithm. The current needs to be free from noise and harmonics for accurate further processing. In this paper, a passive analog filter, as well as a 50th order FIR filter is designed in MATLAB and implemented in Code Composer Studio to remove noise and distortion, and a comparative analysis has been done, for the speed control of an induction motor fed through ZSI, for electric vehicle application.

1. Introduction

For any control technique application to an electrical motor, it is important to remove all the noise and harmonics from the current waveforms, for accurate and correct processing in the controller. That is why a filter block needs to be introduced prior to the signal reaching the controller. In case of this particular application for the electric vehicle, a filter block is added before the Clark macro, in the level 3 of our implementation.

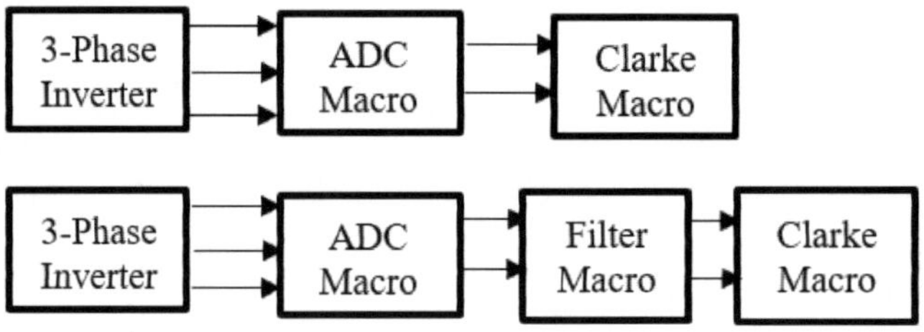

Figure 1: The filter's function in the block diagram of IFOC based Induction Motor drive (EV) fed through ZSI

In digital signal processing, a filter's function is to remove unwanted parts of the signal, such as random noise, or to extract useful parts of the signal, such as components lying within a certain frequency range, as illustrated below:

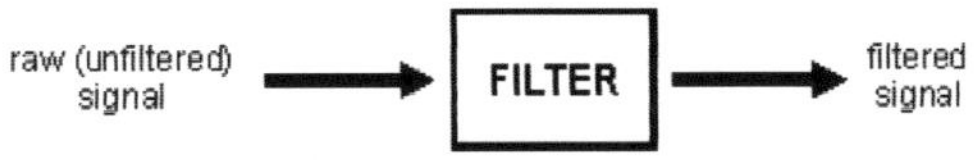

Figure 2: A filter's basic function

There are mainly two kinds of filters – analog and digital.

A digital filter uses a digital processor to perform numerical calculations on sampled values of the signal. The processor is usually a specialized DSP (Digital Signal Processor) chip.

An analog filter uses analog electronic circuits, made from components such as resistors, capacitors and op-amps to deliver the required function of the filter.

Digital filters are programmable, and thus can be easily designed, tested and implemented. Also, fast DSP processors are much more versatile in their ability to process signals in a variety of ways.

2. Digital Filters – FIR and IIR Filters

A finite impulse response (FIR) filter is a filter whose impulse response *(or response to any finite length input)* is of finite duration, because it settles to zero in finite time. In contrast, an infinite impulse response (IIR) filter may continue to respond indefinitely due to internal feedback.

For a causal discrete-time FIR filter of order N, each value of the output sequence is a weighted sum of the most recent input values:

$$y[n] = b_0 x[n] + b_1 x[n-1] + \ldots + b_N x[n-N]$$

(This computation is also known as discrete convolution)

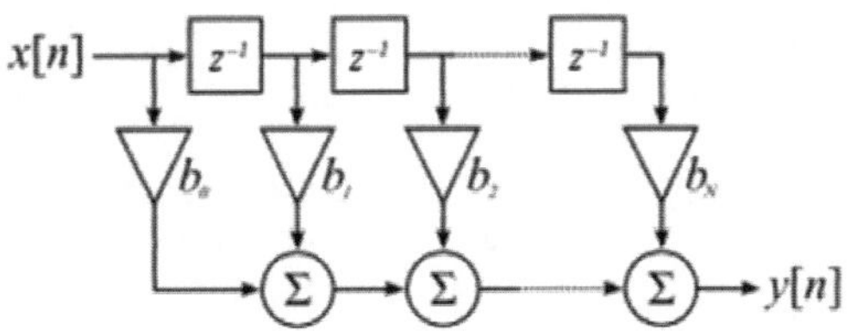

Figure 3: A direct form discrete-time FIR filter of order N

Infinite impulse response (IIR) filters are digital filters with infinite impulse response. Unlike FIR filters, they have the feedback and are therefore known as recursive digital filters.

Because of the feedback being present, IIR filters have much better frequency response than FIR filters of the same order. However, there is one problem with IIR filters, which is potential instability of the filter, because of the presence of feedback.

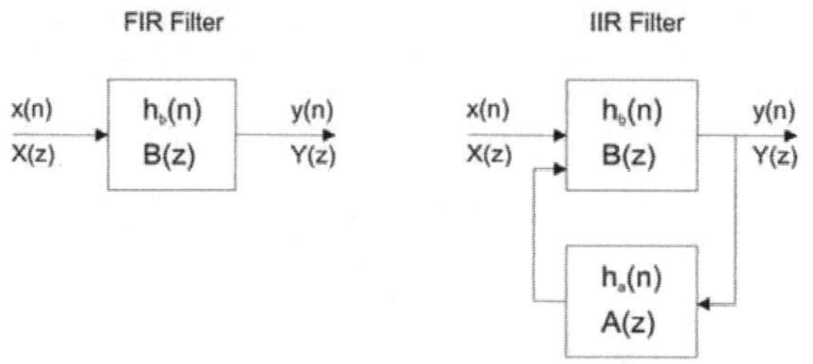

Figure 4: Block diagram representation of FIR and IIR Filters

In this study, we choose an FIR filter for implementation over IIR filters mainly because of two reasons. Firstly, FIR filters are always stable since they contain only zeros and no poles, unlike IIR filters which must be handled with care. Also, FIR filters provide a constant phase difference, thus phase correction is very easy to implement.

3. FIR Filter Design - MATLAB

For this study, the incoming signal has sampling frequency $f_s = 10$ kHz, signal frequency f =25 Hz.

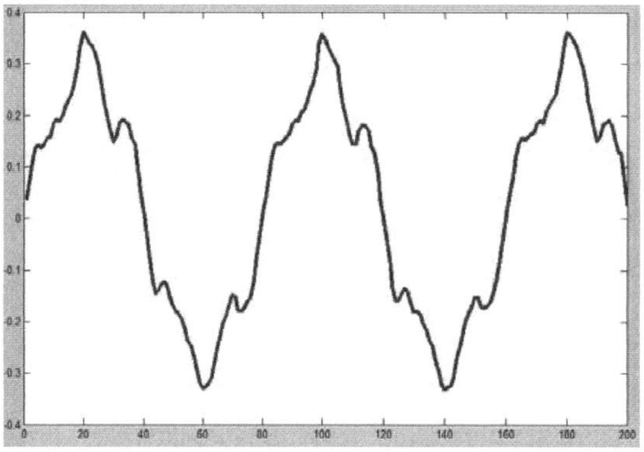

Figure 5: The incoming unfiltered data, with sampling frequency=10kHz, signal frequency=25Hz

For filtering purpose, the coefficients of a 50th order low-pass FIR filter with normalized cutoff frequency $w = 25Hz/10kHz$ are obtained by the following MATLAB command:

$$b = fir1(50,0.0025);$$

The above uses a Hamming Window function, with the following parameters:

Over band width (P/N) = 8

Minimum stopband attenuation = 53 dB

Sidelobe peak amplitude = 41 dB

The resulting waveform, after applying the above described filter in MATLAB, is as under:

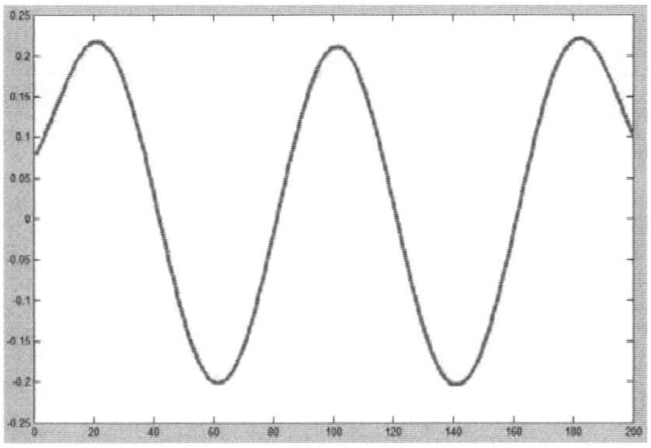

Fig 6: The filtered signal, after applying a 50th order low pass FIR filter with normalized cutoff frequency = 0.0025

The following figures show the frequency content of the unfiltered signal and the filtered signal, using the fft *(fast fourier transform)* command in MATLAB:

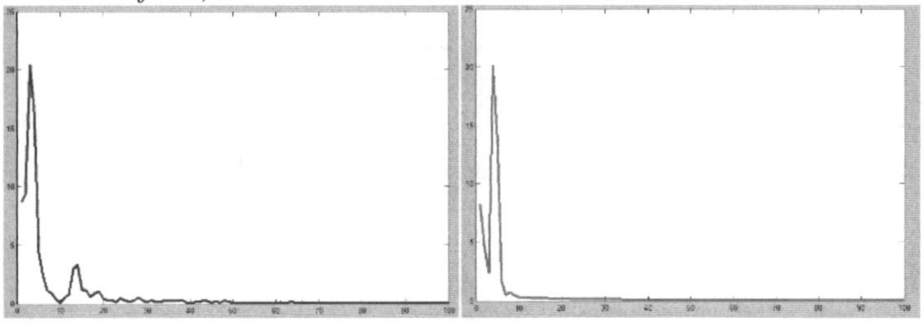

Figure 7: The FFT magnitude plots of the unfiltered signal (on the left) and the FIR filtered signal (on the right)

4. FIR Filter – Realtime Results

The following shows the real time results of the FIR filter, implemented as a C macro using the TI f28335 chip.

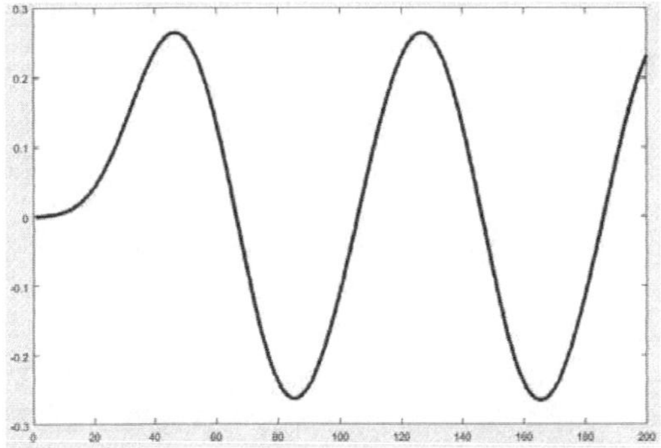

Figure 8: The real-time filtered signal, after applying a 50th order low pass FIR filter with normalized cutoff frequency = 0.0025

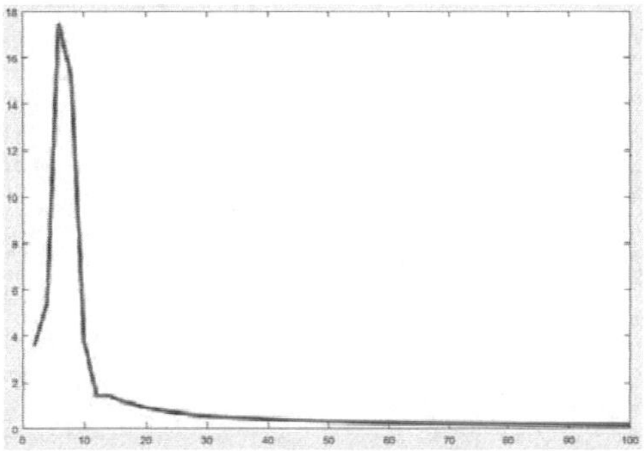

Figure 9: The FFT magnitude plots of the real-time FIR filtered signal

5. Passive Analog Filter

Passive filters are analog filters that do not depend upon an external power supply, i.e. they do not contain active components such as transistors.

The following diagram shows our basic implementation of a passive analog filter. This filter has been used to remove high frequency noise only.

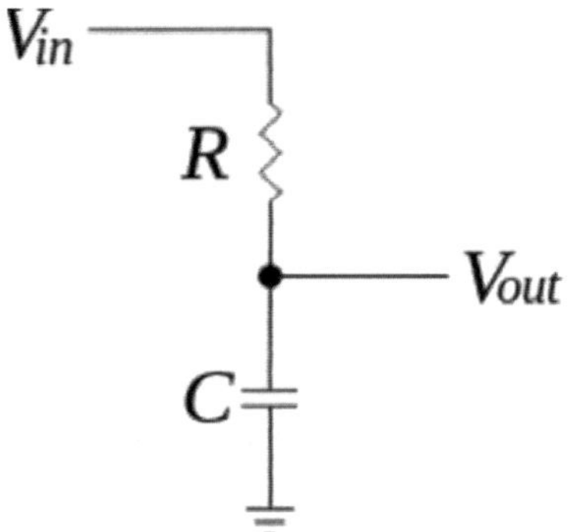

Fig 10: Passive, first-order low-pass RC filter. Here C=10pF

This simple circuit, consisting of a resistor in series with a load, and a capacitor in parallel with the load, acts as a low-pass filter circuit.

The capacitor exhibits reactance, and blocks low-frequency signals, forcing them through the load instead. At higher frequencies, the reactance drops, and the capacitor effectively functions as a short circuit.

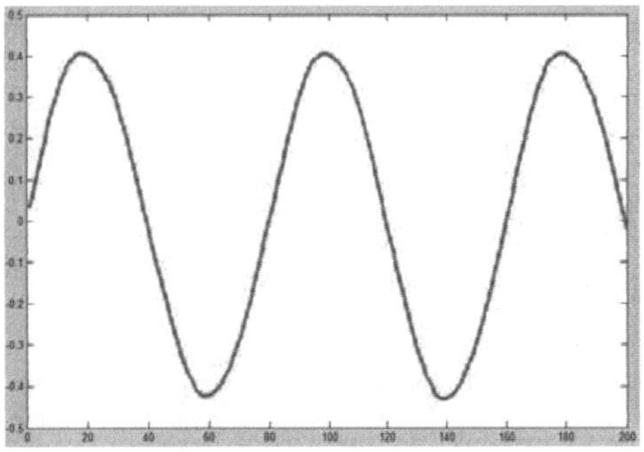

Figure 11: The real-time output obtained from the passive filter

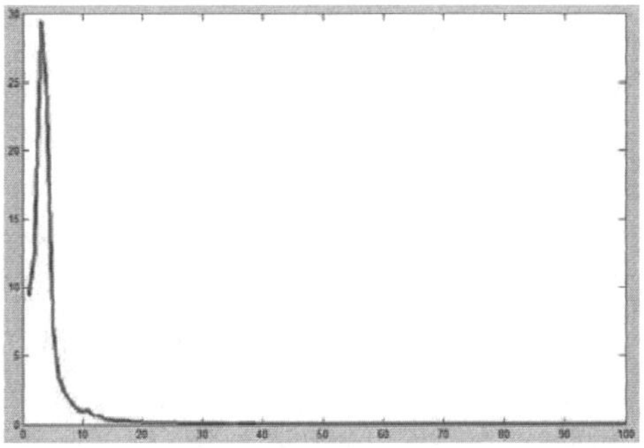

Figure 12: The FFT magnitude plot of the real-time passive filter output

IOP Conf. Series: Materials Science and Engineering **331** (2018) 012002 doi:10.1088/1757-899X/331/1/012002

6. Active Analog Filter

Active filters are analog filters that need an external power supply for operation, i.e. they have active components like transistors, amplifiers in their circuit.

Using amplifiers in a filter design can be beneficial in terms of improving the performance of the filter, while avoiding the need for inductors, which are usually more expensive. An amplifier also prevents the load impedance of the following stage from affecting the characteristics of the filter. Although satisfactory results were obtained from both digital as well as passive filters, active filters could also be used for implementation.

7. Results

Both passive and digital filters were successfully simulated and implemented. From the FFT analysis, it is clear that the FIR filter's performance was marginally better than the passive filter.

Both the filter implementations gave satisfactory results. However, a digital filter is preferred because of ease of implementation, and the ease of modifying the characteristics of the filter.

References

[1] Zhenhe Sun, Wei Yu, Chen He, "FIR Digital Filter Design and MATLAB simulation", 2012 International Conference on Measurement, Information and Control, pp. 677-680.

[2] FIR – MATLAB Documentation, "http://in.mathworks.com/help/signal/ref/fir1.html"

[3] Yagyu M, Yoshida T, Nishihara A, "Design of FIR Digital Filters with Minimum Weight Representation", IEEE Transactions on Circuits and Systems, 1995, 40(3), pp. 227-230.

[4] Shengli Yan, "FIR filter principle and design method", Changchun Engineering college journal, 2003, 4(1), pp. 63-65.

[5] Kaiyu Xu, Zhongming Zhu, Yaling Lu, "Digital Signal Processing", Electronics and Industry Press, 2005, pp. 137-138.

[6] Ananda Kumar A, Dheerendra Singh, "Digital Implementation of DVSPWM Control for EV fed through Impedance Source Inverter", International Journal of Power Electronics and Drive Systems (IJPEDS), vol. 6, no.3, ISSN: 2088-8694.

[7] A. Kulka, T.Undeland, "Voltage harmonic control of z-source inverter for UPS applications", in Power Electronics and Motion Control Conference, 2008. EPE-PEMC 2008. 13th Sept 2008, pp. 657-662.

An exploration of neuromorphic systems and related design issues/challenges in dark silicon era

Mudit Chandaliya, Nitin Chaturvedi, S Gurunarayanan

Birla Institute of Technology and Science
Department of Electrical and Electronics Engineering
Pilani, India, 333031

E-mail: f2015484, nitin80, sguru@pilani.bits-pilani.ac.in

Abstract. The current microprocessors has shown a remarkable performance and memory capacity improvement since its innovation. However, due to power and thermal limitations, only a fraction of cores can operate at full frequency at any instant of time irrespective of the advantages of new technology generation. This phenomenon of under-utilization of microprocessor is called as dark silicon which leads to distraction in innovative computing. To overcome the limitation of utilization wall, IBM technologies explored and invented neurosynaptic system chips. It has opened a wide scope of research in the field of innovative computing, technology, material sciences, machine learning etc. In this paper, we first reviewed the diverse stages of research that have been influential in the innovation of neurosynaptic architectures. These, architectures focuses on the development of brain-like framework which is efficient enough to execute a broad set of computations in real time while maintaining ultra-low power consumption as well as area considerations in mind. We also reveal the inadvertent challenges and the opportunities of designing neuromorphic systems as presented by the existing technologies in the dark silicon era, which constitute the utmost area of research in future.

1. Introduction

The last few years, witnessed successful scaling of MOS dimensions for several technology generations. However, reduction in the MOS dimensions in the present-day fabrication technologies have resulted to unexpected new problem for the chip designers know as dark silicon.

This phenomenon restricts the ever increasing portion of a chip's transistor to remain on, they remain dark in order to keep check on power budget of a chip and restrict the system designers to utilize the extra resources that comes with every new technology generation (Moore's Law). It is probable that we have attained the end of Dennard scaling and furthermore Moore's law. Figure 1 illustrates speed-up projections for last few decades as well as future trends.

In the past, clock frequencies of a single microprocessor have shown significant improvements, whereas subsequent improvement in microprocessors based on von Neumann computer architecture have been achieved by means of multiple parallel cores (parallelization). The current computing systems are based on classical von Neumann Architecture which have one or multiple cores (processing units) and access both data and program using the same shared resources. Since, the processor speed is much faster than memory, the processor stays in the idle state for most of the time. Also, there are many applications in which parallelism is difficult and these architectures are not compatible for it. Moreover, the complexity of the future application continues to increase along with the size of data (distributive), this presents the inconsistency between the massive future real time applications and what conventional von Neumann architecture can offer. The main performance bottleneck of the Von Neumann architecture is mainly due to increased performance gap between processing core and memory also known as memory wall problem, which motivated the researchers from different communities to propose novel computing architectures.

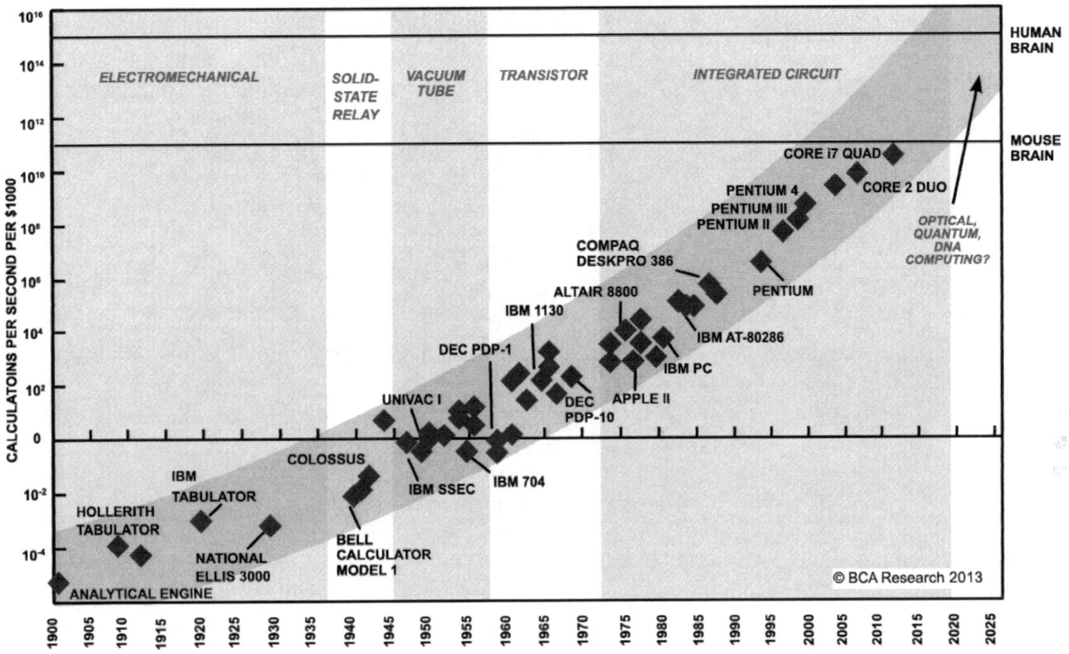

Figure 1. Speedup Projections.

Different research organizations have revealed that valuable insights can be gained from looking into the human brain. Figure 2 illustrates virtual mapping between current state of art technology with brain. This shift in new paradigm for computing is motivated by the structure of human brain and it is efficient in performing complicated tasks rapidly without pre-programming with accuracy and consumes very less energy. This has motivated another field of study focused at the growth and expansion of the neurosynaptic computing systems that could possibly copy the cerebrum's processing proficiency, its size and power. These novel architectures utilizes extensive parallelism and distributed description of data to facilitate novelty to address current unmanageable problem, mainly the reduction of the gap between computational resources and the data storage components present in conventional von Neumann architectures.

In this paper, we present an overview and challenges of the brain-inspired microprocessor architectures and provides a survey of different approaches being followed for developing artificial neural processing systems which exhibits substantial performance observed in biological systems.

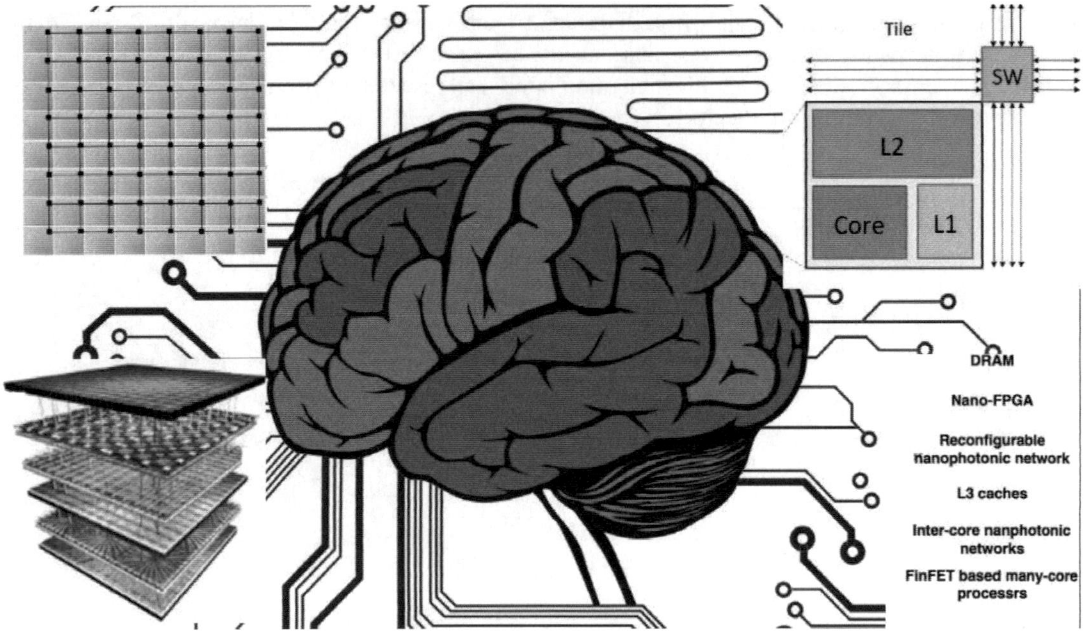

Figure 2. Brain Inspired Computing.

2. Related Work

The interest in neuromorphic computing systems was first seen in late eighties, when Douglas et al. [9] proposed to use the strengths of CMOS VLSI technologies to construct a diverse set of neural analogues, starting from a single synapses, to a set of sensors and simple system. Thereafter, a number of small research labs in the world started working on VLSI model of active cochlea [10], oculomotor system [11]. Recently, an increasing number of researchers from both academia and industry are working on the design of neuromorphic chips [13]. Several researchers reported different new materials for constructing nano device that can mirror some of the properties found in bio-synapses. These neuromorphic computing systems are build using the basic design principles of bio-neurological systems [9].Moreover, these system are designed using advanced mixed analog/digital circuits fabricated using VLSI processes. Analogous to the neurological system, the neuromorphic computing systems utilizes asynchronous, event driven, energy efficient and fault tolerant techniques [5]. The main difference between neuromorphic systems and traditional computing systems (CMP) is in the way, the memory is organized and used. The brain inspired computing presents an interesting solution for the development of modified von Neumann architectures. Merolla et al. [8] designed a digital neurosynaptic processor with the aim of implementing as much as neurons present in a cat brain at IBM SyNAPSE project. A group of researchers works on the comparative studies between neuroscience and neural network based machine learning. For example, Serre et al. presented HMAX, a visual cortex inspired neural network (rate-based, non-spiking, neurons) and demonstrated to be competing with stable machine- learning models. Vogelstein et al. [9] used basic lookup table synapses to reproduce biological neurons in VLSI process.

However, the simple model of synapse was unsuccessful in identifying the plasticity of the bio-synapses, which is key to information storage and brain learning. Bartolozzi et al. [11] build a VLSI plastic synapse model, whereas Jackson et al.[12] did the same using phase change devices with increased synaptic density.

3. Overview of Neurosynaptic System Architecture

The human brain is capable of astonishing feats of perception, action and memory. There are a million neurons in a human body that make multitasking and memory management very efficient and fast. With the present scaling limitations and von Neumann bottleneck, there is a need for more efficient system hence the tremendous amount of research in neuromorphic system.

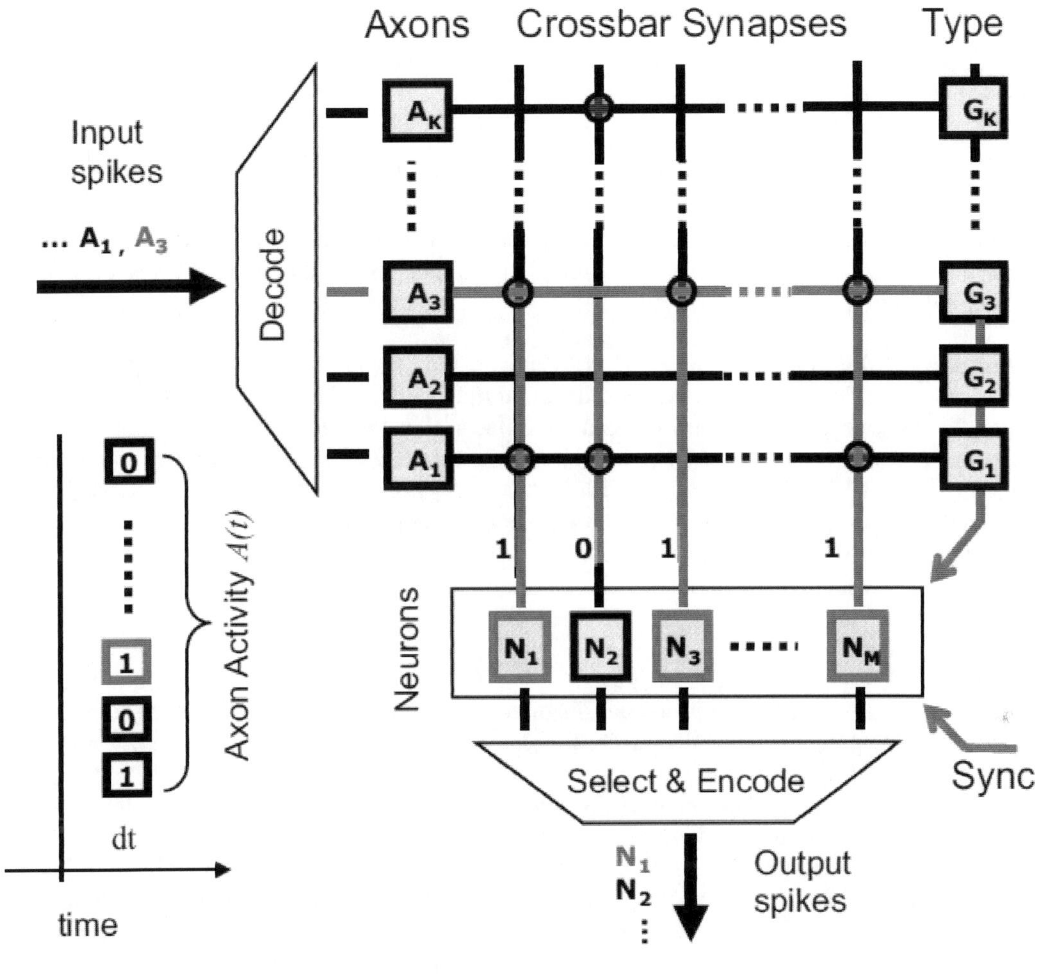

Figure 3. Block Diagram of Neurosynaptic Architecture [13].

3.1. *Architecture*

The idea is to integrate the memory and the computation for quick results. Figure 3 presents the block Diagram of a Neurosynaptic Architecture. The fundamental units of neurosynaptic system are the dendrons, axons and synapses. The dendrons communicate with axons through synapses. The system can have a digital or an analog implementation. The major drawbacks of an analog implementation are its sensitivity to PVT variations and the limited scalability. Hence the digital implementation is preferred. The conventional SRAM can be replaced by memristors for better performance. A very popular implementation is obtained by arranging memory elements in an $M \times K$ crossbar array to store the synaptic weight values. The core contains M axons that are connected to K dendrons through $M \times K$ binary valued synapses. The connection between the jth axon and the ith dendron is given by W_{ji}. The activity bit, $A_j(t)$, for the jth axon represents whether the corresponding neuron was fired in the previous time step. The state of the jth axon is denoted by G_j which can assume a value of 0, 1 or 2. These values distinguish the connection (eg. Excitatory or inhibitory). The synaptic weight between the ith dendron and j^{th} axon is given by

$$S_i^{Gj} \quad (1)$$

Thus the input to the dendron is:

$$A_j(t) \times W_{ji} \times S_i^{Gj} \quad (2)$$

Assuming a membrane voltage of V(t), leak voltage L and threshold voltage, the membrane voltage of a neuron is updated at each time step as:

$$V(t+1) = V(t) + L_i + \sum_{j=1}^{M} A_j(t) \times W_{ji} \times S_i^{Gj} \quad (3)$$

Whenever this membrane voltage of a neuron exceeds the threshold value, it generates a spike and its membrane voltage is reset to 0. The fundamental unit can be arranged in various ways to achieve the required architecture that mimics the brain.

3.2. Deep Neural Networks

Deep Neural Networks (DNN) are composed of multiple layers of neurons. There is a one to one correspondence between the neurons belonging to adjacent layers hence a two way communication is possible. There is a high fan in as well as fan out in this architecture as there is a feedback as well as a feed forward path between neurons from different layers. This architecture is typically used in wide range of classification problems. Convolution network are similar to DNN but they are composed of only feed-forward network. The inputs to a layer from the neurons of previous layer are used to compute the weighted sum which is the input for the subsequent layer. This connectivity pattern forms the convolution kernel.

4. Neuromorphic Engineering Challenges

In order to fully utilize the strength of brain-inspired neuromorphic system. We need to address the following challenges.

4.1. Focused Effort to design Neuromorphic System

Numerous years have passed after the first research publication on neuromorphic system [mead]. In the near past, only a few small groups started working on neuromorphic engineering inorder to utilize huge number of transistors in current technology node also called the dark silicon era, which restricts further performance improvement in chip multiprocessors. In dark silicon based CMPs, all the on-chip cores cannot be switched on together. The progress in neuromorhpic engineering is slow as compared to conventional engineering and in the field of technology (for example: a large number of engineers are assigned to the development of GPUS and CMPS) due to which there is a shortage of technology infrastructure for neuromorphic circuit design.

4.2. Limited Brain Understanding

Another, major hurdle in the rapid development of neuromorphic systems is the limited knowledge and understanding of the functions of brain and neural computing. The complexity of human brain makes it almost impossible to mimic. This complexity comes due to the high number of neurons. The neural engineering research groups much learn and comprehend the art of building real time event-based sensors, motor-reactive systems. The research community should propose novel neural based techniques to link machine learning with neuromorphic systems that models cognitive abilities.

4.3. Multi-disciplinary Research
In order to achieve the aim of developing neuromorphic systems, it is extremely necessary for various researchers from diverse field of engineering to develop a multi-disciplinary groups. These groups together must explore and investigate the main principles of neural computing and develop techniques that serve as a platform to perform complex neural simulations.

4.4. Other Challenges
The existing computing platform are approximately 300 times slower than human brain. Moreover, the existing neural networks are fast and reliable but even a small network grows inevitably in size along with switching time delays. Another, important limitation is the leakage current at the neurosynaptic nodes due to the high density of interconnections which is responsible of transmitting information to the wrong node. Another issue is the interfacing of these neuromorhphic architecture (modified von Neumann) to the existing computing devices with von Neumann architectures.

5. Conclusions
In summary, we presented an overview of a basic brain inspired neuromorphic system architecture. In this work, we provided a survey of few essential challenges in terms of theoretical and experimental neuroscience. For the development of neuromorphic computing systems, inter-disciplinary efforts are required which involve contribution from neurologists/neuroscientists, computer engineers, technology engineers and researchers from material science. There is a strong need to train new generation of scientists and engineers with inter-disciplinary skill and encourage the area specialists on different domains to work together for the deep understanding of brain and to develop brain-inspired high performance system architectures in dark silicon era. As part of our future work, we will use GEM5 full system simulator to model a distributed memory multicore system and to simulate a real heterogeneous system with varying workloads under stringent power-performance constraints. With initial simulation results, we plan to present few fundamental problems that can be solve by the hardware/software research community.

Acknowledgement
We would like to extend our gratitude to the Senior Professors from the Department of Electrical and Electronics Engineering, BITS Pilani for their valuable suggestions.

References
[1] P. S. Magnusson, M. Christensson, J. Eskilson, D. Forsgren, G. Hallberg, J. Hogberg, F. Larsson, A. Moestedt, and B. Werner, *Simics: A Full system Simulator Platform*. Computer, 2002, vol. 35-2, pp. 50-58.
[2] C. Bienia, S. Kumar, J. P. Singh, and K. Li, *The parsec benchmark suite: Characterization and architectural implications*, in Procs. of the International Conference on Parallel Architectures and Compilation Techniques, 2008.
[3] C. Mead, *Analog VLSI and Neural Systems*. Reading, MA, USA: Addison-Wesley, 1989.
[4] E. Fragniere, A. van Schaik, and E. Vittoz, *Design of an analogue VLSI model of an active cochlea*, J. Analog Integr. Circuits Signal Process., vol. 13, no. 1/2, pp. 1935, May 1997.
[5] T. Horiuchi and C. Koch, *Analog VLSI-based modeling of the primate oculomotor system*, Neural Comput., vol. 11, no. 1, pp. 243265, Jan. 1999.
[6] K. Boahen, *Neuromorphic microchips*, Sci. Amer., vol. 292, no. 5, pp. 5663, 2005.
[7] R. Sarpeshkar, *Brain powerVBorrowing from biology makes for low power computing VBionic ear*, IEEE Spectrum, vol. 43, no. 5, pp. 2429, May 2006.
[8] S. H. Jo et al., *Nanoscale memristor device as synapse in neuromorphic systems*, Nano Lett., vol. 10, no. 4, pp. 12971301, 2010.
[9] K. Kim et al., *A functional hybrid memristor crossbar-array/CMOS system for data storage and neuromorphic applications*, Nano Lett., vol. 12, no. 1, pp. 389395, 2011.

IOP Conf. Series: Materials Science and Engineering **331** (2018) 012003 doi:10.1088/1757-899X/331/1/012003

[10] W. Maass, *Noise as a resource for computation and learning in networks of spiking neurons*, Proc. IEEE, vol. 102, no. 5, pp. 860880, May 2014.

[11] P. Merolla, J. Arthur, F. Akopyan, N. Imam, R. Manohar, and D. Modha, *A digital neurosynaptic core using embedded crossbar memory with 45pJ per spike in 45nm*, in IEEE Custom Integrated Circuits Conference, pp. 14, IEEE, Sept. 2011.

[12] T. Serre, L. Wolf, and T. Poggio, *Object Recognition with Features Inspired by Visual Cortex, in Conference on Computer Vision and Pattern Recognition*, pp. 9941000, Ieee, 2005.

[13] Paul Merolla, John Arthur, Filipp Akopyan, Nabil Imam, Rajit Manohar, Dharmendra S. Modha, *A Digital Neurosynaptic Core Using Embedded Crossbar Memory with 45pJ per Spike in 45nm* in IEEE custom integrated circuits conference (CICC), pp.1, 2011.

Applications of Electrical Impedance Tomography (EIT): A Short Review

Tushar Kanti Bera
Department of Instrumentation and Applied Physics, *Indian Institute of Science (IISc)*, Bangalore 560012, INDIA.
Department of Medical Imaging, *University of Arizona (UofA)*, Tucson 85724, Arizona (AZ), USA.

E-mail: tkbera77@gmail.com

Abstract. Electrical Impedance Tomography (EIT) is a tomographic imaging method which solves an ill posed inverse problem using the boundary voltage-current data collected from the surface of the object under test. Though the spatial resolution is comparatively low compared to conventional tomographic imaging modalities, due to several advantages EIT has been studied for a number of applications such as medical imaging, material engineering, civil engineering, biotechnology, chemical engineering, MEMS and other fields of engineering and applied sciences. In this paper, the applications of EIT have been reviewed and presented as a short summary. The working principal, instrumentation and advantages are briefly discussed followed by a detail discussion on the applications of EIT technology in different areas of engineering, technology and applied sciences.

1. Introduction

Electrical impedance tomography (EIT) [1-5] reconstructs the conductivity or resistivity distribution of the object under test within a volume conductor using a voltage-current data measured at the domain boundary (Figure 1). In EIT, a constant sinusoidal current signal is injected to the object and the boundary potentials data are measured through the surface electrodes [1] using an EIT hardware or EIT instrumentation [6-10]. Conductivity or resistivity distribution is reconstructed from the boundary potential data using a computer program known as image reconstruction algorithm [11-13]. Practical phantoms [14-22] which are a model of any targeted object of the real application filed are essentially required in EIT to study, calibrate and evaluate the system before applying it in real or practical applications. In EIT, the practical phantoms are developed with non-biological or biological materials with different conductivities. An electronic instrumentation is required for constant current injection and the boundary voltage data measurements in EIT. A constant current source [7] is used to inject a low frequency constant amplitude sinusoidal current to the phantom boundary and develop boundary potentials are measured. Boundary voltage-current data are used to solve the EIT invers problem and the impedance image reconstruction is conducted to get the tomographic images. The EIT is an ill-posed inverse problem [23-24] has poor signal to noise ratio (SNR) [25-26] and poor spatial resolution [27-28]. Therefore, EIT technique suffers from low spatial resolution compared to conventional tomographic imaging techniques such as [29], CT [29] and MRI [29]. As the quality of the EIT images (especially the special resolution) is not yet competitive with CT and MRI, the EIT has not yet been accepted as gold method in medical and clinical imaging. But, EIT is a low cost, noninvasive, nonionizing, radiation free and portable imaging modality with high temporal resolution, and hence, for these unique advantages [3-6, 11-13], EIT has obtained enormous attention and interest in diversified fields such as medical imaging [30-38], material engineering [39], Nanotechnology and MEMS [40-41], civil engineering [42-43], chemical engineering [44-46], biotechnology [47-49], and other fields of engineering, technology and applied sciences. In this paper the applications of EIT in different fields have been reviewed and a short summary has been presented for the researchers. The working principal, instrumentation and advantages are briefly discussed followed by a detail discussion on the applications in different areas along with their limitations.

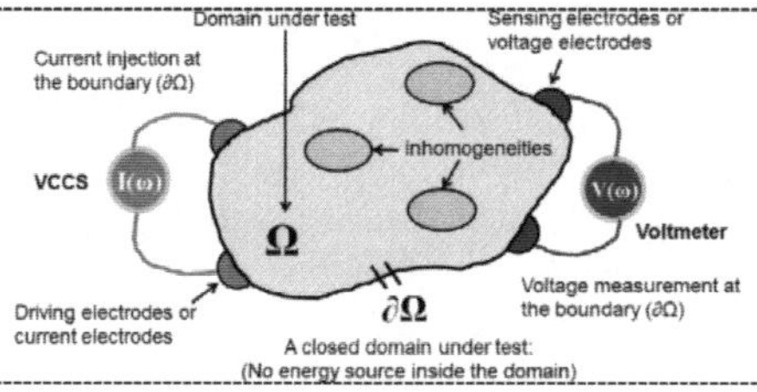

Figure 1. Principle of electrical impedance tomography (EIT) technology.

2. Basics of EIT Technology

In EIT, the electrical impedance of the object under test is reconstructed from a set of voltage current data measured at the boundary by injecting a constant voltage signal or current signal using a constant voltage source [50-51] or constant current source [52-56] using the surface electrodes [1, 57-58]. The surface electrodes are attached to the boundary of the domain under test and are connected to the EIT instrumentation (Figure 2). The boundary potentials are measured by either injecting a constant current signal at a particular frequency (ω) with a single frequency EIT system or at a number of frequencies with a multifrequency EIT system [59-68]. The current injection and the voltage measurement are conducted with, generally, four electrode method [1, 57-58, 69, 75-80] to avoid the electrode-skin interface impedance or contact impedance problem [70-74] in EIT. In four-probe method based EIT data collection scheme, the alternating current signal (I(ω)) is injected through two-electrodes (called driving electrode or current electrodes) as shown by the red electrodes in Fig. 1 and the induced potential (V(ω)) developed is measured across other two-electrodes (called sensing electrodes or voltage electrodes blue electrodes in Fig. 1.

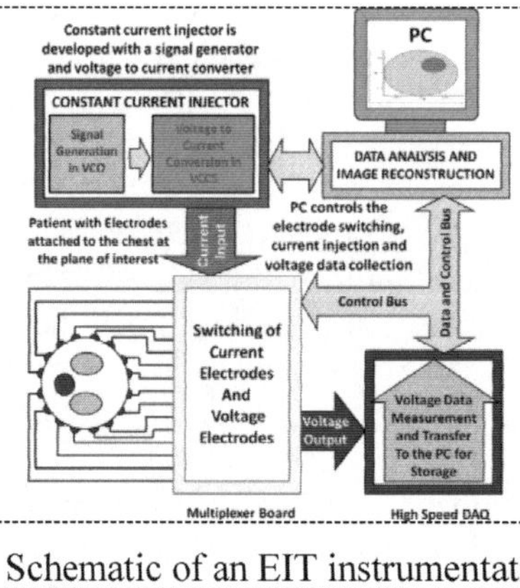

Figure 2. Schematic of an EIT instrumentation system.

3. Applications of EIT Technology

Due to a lot of unique advantages [3-6, 11-13], EIT method has been applied in several fields. The researchers from medical imaging, material engineering, Nanotechnology and MEMS, civil engineering, chemical engineering, biotechnology, and other fields of engineering, technology and applied sciences have tried to solve their numerous research problems using the EIT technology. The applications of EIT in different fields have been reviewed and a short summary is presented below:

3.1. Medical Imaging

EIT has been found suitable for tomographic imaging in medical field for visualizing body anatomy, physiology and several disease conditions. Few of the medical applications of EIT have been summarized below:

3.1.1. Cranial Imaging in the Newborn

EIT has been applied for cranial imaging in the newborn by Tarrasenko and Rolfe, 1984 [30]. They proposed a sensitivity method to obtain images of resistivity distribution within a body segment to image the spatial distribution of electrical resistivity for cranial imaging in the newborn using a weighted backprojection algorithm.

3.1.2. Lung Imaging

A number of studies on lung imaging with EIT technique have been conducted and reported. Meier et al. [31] studied the suitability of EIT on animal subjects to visualize the acute lung injury. They conducted the EIT experimental studies on six pigs with saline-lavage-induced acute lung injury and investigated whether the EIT procedure could be suitable for monitoring the regional lung recruitment and lung collapse. The incremental and decremental positive end-expiratory pressure (PEEP) trials at ten pressure levels were performed and the ventilatory, gas exchange, and hemodynamic parameters were automatically recorded. EIT and computed tomography (CT) scans of the same slice were simultaneously taken at each PEEP level and the results are compared. Authors found EIT is suitable for monitoring the dynamic effects of PEEP variations on the regional change of tidal volume. Pulmonary Edema is also studied by EIT. Harris et al., 1987 [32] reported that applied potential tomography (APT) [1, 5] is able to produce cross-sectional images of the changes in resistivity distribution within the chest. The authors concluded that the APT may be used to detect ventilatory defects in certain types of lung disease to detect the onset of pulmonary edema in high risk patients and to detect and monitor the edema following treatment. Adler et al., 1995 [33] have conducted a series of EIT experiments on dogs to quantify the capacity of this technology to measure lung tidal volume and fluid content. The authors reported that their results obtained from the experimentations conducted on dogs indicated good correlation with standard reference techniques. They also reported that it was difficult to obtain the long term stability of electrodes on the subject and instrumentation.

3.1.3. Hyperthermia Treatment

EIT has been also used for thermal monitoring of hyperthermia treatment. Conway et al. 1985 [34] and Conway, 1987 [35] presented an impedance imaging method to display the induced thermal gradients during hyperthermia treatment. They presented the results obtained with EIT method applied with the experimentations both in vitro and in vivo. A time series of conductivity images showed the distribution of thermal changes and suggested a temperature resolution better than 1 °C. They reported that the spatial resolution was limited to approximately 10% of the circular array diameter by the 104 independent measurements from 16 electrodes used in the EIT system.

3.1.4. Paediatric Lung Disease

EIT has also been applied for paediatric lung disease. Pham et al., 1994 [36] used the EIT technique to study the regional ventilation distribution in neonatal and paediatric lung disease. EIT has been applied on spontaneously breathing infants in their neonatal period, at 3 and 6 months of age during their non-rapid eye movement sleep and the data on regional ventilation distribution and regional filling characteristics were obtained. It is reported that the regional impedance amplitudes increased with the age but regional ventilation distribution remained unchanged in all infants at any age, with the dependent (posterior) lung always better ventilated. Authors also reported that the regional ventilation distribution and regional filling characteristics remained unchanged over the first 6 months of life, and the results obtained on regional ventilation distribution with EIT were found very similar to those in adult subjects.

3.1.5. Breast Imaging

Breast Imaging has been studied with EIT by several research groups. Zain and Chelliah, 2014 [37], applied EIT technique on one hundred and fifty mammography patients above 40 years. Visual interpretation of the images was conducted using the breast imaging - electrical impedance (BI-EIM) classification for detection of abnormalities. Quantitative assessment was carried out by comparison of the breast average electric conductivity with the norm and correlations with visual interpretation of the images, made by a radiologist, were determined using Chi-square. The one-way analysis of variance (ANOVA) was used to compare the mean electrical conductivity between groups and t-test was used for comparisons with pre-existing Caucasians statistics. Authors reported that the Quantitative assessment of electrical impedance tomography was significantly related with visual interpretation of images of the breast.

3.1.6. Brain Imaging

Holder et al., 1996 [38] reported their studies on EIT for imaging the impedance changes of several per cent over tens of seconds, known to occur during evoked activity of the cerebral cortex. The results demonstrated that, for the first time, reproducible impedance changes of few per cent can be seen in EIT images during functional cerebral stimulation and they concluded that EIT may eventually be suitable for imaging evoked responses or epilepsy in human subjects.

3.2. Material Engineering

EIT has been also studied in material engineering and the manufacturing technology such as semiconductor manufacturing. Michiel and Kruger, 2003 [39] applied EIT procedure to study the semiconductor manufacturing. The author used the standard EIT technique to estimate the conductivity distribution of the thin film of a conductive polysilicon across a wafer and the estimated conductivity distribution is related to the thickness of the polysilicon film. Results showed that the EIT is suitable to visualize the conductivity changes occurred due to the physical and chemical effects inside the interior of a wafer during semiconductor manufacturing. Author reported that the differential thickness measurements from the EIT based prototype etch rate sensor correlated very well with optical thickness measurements. Author also proposed a novel EIT based sensor which was capable of providing temporal and spatial wafer surface potential information during plasma processing. Promising results are obtained from the simulation studies conducted on the prototype sensor using the depletion mode NMOSFETs as transduction elements and justified the suitability of the EIT based metrology technique.

3.3. Nanotechnology and MEMS

EIT has been studied in thin film technology and nanotechnology such as imaging of Carbon Nanotube (CNT) composite thin films. Hou et al. 2007 [40-41] applied EIT to reconstruct the spatial conductivity distribution to visualize the structural defects in CNT thin films and its response to various pH environments. The ability of the EIT to image the spatial distribution of conductivity of CNT thin films was found advantageous to develop the multifunctional CNT-based sensing skins.

3.4. Civil Engineering

EIT has been also studied in several civil engineering applications and a number of research works have been reported such as imaging leaks from buried pipes by Jordana et al. 2001 [42], brick walls imaging by Hola et al., 2008 [43]. Jordana et al. 2001 [42] applied EIT to detect the leaks in buried pipes using a surface linear electrode array perpendicular to the axis of the pipe. The author tested their proposed system in the laboratory using a stainless steel tube immersed in water and covered by a rubber sleeve to simulate a non-conductive leak. The system has also been studied with the field measurements involved with the simulated leaks of water from a plastic tube buried in a farm field. Author reported that the simulated leak was successfully distinguished with EIT based pipe leakage detection system. Hola et al., 2008 [43] studied the applicability of EIT for detection of brick wall dampness in real building structures. The EIT setup was applied to determine the dampness of test brick walls on a specially built laboratory test rig and the results obtained from the EIT based studies are compared with the results obtained from the conventional destructive dry-weight method. Author reported that, a satisfactory agreement was found between the results and the further research aimed to apply the EIT method in building engineering practice.

3.5. Chemical Engineering

EIT has large potential in the chemical engineering applications and hence EIT has been also studied in this area. Dickin and Wang, 1996 [44] presented an overview of the important aspects of an ERT system for use in process applications by focusing on the design and operation of flexible ERT instrumentation. A few of the research works reported so far could be summarized as visualizing combustion by Waterfall et al., 1997 [45] and Tapp et al. 2003 [46].

3.6. Biotechnology

Cell culture imaging has been studied by several research groups such as Linderholm et al. 2008 [47], Sun et al. 2008 [48] and Sun et al. 2010 [49]. In EIT based cell culture imaging, the microtechnology is used to develop the microelectrodes which are housed in a small domain where the cell culture is process and monitoring is conducted. A biological culture media of living cells or tissue is used as the object and small amount of electrical current has been injected and the boundary data are collected. The collected boundary data are fed to the computer and the impedance images are reconstructed using a reconstruction algorithm. Linderholm et al. 2008 [47], presented a novel, inexpensive, and fast microimpedance tomography system capable of four-electrode measurements using 16 planar microelectrodes integrated into a culture chamber for two-dimensional imaging of cell and tissue cultures. Impedance measurements are conducted with an Agilent 4294A impedance analyzer combined with a front-end amplifier and 2D images are obtained using a reconstruction algorithm. The shape and position of a human hair, yielding vertical cross sections of the object and the human epithelial stem cells (YF 29) grown directly on the device surface are studied. The rapid resistivity decrease caused by permeabilized cell membranes is monitored by EIT system which suggested that the technique can be used in electroporation studies. Sun et al. 2008 [48] and Sun et al. 2010 [49] presented a miniaturized EIT system that can image the electrical conductivity distribution within a two-dimensional cell culture. A chip containing a circular 16-electrode array is fabricated with PCB technology and used to inject current and to measure spatial voltage across a multi-nuclear single cellular organism, Physarum Polycephalum. An impedance analyzer operating in four-electrode mode is used to collect the voltage –current data and the impedance images are reconstructed with the open source software, EIDORS. Author claimed that the developed system providing a non-invasive lab-on-a-chip (LOC) technology for mapping the spatially distribution of electrical properties of single cells will be significantly useful for diagnostic and clinical applications.

4. Discussion

Due to a number of advantages, the researchers from a number of research fields are applying EIT to solve their problems. Though the EIT technology has number advantages, but the technology suffers from poor SNR and poor spatial resolution. Increasing the number of electrodes increases the spatial resolution but housing a lot of electrodes is not always possible in all the real applications. Increasing FEM mesh with large number of elements also improves the image quality but computation time and cost may increase. Electrodes used in medical imaging are generally found as the surface electrodes in most of the cases. In other field the electrodes can be soldered or penetrated to the object. The injected current amplitude and frequency are required to be chosen very carefully for the medical EIT whereas the amplitude could be increased more for the non-biological sample imaging with EIT. Contact impedance is found a crucial issue in EIT in most of the applications and hence the four electrode method is preferred. Being an ill-posed inverse problem the EIT produces a large noise at the output in response to the small error or noise at the input. Also, the solution process of the inverse problem needs a suitable regularization technique to get the better solution. Therefore, we need to carefully design the system to get the optimum images in a particular application. Therefore, a lot of studies are still required to be conducted to solve all the challenges in sensors design, instrumentation developments, and the reconstruction process.

5. Conclusions

EIT has been applied in different filed of engineering, technology and applied sciences. The research works conducted on the application of EIT in medical imaging, material engineering, Micro and Nanotechnology, civil engineering, chemical engineering, biotechnology have been reviewed and presented as a short summary. In all the fields the EIT technology is found as a low cost portable and fast tomographic imaging modality along with some other advantages and limitations. Due to number of advantages, EIT is possible to apply in several research problems. In this paper a number of research works conducted on the application of EIT in different areas have been discussed and summarized. The paper provides a short summary of the application of EIT and the summary of the EIT applications may help to cultivate the new ideas to apply the EIT technology in new fields where the technology has not yet been applied or studied so far.

6. References

[1] J.G. Webster, Electrical impedance tomography. Adam Hilger Series of Biomedical Engineering, Adam Hilger, New York, USA. 1990.
[2] M. Cheney, D. Isaacson, J.C. Newell, Electrical Impedance Tomography, SIAM Review, 1999; 41(1): 85–101.
[3] R.H. Bayford, Bioimpedance Tomography (Electrical Impedance Tomography), Annual Review of Biomedical Engineering. 2006; 8: 63-91.
[4] C.W.L. Denyer, Electronics for Real-Time and Three-Dimensional Electrical Impedance Tomographs, PhD Thesis, Oxford Brookes University, 1996.
[5] D.S. Holder, Electrical impedance tomography: methods, history and applications (Series in Medical Physics and Biomedical Engineering), 1st Ed. Institute of Physics Publishing Ltd., UK. 2005.
[6] Bera TK and Nagaraju J, A Multifrequency Electrical Impedance Tomography (EIT) System for Biomedical Imaging, IEEE International Conference on Signal Processing and Communications (SPCOM 2012), India, IISc - Bangalore, India pp 1-5.
[7] Bera TK and Nagaraju J, A Multifrequency Constant Current Source for Medical Electrical Impedance Tomography, Proceedings of the IEEE International Conference on Systems in Medicine and Biology 2010 (IEEE ICSMB 2010), India, 2010, pp 278-283.
[8] Riu PJ, Rosell J, Lozano A, Pallàs-Areny R, Multi-frequency static imaging in electrical impedance tomography: Part 1. Instrumentation requirement, Med Biol Eng Comput. 1995 Nov;33(6):784-92.
[9] Oh TI, Wi H, Kim do Y, Yoo PJ, Woo EJ, A fully parallel multi-frequency EIT system with flexible electrode configuration: KHU Mark2, Physiol Meas. 2011 Jul;32(7):835-49. doi: 10.1088/0967-3334/32/7/S08. Epub 2011 Jun 7.
[10] 13K G Boone and D S Holder, Current approaches to analogue instrumentation design in electrical impedance tomography, TOPICAL REVIEW, Physiol. Meas. 17 (1996) 229–247. Printed in the UK
[11] W.R.B. Lionheart, EIT reconstruction algorithms: pitfalls, challenges and recent developments, Review Article, Physiol. Meas. 2004; 25: 125–142.
[12] T.K. Bera, S.K. Biswas, K. Rajan and Nagaraju J, Improving Conductivity Image Quality Using Block Matrix-based Multiple Regularization (BMMR) Technique in EIT: A Simulation Study, Journal of Electrical Bioimpedance, vol. 2, pp. 33–47, 2011.
[13] T.K. Bera, S.K. Biswas, K. Rajan and Nagaraju J, Improving Image Quality in Electrical Impedance Tomography (EIT) Using Projection Error Propagation-Based Regularization (PEPR) Technique: A Simulation Study, Journal of Electrical Bioimpedance, vol. 2, pp. 2–12, 2011. DOI:10.5617/jeb.158
[14] Holder, D. S., Hanquan, Y., & Rao, A. (1996). Some practical biological phantoms for calibrating multifrequency electrical impedance tomography. Physiological measurement, 17(4A), A167.
[15] P.J. Vauhkonen, M. Vauhkonen, T. Makinen, P.A. Karjalainen and J.P. Kaipio, Dynamic electrical impedance tomography—phantom studies, Inverse Problems in Engineering, 2000, Vol. 8, pp. 495-510.
[16] T.K. Bera and Nagaraju J, Resistivity Imaging of A Reconfigurable Phantom With Circular Inhomogeneities in 2D-Electrical Impedance Tomography, Measurement, Volume 44, Issue 3, March 2011, Pages 518-526. DOI:10.1016/j.measurement.2010.11.015
[17] H. Griffiths, Z. Zhang and M. Watts, A constant-perturbation saline phantom for electrical impedance tomography, Phys. Med. Biol., 1989, Vol. 34, No 8, 1063-1071. Printed in the UK
[18] T.K. Bera, J. Nagaraju and G. Lubineau, Electrical impedance spectroscopy (EIS)-based evaluation of biological tissue phantoms to study multifrequency electrical impedance tomography (Mf-EIT) systems. Journal of Visualization, Vol. 19(4), pp 691-713, 2016

[19] B.S. Kim, K.Y. Kim, T.J. Kao, J.C. Newell, D. Isaacson, and G.J. Saulnier, Dynamic electrical impedance imaging of a chest phantom using the Kalman filter, Physiol Meas. 2006, 27(5): S81–S91.
[20] A. Kimoto and K. Shida, Imaging of Temperature-Change Distribution in the Brain Phantom by Means of Capacitance Measurement, IEEE Trans. on Instrumentation and Measurement, Vol. 49, No. 3, 2000
[21] Bera, T. K., Bera, S., Nagaraju, J., & Chakraborty, B. (2017, March). Multiple inhomogeneity phantom imaging with a LabVIEW-based Electrical Impedance Tomography (LV-EIT) System. In Computer, Communication and Electrical Technology: Proceedings of the Int. Conf. on Advancement of Computer Comm. and Electrical Technology (ACCET 2016), India, 2016 (p. 273). CRC Press.
[22] Bera, T. K., Chowdhury, A., Mandai, H., Kar, K., Haider, A., & Nagaraju, J. (2015, February). Thin domain wide electrode (TDWE) phantoms for Electrical Impedance Tomography (EIT). In Computer, Communication, Control and Information Technology (C3IT), 2015 Third International Conference on (pp. 1-5). IEEE.
[23] O'Sullivan, F. (1986). A statistical perspective on ill-posed inverse problems. Statistical science, 502-518.
[24] Tikhonov, A. N., & Goncharsky, A. V. (Eds.). (1987). Ill-posed problems in the natural sciences (p. 344). Moscow: Mir.
[25] A T Tidswell, A Gibson, R H Bayford, and D S Holder, Electrical impedance tomography of human brain activity with a two-dimensional ring of scalp electrodes, Physiol. Meas. 22 (2001) 167–175.
[26] Andy Adler and Robert Guardo, A Neural Network Image Reconstruction Technique for Electrical Impedance Tomography, IEEE Trans Med Imaging, Vol. 13, No. 4, December 1994, pp 594-600.
[27] A M Sinton, B H Brown, D C Barber, F J McArdle and A D Leathard, Noise and spatial resolution of a real-time electrical impedance tomograph, Phhysiol. Meas.. 1992 Vol. 13, Suppl. A, 125-130.
[28] James L Wheeler, WeiWang and Mengxing Tang, A comparison of methods for measurement of spatial resolution in two-dimensional circular EIT images, Physiol. Meas. 23 (2002) 169–176.
[29] Bushberg, J.T., Seibert, J.A., Leidholdt, Jr.E.M., Boone, J.M.: The Essential Physics of Medical Imaging. 2nd Edition, Lippincott Williams & Wilkins, USA (2001).
[30] Tarrasenko L and Rolfe P, Imaging spacial distributions of resistivity - an alternative approach, Electronics Letters, 20, pp574-5, 1984.
[31] Meier, T., Luepschen, H., Karsten, J., Leibecke, T., Großherr, M., Gehring, H., & Leonhardt, S. (2008). Assessment of regional lung recruitment and derecruitment during a PEEP trial based on electrical impedance tomography. Intensive care medicine, 34(3), 543-550.
[32] Harris ND, Suggett AJ, Barber DC and Brown BH, Applications of applied potential tomography (APT) in respiratory medicine, Clin Phys. Physiol. Meas., 8 Suppl. A, pp 155-65, 1987
[33] Adler, A., Berthiaume, Y., Guardo, R., & Amyot, R. (1995, September). Imaging of pulmonary edema with electrical impedance tomography. In Engineering in Medicine and Biology Society, 1995., IEEE 17th Annual Conference (Vol. 1, pp. 557-558). IEEE.
[34] J. Conway, M. S. Hawley, A. D. Seagar, B. H. Brown, D. C. Barber, Applied Potential Tomography (APT) For Noninvasive Thermal Imaging During Hyperthermia Treatment, Electronics Letters, 1985, Vol. 21 No. 19, 836-838.
[35] 34Conway J, Electrical impedance tomography for thermal monitoring of hyperthermia treatment: an assessment using in vitro and in vivo measurements, Clin. Phys, Physiol. Meas., 8, Suppl. A, pp 141-6, 1987
[36] T.M.T. Pham, M. Yuill, C. Dakin and A. Schibler, Regional ventilation distribution in the first 6 months of life, European Respiratory Journal (ERJ) April 1, 2011 vol. 37 no. 4 919-924.
[37] Zain, N. M., & Chelliah, K. K. (2014). Breast imaging using electrical impedance tomography: correlation of quantitative assessment with visual interpretation. Asian Pacific Journal of Cancer Prevention, 15(3), 1327-1331..
[38] D S Holder, A Rao and Y Hanquan, Imaging of physiologically evoked responses by electrical impedance tomography with cortical electrodes in the anaesthetized rabbit, Physiol. Meas. 17 (1996) A179–A186.
[39] Michiel Victor Paul Kruger, Tomography as a Metrology Technique for Semiconductor Manufacturing, PhD Thesis, University of California, Berkeley, 2003.
[40] Tsung-Chin Hou, Kenneth J Loh and Jerome P Lynch, Spatial conductivity mapping of carbon nanotube composite thin films by electrical impedance tomography for sensing applications, Nanotechnology 18 (2007) 315501 (9pp).
[41] Tsung-Chin Hou, Kenneth J. Loh, Jerome P. Lynch, Electrical impedance tomography of carbon nanotube composite materials, Sensors and Smart Structures Technologies for Civil, Mechanical, and Aerospace Systems 2007, edited by Masayoshi Tomizuka, Chung-Bang Yun, Victor Giurgiutiu, Proc. of SPIE Vol. 6529, 652926, (2007).
[42] Josep Jordana, Manel Gasulla and Ramon Pall`as-Areny, Electrical resistance tomography to detect leaks from buried pipes, Meas. Sci. Technol. 12 (2001) 1061–1068.
[43] Jerzy Hola, Zygmunt Matkowski, Krzysztof Schabowicz, Jan Sikora, Stefan Wójtowicz, New Method of Investigation of Rising Damp in Brick Walls by means of Impedance Tomography, 17th World Conference on Nondestructive Testing, 25-28 Oct 2008, China.
[44] Fraser Dickin and Mi Wang, Electrical resistance tomography for process applications, Meas. Sci. Technol. 7 (1996) 247–260.
[45] Roger C. Waterfall, Ruhua He and Christopher M. Beck, Visualizing combustion using electrical impedance tomography, Chemicnl Engineering Science, Vol. 52, No 13, pp. 2129-2138, 1997.
[46] H.S. Tapp, A.J. Peyton, E.K. Kemsley, R.H. Wilson, Chemical engineering applications of electrical process tomography, Sensors and Actuators B 92 (2003) 17–24.
[47] Pontus Linderholm, Laurent Marescot, Meng Heng Loke, and Philippe Renaud, Cell Culture Imaging Using Microimpedance Tomography, 138 IEEE Trans Biomed Eng., Vol. 55, No. 1, 2008, pp 138-146.
[48] T. Sun, S. Tsuda, N.G. Green, K.P. Zauner and H. Morgan, On-Chip Electrical Impedance Tomography for Moniroting The Kinetics in The Cell Culture, Twelfth International Conference on Miniaturized Systems for Chemistry and Life Sciences, 2008, USA, pp 973-975.
[49] Tao Sun, Soichiro Tsuda, Klaus-Peter Zauner, Hywel Morgan, On-chip electrical impedance tomography for imaging biological cells, Biosensors and Bioelectronics 25 (2010) 1109–1115.

[50] T R Qureshi, C R Chatwin and W Wang, Design of Wideband Voltage Source Having Low Output Impedance, Flexible Gain and Controllable Feedback Current for EIT Systems, 2012 2nd International Conference on Biomedical Engineering and Technology, IPCBEE vol. 34 (2012), IACSIT Press, Singapore, 45-50.
[51] Youssoufa Mohamadou, Tong In Oh, HunWi, Harsh Sohal, Adnan Farooq, Eung Je Woo and Alistair Lee McEwan, Performance evaluation of wideband bio-impedance spectroscopy using constant voltage source and constant current source, Meas. Sci. Technol. 23 (2012) 105703 (10pp).
[52] Alexander S Ross, G J Saulnier, J C Newell and D Isaacson, Current source design for electrical impedance tomography, Physiol. Meas. 24 (2003) 509–516.
[53] Jeong Whan Lee, Tong In Oh, Sang Min Paek, Jae Sang Lee, and Eung Je Woo, Precision Constant Current Source for Electrical Impedance Tomography, Proceedings of the 25" Annual International Conference of the IEEE EMBS, Mexico, 2003, 1066-1069.
[54] José Antonio Gutiérrez Gnecchi, Voltage Controlled Current Source (VCCCS) for Electrical Impedance Tomography (EIT) Measurements in the α and β Dispersion Frequency Ranges, 2010 Electronics, Robotics and Automotive Mechanics Conf., 677-681.
[55] Bera TK and Nagaraju J, A Multifrequency Constant Current Source Suitable for Electrical Impedance Tomography (EIT), Proceedings of 2010 International Conference on Systems in Medicine and Biology, 2010, India, pp 278-283.
[56] Engr. Angelito A. Silverio, M.S., Engr. Angelina A. Silverio, M.A., M.S., A High Output Impedance Current Source for Wideband Bioimpedance Spectroscopy Using 0.35μm Tsmc Cmos Technology, International Journal of Engineering and Applied Sciences, 2012, December 2012. Vol. 1, No.2. 68-75.
[57] Malmivuo, J. & Plonsey, R. (1995). Bioelectromagnetism: Principles and Applications of Bioelectric and Biomagnetic Fields, New York, Oxford: Oxford University Press.
[58] J. G. Webster, Medical Instrumentation: Principle and design, John Wiley & Sons, Hoboken, NJ, USA, 1995
[59] Romsauerova, A., McEwan, A., Horesh, L., Yerworth, R., Bayford, R. H. & Holder, D. S. (2006). Multi-frequency electrical impedance tomography (EIT) of the adult human head: initial findings in brain tumours, arteriovenous malformations and chronic stroke, development of an analysis method and calibration, Physiol. Meas. 27, S147–S161.
[60] Soni, N. K., Hartov, A., Kogel, C., Poplack C. S. & Paulsen1, K. D. (2004). Multi-frequency electrical impedance tomography of the breast: new clinical results, Physiol. Meas. 25, 301–314.
[61] Bera TK and Nagaraju J, A Battery Based Multifrequency Electrical Impedance Tomography (BbMf-EIT) System for Impedance, Imaging of Human Anatomy, 15th International Conference on Electrical Bio-Impedance (ICEBI) and 14th Conference on Electrical Impedance Tomography (EIT), Germany, April 22–25, pp 217.
[62] Wi, H., Kim, T. E., Oh, T. I. & Woo, E. J. (2012). Expandable Multi-frequency EIT System for Clinical Applications, In, Progress In Electromagnetics Research Symposium Proceedings (49-52). KL, MALAYSIA.
[63] Bera TK and Nagaraju J, A LabVIEW Based Multifunction Multifrequency Electrical Impedance Tomography (MfMf-EIT) Instrumentation for Flexible and Versatile Impedance Imaging, 15th International Conference on Electrical Bio-Impedance (ICEBI) and 14th Conference on Electrical Impedance Tomography (EIT), Germany, April 22–25, 2013, 216.
[64] Oh, T. I., Koo, H., Lee, K. H., Kim, S. M., Lee, J., Kim, S. W., Seo, J. K. and Woo, E. J. (2008). Validation of a multi-frequency electrical impedance tomography (mfEIT) system KHU Mark1: impedance spectroscopy and time-difference imaging, Physiol. Meas. 29, 295–307.
[65] Griffiths, H. (1988). A phantom for electrical impedance tomography, Clin. Phys. Physiol. Meas., 9 (Suppl. A), 15-20.
[66] S. Kaufmann, A. Latif, W. C. Saputra, T. Moray, J. Henschel, U. G. Hofmann, M. Ryschka, Multi-frequency Electrical Impedance Tomography for Intracranial Applications, World Congress on Medical Physics and Biomedical Engineering May 26-31, 2012, Beijing, China, IFMBE Proceedings Volume 39, 2013, pp 961-963.
[67] Robitaille, N., Guardo, R., Maurice, I., Hartinger, A. E. & Gagnon, H. (2009). A multi-frequency EIT system design based on telecommunication signal processors, Physiol. Meas. 30, S57–S71.
[68] Kerner, T. E., Williams, D. B., Osterman, K. S., Reiss, F. R., Hartov, A. & Paulsen, K. D. (2000). Electrical impedance imaging at multiple frequencies in phantoms, Physiol. Meas. 21, 67–77.
[69] Bera TK and Nagaraju J, Electrical Impedance Spectroscopic Study of Broiler Chicken Tissues Suitable for The Development of Practical Phantoms in Multifrequency EIT, Journal of Electrical Bioimpedance, vol. 2, pp. 48–63, 2011. DOI:10.5617/jeb.174
[70] E. J. Woo, P. Hua, Dr J. G. Webster, W. J. Tompkins, R. Pallás-Areny, Skin impedance measurements using simple and compound electrodes, Medical and Biological Engineering and Computing, January 1992, Volume 30, Issue 1, pp 97-102
[71] Hua, P., Woo E.J., Webster J.G., Tompkins Willis J., An electrical impedance tomograph using compound electrodes, An electrical impedance tomograph using compound electrodes, Date of Conference: 9-12 Nov 1989, vol.2, Page(s): 472 – 473.
[72] Ping Hua, Eung Je Woo, John G. Webster, and Willis J. Tompkins, Using Compound Electrodes In Electrical Impedance Tomography, IEEE TRANSACTIONS ON BIOMEDICAL ENGINEERING, VOL. 40, NO.1, JANUARY 1993, 29-34.
[73] B H Cornish, A Jacobs, B J Thomas and L C Ward, Optimizing electrode sites for segmental bioimpedance measurements, Physiol. Meas. 20 (1999) 241–250.
[74] A. Hinton and B. Sayers. (1998). "Advanced Instrumentation for Bioimpedance Measurements," Solartron Analytical, Farnborough, U.K., Technical Note. [Online]. Available: http://www.solartronanalytical.com/downloads/technotes/ap04.pdf
[75] Zhang WB, Jeong DM, Lee YH, Lee MS., Measurement of subcutaneous impedance by four-electrode method at acupoints located with single-power alternative current, Am J Chin Med. 2004;32(5):779-88.
[76] Alberto Yúfera and Adoración Rueda, A Method for Bioimpedance Measure With Four- and Two-Electrode Sensor Systems, 30th Annual International IEEE EMBS Conference, Vancouver, British Columbia, Canada, August 20-24, 2008, pp 2318-2321.
[77] Padma Batra1 , Sukwinder Singh1, Satyendra Pratap Singh1, A New Method for Measurement of Bioelectrical Impedance, 2011 International Conference on Communication Systems and Network Technologies, pp 740-743.
[78] Fernando Seoane1,2, Javier Ferreira1, Juan Jos´e Sanch´ez1 and Ramon Brag´os3, An analog front-end enables electrical impedance spectroscopy system on-chip for biomedical applications, Physiol. Meas. 29 (2008) S267–S278 doi:10.1088/0967-3334/29/6/S23.
[79] A. Kežionis, P. Butvilas, T. Šalkus, S. Kazlauskas, D. Petrulionis, T. Žukauskas, E. Kazakevičius, and A. F. Orliukas, Four-electrode impedance spectrometer for investigation of solid ion conductors, Rev. Sci. Instrum. 84, 013902 (2013).
[80] H Asfour, W Soller, NG Posnack, AE Pollard and MW Kay, Low frequency impedance spectroscopy of cell monolayers using the four-electrode method, Journal of Physics: Conference Series 224 (2010) 012085 doi:10.1088/1742-6596/224/1/012085.

IOP Conf. Series: Materials Science and Engineering **331** (2018) 012005 doi:10.1088/1757-899X/331/1/012005

Bioelectrical Impedance and The Frequency Dependent Current Conduction Through Biological Tissues: A Short Review

Tushar Kanti Bera
Department of Instrumentation and Applied Physics, Indian Institute of Science (IISc), Bangalore – 560012, INDIA.
Department of Medical Imaging, *University of Arizona (UofA),*
Tucson 85724, Arizona (AZ), USA.

E-mail: tkbera77@gmail.com

Abstract. Biological tissues are developed with biological cells which exhibit complex electrical impedance called electrical bioimpedance. Under an alternating electrical excitation the bioimpedance varies with the tissue anatomy, composition and the signal frequency. The current penetration and conduction paths vary with frequency of the applied signal. Bioimpedance spectroscopy is used to study the frequency response of the electrical impedance of biological materials noninvasively. In bioimpedance spectroscopy, a low amplitude electrical signal is injected to the tissue sample or body parts to characterization the sample in terms of its bioimpedance. The electrical current conduction phenomena, which is highly influenced by the tissue impedance and the signal frequency, is an important phenomena which should be studied to understand the bioimpedance techniques like bioelectrical impedance analysis (BIA), EIS, or else. In this paper the origin of bioelectrical impedance and current conduction phenomena has been reviewed to present a brief summary of bioelectrical impedance and the frequency dependent current conduction through biological tissues. Simulation studies are conducted with alternation current injection through a two dimensional model of biological tissues containing finite number of biological cells suspended in extracellular fluid. The paper demonstrates the simulation of alternating current conduction through biological tissues conducted by COMSOL Multiphysics. Simulation studies also show the frequency response of the tissue impedance for different tissue compositions.

1. Introduction

Biological tissues are three dimensional (3D) materials which are developed with biological cells. The cells are developed with intracellular fluid (ICF) [1-4] surrounded by the cell envelop. Cells are arranged in a 3D structure and suspended in an extracellular matrix called extracellular fluid (ECF) [1-4]. ICF and ECF provide low resistive properties to an alternating electrical excitation [2, 5] whereas the cell membrane produces some capacitance [2, 5]. Therefore, a biological cell surrounded by ECF collectively produces complex electrical impedance under an alternating electrical excitation [2, 5]. The electrical impedance of biological tissues is termed as bioelectrical impedance or electrical bioimpedance or simply bioimpedance [2, 5-8]. The electrical impedance of the biological tissue depends on the tissue composition, tissue anatomy and frequency of the applied signal [2]. Also, as the capacitive reactance of the cell membrane changes with the signal frequency, the bioimpedance also varies with the frequency of the applied electrical signal. Therefore the current penetration and conduction paths significantly changes with the frequency. Electrical impedance spectroscopy (EIS)

[2, 9-17] studies the frequency response of the electrical impedance of a material and hence it has been applied as an impedance based non-invasive material characterization technique in several fields of engineering, technology and applied sciences [18-27]. As shown in Fig. 1, the EIS applies a constant amplitude electrical signal (current or voltage) to a SUT at different frequency points and the developed signal on the sample surface is measured at all the frequencies using, generally, a linear array of surface electrodes [28-30]. The bioimpedance spectroscopy, which is an EIS technique applied for the biological materials, is used to study and characterize the biological materials for non-invasive analysis of anatomy, physiology and pathology. In bioimpedance spectroscopy, a low amplitude electrical signal is injected to the biological sample (cells, tissues or boy parts) and the surface potentials are measured to calculate the bioimpedance for their non-invasive characterization. As the current conduction phenomena is highly influenced by signal frequency, tissue composition and structure, it is important to study the bioimpedance and its frequency response to understand the bioelectrical phenomena as well as to obtain the information required for tissue characterization. Also, the current conduction paths through the tissue can be controlled by suitably changing the frequency which helps us to calculate the impedance of the tissue for different current paths. Therefore, the studies on the current conduction through biological tissue are very important for understanding the tissue behaviour and tissue composition analysis techniques such as bioelectrical impedance analysis (BIA) [7, 31-33]. Visualization of current paths in real materials is very difficult and hence the simulated model analysis with modern softwares helps us to virtually develop the tissue structure and understand the tissue behaviour under electrical excitations and other bioelectrical phenomena. In this direction, the origin of bioelectrical impedance and isolated cell model are discussed and current conduction through biological tissue has been reviewed to present a brief summary of bioelectrical impedance phenomena. The frequency dependent current conduction through biological tissues is also discussed and the simulation of alternating current conduction through biological tissues is studied by COMSOL Multiphysics [34]. The simulation studies are conducted with alternation current injection through a two dimensional model of biological tissues containing finite number of biological cells suspended in extracellular fluid.

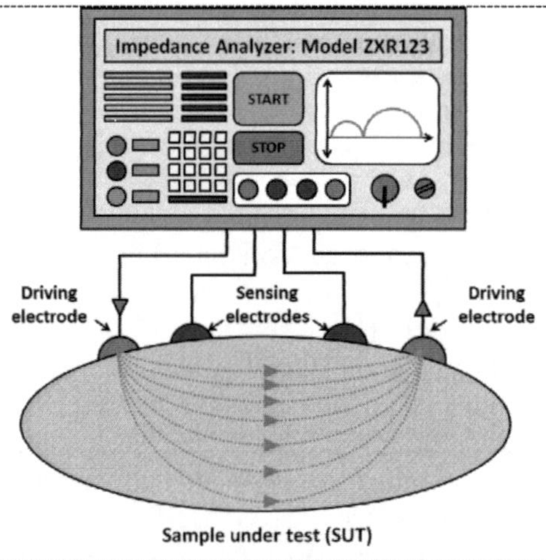

Figure 1. Schematic of the electrical impedance spectroscopy (EIS) of a material.

2. Materials and Methods

2.1. Bioelectrical Impedance

ICF is developed with protoplasm which is composed of nucleus and cytoplasm [4, 35]. Cytoplasm is developed with cell organelles, cytoskeleton, and the cytosol which is made up of cell organelles. ICF is composed of water, salts, and proteins [4, 35] and hence it is found as the good conductor electricity

[4, 35] and produces a resistance (R_{ICF}) to the electrical signal applied. The extracellular fluid (ECF) is a collection of extracellular molecules secreted by cells that provides structural and biochemical support to the cells [35]. ECF is developed with the intricate network of macromolecules [35] which are composed of a number of proteins and polysaccharides [35] and produces an extracellular resistance (R_{ECF}). The cell membranes are developed with protein-lipid-protein (P-L-P) sandwich structure. Protein is conductor of electricity while the lipid is a poor conductor. Therefore, the P-L-P structured cell membrane produces capacitance called membrane capacitance (C_M) under an alternating excitation. If the frequency of the alternating electrical excitation is f, the membrane capacitance (C_M) produces a capacitive reactance ($X_M = 1/\omega C_M = 1/2\pi f C_M$) [4] which depends on the membrane structure, membrane composition and the signal frequency. Therefore, a cell surrounded by ECF produces an electrical impedance which is function of R_{ICF}, R_{ECF}, C_M and f could be modelled in several ways. A popular model has been shown in the Fig. 2b.

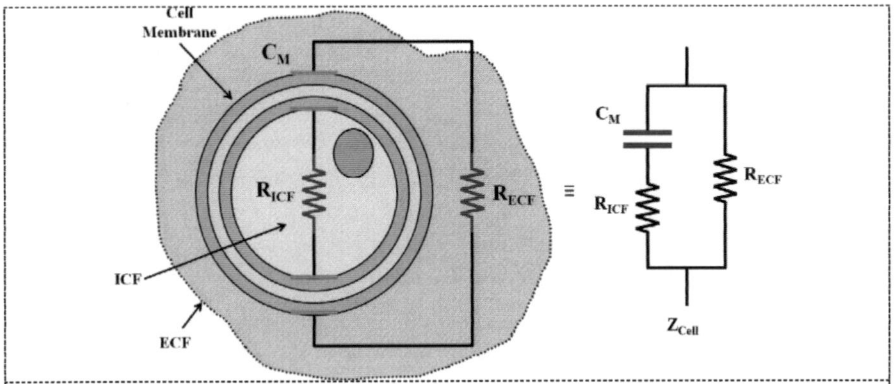

Figure 2. Electrical impedance modelling of a biological cell (a) a cell suspended in an extracellular fluid (ECF), (b) approximate equivalent electrical model of the cell impedance.

The approximate equivalent electrical model of the cell impedance as shown in Fig. 2b is given by the Eq. 1 which can be resolved to get the real and imaginary terms as described below:

$$Z_{Cell} = \left(R_{ICF} - jX_M\right) \| R_{ECF} \tag{1}$$

$$Z_{Cell} = \frac{\left(R_{ICF} - jX_M\right) \times R_{ECF}}{\left(R_{ICF} + R_{ECF}\right) - jX_M}$$

$$Z_{Cell} = \frac{\left(R_{ICF}R_{ECF} - jX_M R_{ECF}\right)\left(\left(R_{ICF} + R_{ECF}\right) + jX_M\right)}{\left(R_{ICF} + R_{ECF}\right)^2 + \left(X_M\right)^2}$$

$$Z_{Cell} = \frac{R_{ICF}R_{ECF}\left(R_{ICF} + R_{ECF}\right) - jX_M R_{ECF}\left(R_{ICF} + R_{ECF}\right) + R_{ICF}R_{ECF}\left(jX_M\right) + \left(X_M\right)^2 R_{ECF}}{\left(R_{ICF} + R_{ECF}\right)^2 + \left(X_M\right)^2}$$

$$Z_{Cell} = \frac{R_{ICF}R_{ECF}\left(R_{ICF} + R_{ECF}\right) + \left(X_M\right)^2 R_{ECF} - jX_M\left(R_{ECF}\left(R_{ICF} + R_{ECF}\right) - R_{ICF}R_{ECF}\right)}{\left(R_{ICF} + R_{ECF}\right)^2 + \left(X_M\right)^2}$$

$$Z_{Cell} = \frac{R_{ICF}R_{ECF}(R_{ICF}+R_{ECF})+(X_M)^2 R_{ECF}}{(R_{ICF}+R_{ECF})^2+(X_M)^2} - j\frac{X_M(R_{ECF}(R_{ICF}+R_{ECF})-R_{ICF}R_{ECF})}{(R_{ICF}+R_{ECF})^2+(X_M)^2} \quad (2)$$

It is observed from the Eq. 1 through Eq. 2 that, as both the real and imaginary terms of cell impedance (Eq. 2) contain capacitive reactance (X_M), they both are found frequency dependent. Also, both the real and imaginary terms are found as the function of R_{ICF} and R_{ECF} and C_M.

2.2. Electrical Current conduction Through Biological Tissues

Electrical current can flow through the biological tissues depending on the electrical properties of the biological tissues and the property of the electrical signal applied. Bioelectrical impedance is developed with the resistance provided by the ICF (R_{ICF}) and ECF (R_{ECF}) and the capacitance produced by the cell envelop (C_M). The capacitive reactance (X_M) depends on C_M as well as the frequency (f) of the applied signal. At low frequency, the magnitude of the capacitive reactance ($X_M = 1/\omega C_M = 1/2\pi f C_M$) becomes large which does not allow the current to pass the through the cell membrane and hence the major part of the electrical current is conducted through the low resistive medium i.e. ECF. But when the signal frequency increases, the X_M reduces gradually and current start penetrating the cell membrane and passes through the cells. At sufficiently high frequency, the X_M becomes significantly low in amplitude and hence the current signal easily passes through the cells. Therefore, the impedance measured with low frequency signal provided the information about the extracellular medium only whereas the impedance measured with high frequency signal gives the information of ICF and ECF both. Also, using the low frequency impedance and high frequency impedance we can calculate the compositions of the tissue under test which is the basis of the BIA technique.

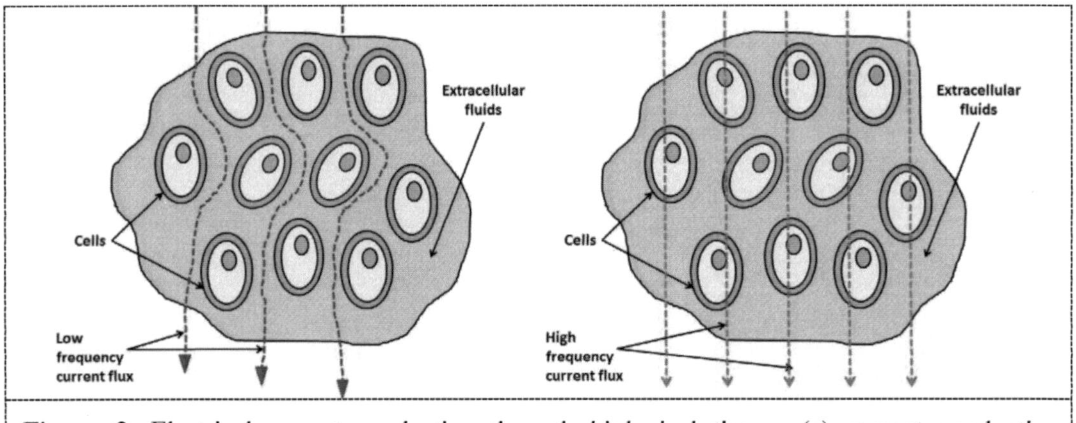

Figure 3. Electrical current conduction through biological tissues (a) current conduction through biological tissue for low frequency signal, (b) current conduction through biological tissue for high frequency signal.

2.3. EIS

Electrical impedance spectroscopy (EIS) provides the frequency response of the material to study the material properties noninvasively. In EIS, a low amplitude electrical signal is injected to the sample under test (SUT) at different frequencies and the electrical impedance is calculated at each frequency points using Ohms Law. Generally a low amplitude current signal (or voltage signal) is applied to a SUT and the developed surface potential (or current) is measured across (or through) a material segment on the SUT surface using either by two electrode method (Fig. 3a) or four electrode method (Fig. 3b). In two electrode method, current injection and voltage measurement are conducted using the same electrode pairs (Fig. 3a) whereas the four electrode method uses two different electrodes pairs for voltage-current data collection (Fig. 3b). As the material impedance depends on the material composition and structure, the electrical impedance could be used as the signature of material information for material characterization. As the electrical impedance is a function of frequency also, the frequency response of the electrical impedance provides more information about the material under test which helps us for better characterization process.

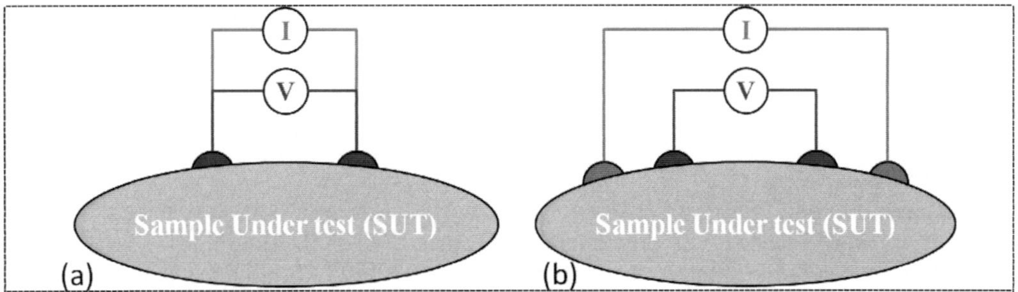

Figure 4. Schematic of impedance measurement with two-electrode and four-electrode methods (a) two-electrode method, (b) four-electrode method.

2.4. Bioelectrical Impedance Spectroscopy (BIS)

Bioelectrical impedance spectroscopy (BIS) [36-42] is the impedance spectroscopic technique which is applied for biological tissues. The BIS can be found useful for characterizing the biological mediums such as cells, tissues and biological body parts. Even the BIS can be utilized for single cell spectroscopy also [43-44]. BIS has been found suitable for food analysis [9, 19, 21, 23, 42, 45-50], cell or tissue culture studies [51-55] and the disease detection [56-67] and so on.

2.5. Bioimpedance Studies in COMSOL

To visualize the frequency dependent of bioelectrical impedance and its response to alternating current excitation, the electrical conduction through biological tissues electrical, current conduction studies are conducted with 2 two dimensional (2D) biological tissue model (Fig. 4) using COMSOL Multiphysics software [34].

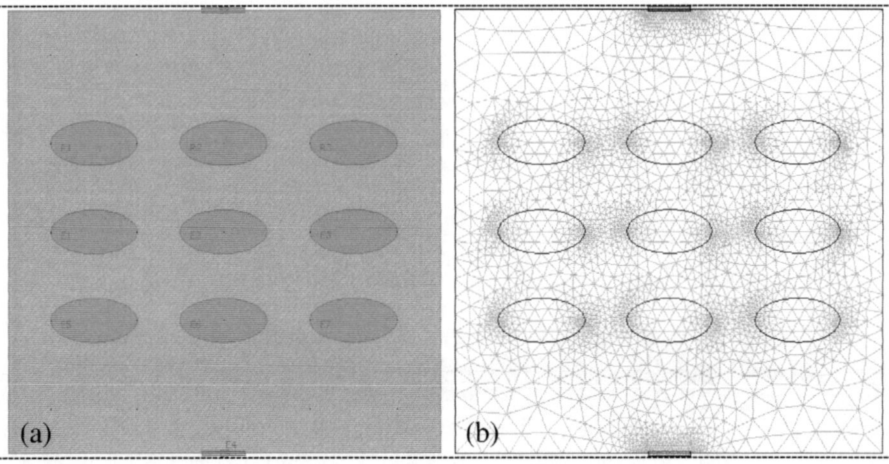

Figure 5. Simulated model of the biological tissue and the bioimpedance simulation in COMSOL Multiphysics (a) 2D model of the biological tissue with electrodes (b) tissue model discretised with the FEM mesh.

A 2D model of a square geometry tissue sample containing nine cells has been developed and the tissue model is equipped with two high conductive electrodes positioned on the two horizontal sides of the geometry as shown in Fig. 4a. The tissue sample is discretized with triangular elements mesh (containing 5528 elements) using the finite element method (FEM) [68-72].

The FEM mesh is developed with non-uniform refinement of the elements to obtained optimum results. As shown in the Fig. 4b, the meshing is also made more refined near the electrode region and cell-ECF interfaces with large curvature. The EIS studies are conducted by injecting an alternating sinusoidal current (1 mA amplitude) from 1 Hz to 1 GHz and the current paths through the tissue sample at different frequencies are studied.

3. Results and Discussion

The biological cell model and current conduction has been discussed in this paper. The current conduction paths are visualized with the simulation studies COMSOL Multiphysics software. Fig. 5 shows the current conduction paths obtained from the COMSOL Multiphysics based simulation studies on the multifrequency current injection.

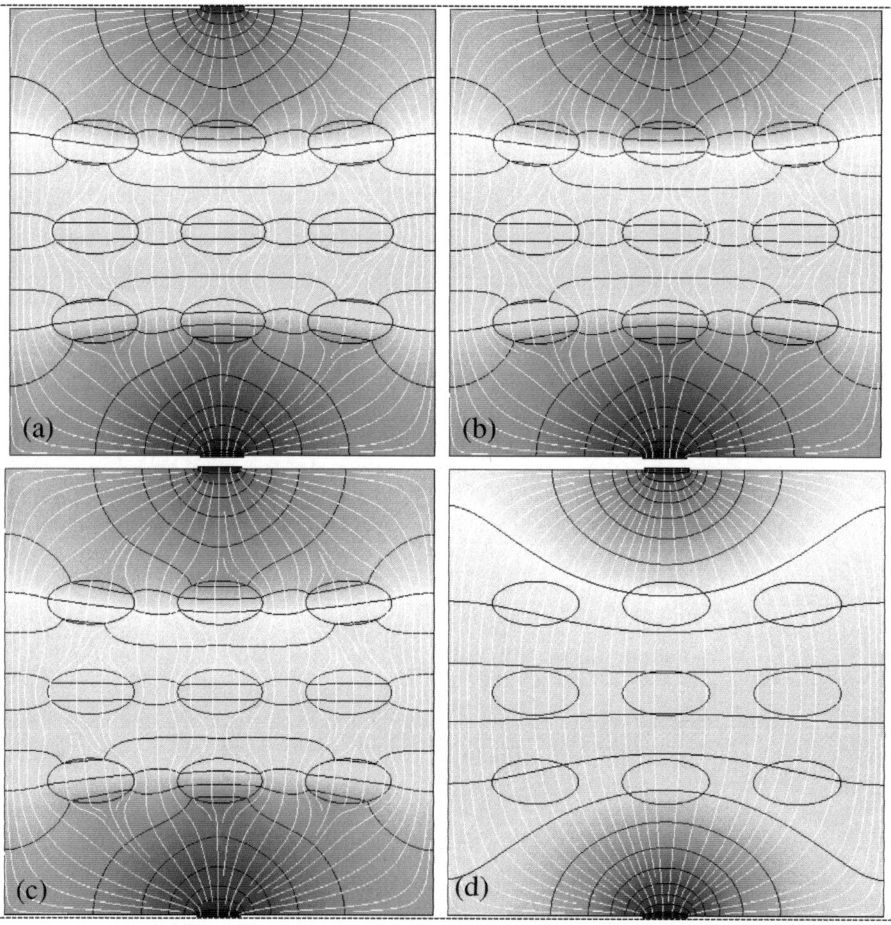

Figure 6. Current conduction paths through the tissue sample at different frequency of the applied current signal (a) 1 Hz, (b) 1 kHz, (c) 1 MHz, (d) 1 GHz.

The Fig. 5a, Fig. 5b, Fig. 5c and Fig. 5d show the current conduction paths at 1 Hz, 1 kHz, 1 MHz and 1 GHz respectively. As shown in the figures from Fig. 5a, through Fig. 5d, it is clearly observed that at the low frequency the current signal tries to avoid conducting through the cells and majorly conducts through the ECF whereas, the penetration of electrical current through the cells increases as the frequency increases. For the current injection with the signal of 1 kHz (Fig. 5b), it is observed that the major part of the current flux lines are trying to avoid the cells whereas the penetration through the cells increases with the increase in signal frequency (Fig. 5b, Fig. 5c and Fig. 5d). At very high frequency the current penetration become maximum and current flux lines become almost linear through the ECF and ICF as shown in Fig. 5d. In the work presented in this paper is the preliminary study on a basic tissue model where the ECF and ICF of a basic animal tissue model have been studied. In the future studies, the current conduction studies with 3D anatomy could be conducted.

4. Conclusions

The origin of the electrical bioimpedance has been discussed from the single cell model and the cell impedance has been deduced to show the frequency dependent behaviour of the impedance and its real and imaginary components. With the mathematical model, it has also been discussed that the bioimpedance and their real and imaginary components depend on the cell composition i.e. ICF, ECF and cell membrane. The COMSOL simulation demonstrates the frequency dependent current conduction through biological tissues. It is observed that the at low frequency, the major part of the current signal avoid to conduct through the cells due to the large capacitive reactance of the cell membrane and prefers to pass through the extracellular fluid region. On the contrary, at high frequency the capacitive reactance reduces and the electrical current penetrates the cell membranes and therefore, it conducts through both the ECF and ICF. The present paper presents a brief review on the bioelectrical impedance and its frequency dependent behaviour to the alternation current. The discussion about the origin of the bioimpedance, equivalent circuit modelling of biological cells and the mathematical representation of the bioimpedance, collectively helps to understand the bioimpedance and bioelectricity basics.

5. References

[1] Van Loan, M. D., Withers, P., Matthie, J., & Mayclin, P. L. (1993). Use of bioimpedance spectroscopy to determine extracellular fluid, intracellular fluid, total body water, and fat-free mass. In Human body composition (pp. 67-70). Springer US.

[2] Bera, T. K. (2014). Bioelectrical impedance methods for noninvasive health monitoring: a review. Journal of medical engineering, 2014.

[3] Lin, Y. P., Yu, W. C., Hsu, T. L., Ding, P. Y. A., Yang, W. C., & Chen, C. H. (2003). The extracellular fluid—to—intracellular fluid volume ratio is associated with large-artery structure and function in hemodialysis patients. American journal of kidney diseases, 42(5), 990-999.

[4] Bera, T. K., Jampana, N., & Lubineau, G. (2016). A LabVIEW-based electrical bioimpedance spectroscopic data interpreter (LEBISDI) for biological tissue impedance analysis and equivalent circuit modelling. Journal of Electrical Bioimpedance.

[5] Hayden, R. I., Moyse, C. A., Calder, F. W., Crawford, D. P., & Fensom, D. S. (1969). Electrical impedance studies on potato and alfalfa tissue. Journal of Experimental Botany, 20(2), 177-200.

[6] Martinsen, O. G., & Grimnes, S. (2011). Bioimpedance and bioelectricity basics. Academic press.

[7] Kyle, U. G., Bosaeus, I., De Lorenzo, A. D., Deurenberg, P., Elia, M., Gómez, J. M., ... & Scharfetter, H. (2004). Bioelectrical impedance analysis—part I: review of principles and methods. Clinical nutrition, 23(5), 1226-1243.

[8] Clarke, D. E., & Raffin, T. A. (1993). Thoracic electrical bioimpedance measurement of cardiac output-Not ready for prime time. Critical care medicine, 21(8), 1111.

[9] Bera, T. K., Bera, S., Chowdhury, A., Ghoshal, D., & Chakraborty, B. (2017, March). Electrical Impedance Spectroscopy (EIS) based fruit characterization: A technical review. In Computer, Communication and Electrical Technology: Proceedings of the International Conference on Advancement of Computer Communication and Electrical Technology (ACCET 2016), West Bengal, India, 21-22 October 2016 (p. 279). CRC Press.

[10] Zhao, X., Zhuang, H., Yoon, S. C., Dong, Y., Wang, W., & Zhao, W. (2017). Electrical Impedance Spectroscopy for Quality Assessment of Meat and Fish: A Review on Basic Principles, Measurement Methods, and Recent Advances. Journal of Food Quality, 2017.

[11] Grossi, M., & Riccò, B. (2017). Electrical impedance spectroscopy (EIS) for biological analysis and food characterization: a review. Journal of Sensors and Sensor Systems, 6(2), 303.

[12] Zarafshani, A., Bach, T., Chatwin, C., Xiang, L., & Zheng, B. (2017). Current source enhancements in Electrical Impedance Spectroscopy (EIS) to cancel unwanted capacitive effects. Proceedings of SPIE, Medical Imaging 2017: Biomedical Applications in Molecular, Structural, and Functional Imaging, 10137, 101371X.

[13] Pathiraja, A., Ziprin, P., Shiraz, A., Mirnezami, R., Tizzard, A., Brown, B., ... & Bayford, R. (2017). Detecting colorectal cancer using electrical impedance spectroscopy: an ex vivo feasibility study. Physiological measurement, 38(6), 1278.

[14] Fortin, J., Habenbacher, W., Heller, A., Hacker, A., Grüllenberger, R., Innerhofer, J., ... & Pacher, R. (2006). Non-invasive beat-to-beat cardiac output monitoring by an improved method of transthoracic bioimpedance measurement. Computers in biology and medicine, 36(11), 1185-1203.
[15] Nuñez, C., Gallagher, D., Visser, M., Pi-Sunyer, F. X., Wang, Z. I. M. I. A. N., & Heymsfield, S. B. (1997). Bioimpedance analysis: evaluation of leg-to-leg system based on pressure contact footpad electrodes. Medicine and Science in Sports and Exercise, 29(4), 524-531.
[16] Kerner, T. E., Paulsen, K. D., Hartov, A., Soho, S. K., & Poplack, S. P. (2002). Electrical impedance spectroscopy of the breast: clinical imaging results in 26 subjects. IEEE transactions on medical imaging, 21(6), 638-645.
[17] He, Z., & Mansfeld, F. (2009). Exploring the use of electrochemical impedance spectroscopy (EIS) in microbial fuel cell studies. Energy & Environmental Science, 2(2), 215-219.
[18] Bera, T. K., Mohamadou, Y., Lee, K., Wi, H., Oh, T. I., Woo, E. J., ... & Seo, J. K. (2014). Electrical impedance spectroscopy for electro-mechanical characterization of conductive fabrics. Sensors, 14(6), 9738-9754.
[19] Chowdhury, A., Bera, T. K., Ghoshal, D., & Chakraborty, B. (2015, February). Studying the electrical impedance variations in banana ripening using electrical impedance spectroscopy (EIS). In Computer, Communication, Control and Information Technology (C3IT), 2015 Third International Conference on (pp. 1-4). IEEE.
[20] Bera, T. K., Nagaraju, J., & Lubineau, G. (2016). Electrical impedance spectroscopy (EIS)-based evaluation of biological tissue phantoms to study multifrequency electrical impedance tomography (Mf-EIT) systems. Journal of Visualization, 19(4), 691-713.
[21] Chowdhury, A., Kanti Bera, T., Ghoshal, D., & Chakraborty, B. (2017). Electrical impedance variations in banana ripening: an analytical study with electrical impedance spectroscopy. Journal of Food Process Engineering, 40(2).
[22] Almuhammadi, K., Bera, T. K., & Lubineau, G. (2017). Electrical impedance spectroscopy for measuring the impedance response of carbon-fiber-reinforced polymer composite laminates. Composite Structures, 168, 510-521.
[23] Chowdhury, A., Singh, P., Bera, T. K., Ghoshal, D., & Chakraborty, B. (2017). Electrical impedance spectroscopic study of mandarin orange during ripening. Journal of Food Measurement and Characterization, 1-11.
[24] Bera, T. K., & Nagaraju, J. (2011). Electrical impedance spectroscopic studies of the electronic connectors of DIP switch based multiplexers suitable for multifrequency electrical impedance tomography, biomedical engineering, Narosa Publishing House. In Proceeding of the International Conference on Biomedical Engineering (ICBME'11) (pp. 58-65).
[25] Bera, T. K., & Nagaraju, J. (2011). Electrical impedance spectroscopic studies on broiler chicken tissue suitable for the development of practical phantoms in multifrequency EIT. Journal of Electrical Bioimpedance, 2(1), 48-63.
[26] Tai, Y., Bera, T. K., Lubineau, G., & Yang, Z. (2017). Combining the converse humidity/resistance response behaviors of rGO films for flexible logic devices. Journal of Materials Chemistry C, 5(15), 3848-3854.
[27] Tai, Y., Bera, T. K., Yang, Z., & Lubineau, G. (2017). Leveraging a temperature-tunable, scale-like microstructure to produce multimodal, supersensitive sensors. Nanoscale.
[28] Webster, J. G. (1978). Medical Instrumentation-Application and Design. Journal of Clinical Engineering, 3(3), 306.
[29] Rush, S., & Driscoll, D. A. (1968). Current distribution in the brain from surface electrodes. Anesthesia & Analgesia, 47(6), 717-723.
[30] Ishijima, M. (1993). Monitoring of electrocardiograms in bed without utilizing body surface electrodes. IEEE Transactions on Biomedical Engineering, 40(6), 593-594.
[31] Kyle, U. G., Bosaeus, I., De Lorenzo, A. D., Deurenberg, P., Elia, M., Gómez, J. M., ... & Scharfetter, H. (2004). Bioelectrical impedance analysis—part II: utilization in clinical practice. Clinical nutrition, 23(6), 1430-1453.
[32] Kushner, R. F., & Schoeller, D. A. (1986). Estimation of total body water by bioelectrical impedance analysis. The American journal of clinical nutrition, 44(3), 417-424.
[33] Segal, K. R., Van Loan, M., Fitzgerald, P. I., Hodgdon, J. A., & Van Itallie, T. B. (1988). Lean body mass estimation by bioelectrical impedance analysis: a four-site cross-validation study. The American journal of clinical nutrition, 47(1), 7-14.
[34] COMSOL Multiphysics® Modeling Software, COMSOL Inc., USA.
[35] Lodish, H., Berk, A., Zipursky, S. L., Matsudaira, P., Baltimore, D., & Darnell, J. (2000). Molecular cell biology 4th edition. National Center for Biotechnology InformationÕs Bookshelf.
[36] Lukaski, H. C., Bolonchuk, W. W., Hall, C. B., & Siders, W. A. (1986). Validation of tetrapolar bioelectrical impedance method to assess human body composition. Journal of applied physiology, 60(4), 1327-1332.
[37] Jaffrin, M. Y., & Morel, H. (2008). Body fluid volumes measurements by impedance: A review of bioimpedance spectroscopy (BIS) and bioimpedance analysis (BIA) methods. Medical engineering & physics, 30(10), 1257-1269.
[38] Tengvall, M., Ellegård, L., Malmros, V., Bosaeus, N., Lissner, L., & Bosaeus, I. (2009). Body composition in the elderly: reference values and bioelectrical impedance spectroscopy to predict total body skeletal muscle mass. Clinical nutrition, 28(1), 52-58.
[39] Cox-Reijven, P. L., van Kreel, B., & Soeters, P. B. (2002). Accuracy of bioelectrical impedance spectroscopy in measuring changes in body composition during severe weight loss. Journal of Parenteral and Enteral Nutrition, 26(2), 120-127.
[40] Cox-Reijven, P. L., & Soeters, P. B. (2000). Validation of bio-impedance spectroscopy: effects of degree of obesity and ways of calculating volumes from measured resistance values. International journal of obesity, 24(3), 271.
[41] Bartok, C., & Schoeller, D. A. (2004). Estimation of segmental muscle volume by bioelectrical impedance spectroscopy. Journal of Applied Physiology, 96(1), 161-166.
[42] Toyoda, K., Tsenkova, R. N., & Nakamura, M. (2001). Characterization of osmotic dehydration and swelling of apple tissues by bioelectrical impedance spectroscopy. Drying Technology, 19(8), 1683-1695.
[43] Sun, T. (2007). Single cell impedance spectroscopy (Doctoral dissertation, University of Southampton).
[44] Sun, T., Gawad, S., Bernabini, C., Green, N. G., & Morgan, H. (2007). Broadband single cell impedance spectroscopy using maximum length sequences: theoretical analysis and practical considerations. Measurement Science and Technology, 18(9), 2859.
[45] Bauchot, A. D., Harker, F. R., & Arnold, W. M. (2000). The use of electrical impedance spectroscopy to assess the physiological condition of kiwifruit. Postharvest Biology and technology, 18(1), 9-18.
[46] Masot, R., Alcañiz, M., Fuentes, A., Schmidt, F. C., Barat, J. M., Gil, L., ... & Soto, J. (2010). Design of a low-cost non-destructive system for punctual measurements of salt levels in food products using impedance spectroscopy. Sensors and Actuators A: Physical, 158(2), 217-223.
[47] Niu, J., & Lee, J. Y. (2000). A new approach for the determination of fish freshness by electrochemical impedance spectroscopy. Journal of food science, 65(5), 780-785.

[48] Hollaender, J. (1997). Rapid assessment of food/package interactions by electrochemical impedance spectroscopy (EIS). Food Additives & Contaminants, 14(6-7), 617-626.
[49] Li, X., Toyoda, K., & Ihara, I. (2011). Coagulation process of soymilk characterized by electrical impedance spectroscopy. Journal of Food Engineering, 105(3), 563-568.
[50] Scandurra, G., Tripodi, G., & Verzera, A. (2013). Impedance spectroscopy for rapid determination of honey floral origin. Journal of Food Engineering, 119(4), 738-743.
[51] K'Owino, I. O., & Sadik, O. A. (2005). Impedance spectroscopy: a powerful tool for rapid biomolecular screening and cell culture monitoring. Electroanalysis, 17(23), 2101-2113.
[52] Rahman, A. R. A., Register, J., Vuppala, G., & Bhansali, S. (2008). Cell culture monitoring by impedance mapping using a multielectrode scanning impedance spectroscopy system (CellMap). Physiological measurement, 29(6), S227.
[53] Soley, A., Lecina, M., Gámez, X., Cairo, J. J., Riu, P., Rosell, X., ... & Godia, F. (2005). On-line monitoring of yeast cell growth by impedance spectroscopy. Journal of biotechnology, 118(4), 398-405.
[54] Ehret, R., Baumann, W., Brischwein, M., Schwinde, A., Stegbauer, K., & Wolf, B. (1997). Monitoring of cellular behaviour by impedance measurements on interdigitated electrode structures. Biosensors and Bioelectronics, 12(1), 29-41.
[55] Kyle, A. H., Chan, C. T., & Minchinton, A. I. (1999). Characterization of three-dimensional tissue cultures using electrical impedance spectroscopy. Biophysical journal, 76(5), 2640-2648.
[56] Han, A., Yang, L., & Frazier, A. B. (2007). Quantification of the heterogeneity in breast cancer cell lines using whole-cell impedance spectroscopy. Clinical cancer research, 13(1), 139-143.
[57] Wilkinson, B. A., Smallwood, R. H., Keshtar, A., Lee, J. A., & Hamdy, F. C. (2002). Electrical impedance spectroscopy and the diagnosis of bladder pathology: a pilot study. The Journal of urology, 168(4), 1563-1567.
[58] Keshtkar, A., Keshtkar, A., & Smallwood, R. H. (2006). Electrical impedance spectroscopy and the diagnosis of bladder pathology. Physiological Measurement, 27(7), 585.
[59] Baarends, E. M., Lichtenbelt, W. V. M., Wouters, E. F. M., & Schols, A. M. W. J. (1998). Body-water compartments measured by bio-electrical impedance spectroscopy in patients with chronic obstructive pulmonary disease. Clinical Nutrition, 17(1), 15-22.
[60] Smallwood, R. H., Keshtkar, A., Wilkinson, B. A., Lee, J. A., & Hamdy, F. C. (2002). Electrical impedance spectroscopy (EIS) in the urinary bladder: the effect of inflammation and edema on identification of malignancy. IEEE transactions on medical imaging, 21(6), 708-710.
[61] Halter, R. J., Hartov, A., Heaney, J. A., Paulsen, K. D., & Schned, A. R. (2007). Electrical impedance spectroscopy of the human prostate. IEEE Transactions on Biomedical Engineering, 54(7), 1321-1327.
[62] Bhavsar, K., Fairchild, A., Alonas, E., Bishop, D. K., La Belle, J. T., Sweeney, J., ... & Joshi, L. (2009). A cytokine immunosensor for Multiple Sclerosis detection based upon label-free electrochemical impedance spectroscopy using electroplated printed circuit board electrodes. Biosensors and Bioelectronics, 25(2), 506-509.
[63] Kun, S., Peura, R. A., & Ristic, B. (1998). U.S. Patent No. 5,807,272. Washington, DC: U.S. Patent and Trademark Office.
[64] Li, N., Brahmendra, A., Veloso, A. J., Prashar, A., Cheng, X. R., Hung, V. W., ... & Kerman, K. (2012). Disposable immunochips for the detection of Legionella pneumophila using electrochemical impedance spectroscopy. Analytical chemistry, 84(8), 3485-3488.
[65] Küttel, C., Nascimento, E., Demierre, N., Silva, T., Braschler, T., Renaud, P., & Oliva, A. G. (2007). Label-free detection of Babesia bovis infected red blood cells using impedance spectroscopy on a microfabricated flow cytometer. Acta tropica, 102(1), 63-68.
[66] Derkus, B., Emregul, E., Yucesan, C., & Emregul, K. C. (2013). Myelin basic protein immunosensor for multiple sclerosis detection based upon label-free electrochemical impedance spectroscopy. Biosensors and Bioelectronics, 46, 53-60.
[67] Mohr, P., Birgersson, U., Berking, C., Henderson, C., Trefzer, U., Kemeny, L., ... & Reinhold, U. (2013). Electrical impedance spectroscopy as a potential adjunct diagnostic tool for cutaneous melanoma. Skin Research and Technology, 19(2), 75-83.
[68] Reddy, J. N. (1993). An introduction to the finite element method (Vol. 2, No. 2.2). New York: McGraw-Hill.
[69] Strang, G., & Fix, G. J. (1973). An analysis of the finite element method (Vol. 212). Englewood Cliffs, NJ: Prentice-hall.
[70] Bera, T. K., & Nagaraju, J. (2009, March). A FEM-based forward solver for studying the forward problem of electrical impedance tomography (EIT) with a practical biological phantom. In Advance Computing Conference, 2009. IACC 2009. IEEE International (pp. 1375-1381). IEEE.
[71] Šolin, P. (2006). Partial differential equations and the finite element method. Wiley-Interscience.
[72] Bera, T. K., & Nagaraju, J. (2009, June). A reconfigurable practical phantom for studying the 2-D electrical impedance tomography (EIT) using a FEM based forward solver. In International Conference on Biomedical Applications of Electrical Impedance Tomography.

IOP Conf. Series: Materials Science and Engineering **331** (2018) 012006 doi:10.1088/1757-899X/331/1/012006

ComVisMD - compact visualization of multidimensional data: experimenting with cricket players data

Shridhar B Dandin[1] and Mireille Ducassé[2]

[1]BKBIET-Pilani 333031, Rajasthan, INDIA.
[2]IRISA-INSA Rennes, FRANCE

[1]shridhar.dandin@bkbiet.ac.in
[2]mireille.ducasse@irisa.fr

Abstract. Database information is multidimensional and often displayed in tabular format (row/column display). Presented in aggregated form, multidimensional data can be used to analyze the records or objects. Online Analytical database Processing (OLAP) proposes mechanisms to display multidimensional data in aggregated forms. A choropleth map is a thematic map in which areas are colored in proportion to the measurement of a statistical variable being displayed, such as population density. They are used mostly for compact graphical representation of geographical information. We propose a system, ComVisMD inspired by choropleth map and the OLAP cube to visualize multidimensional data in a compact way. ComVisMD displays multidimensional data like OLAP Cube, where we are mapping an attribute *a* (first dimension, e.g. year started playing cricket) in vertical direction, object coloring based on *b* (second dimension, e.g. batting average), mapping varying-size circles based on attribute *c* (third dimension, e.g. highest score), mapping numbers based on attribute *d* (fourth dimension, e.g. matches played). We illustrate our approach on cricket players data, namely on two tables *Country* and *Player*. They have a large number of rows and columns: 246 rows and 17 columns for players of one country. ComVisMD's visualization reduces the size of the tabular display by a factor of about 4, allowing users to grasp more information at a time than the bare table display.

1. Introduction

In present age, vast amount of data is available at our disposal by the introduction of various Information Systems or Information Processing Devices. In general, the information in a database is in tabular format, in which values are provided by means of rows (records) and columns (fields). A data visualization system can be developed augmenting the tabular representation of information as a graphical representation of results to give or reveal more meaning from the data stored in a database of an Information System.

Map can also be used as information visualization tool for presenting final results. Mapping data enables to notice new trends. For example, in [1] readers can detect *scattergram* (a classification method) to map *percentage of rural population* and *percentage of population under the age of 18*. A type of map called choropleth map presents geographical data in numerical values using colors. It is popular for cartographic presentations.

IOP Conf. Series: Materials Science and Engineering **331** (2018) 012006 doi:10.1088/1757-899X/331/1/012006

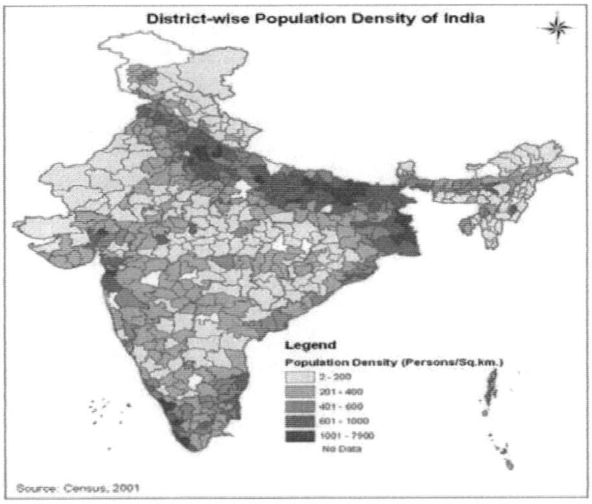

Figure 1: Choropleth Map, an example

OLAP is an acronym for OnLine Analytical Processing, which is a computer-based technique of analyzing data to look for insights[1]. The term cube refers to a multidimensional data set, which is also sometimes called a hyper cube if the number of dimensions is greater than 3. Each cell of the cube holds a number that represents some measure such as sales, profits, expenses, budget, etc.

We propose a compact data visualization system *ComVisMD* inspired by choropleth map and the OLAP cube. The choropleth maps we are using for visualization have dynamic visualization of multidimensional data. The data for illustrating our mapping are taken from the tables of the database of the sports domain **cricket**.

2. Choropleth Map

A type of map called choropleth map presents geographical data using colors. A choropleth map is a thematic map in which areas are shaded or patterned in proportion to the measurement of the statistical variable being displayed on the map, such as population density or per-capita income[2]. The data used in choropleth map can be classified as equal intervals or geometric progressions or standard deviation. When using color progressions, generally, five to seven color categories are recommended. The map user should be able to easily identify the implied magnitude of the hue and to match it with the legend.

A choropleth map example, shown in Figure 1, represents population density i.e. persons per square kilometre of India, based on 2001 census, district-wise. Straightaway one can see highly populated regions just by observing the map, they are either near sea or hill-regions. As identified in [2] choropleth maps are extensively used for data presentation. Any of the patterns of the map can expose some important feature of the data.

3. Cricket players data

For experimenting in *ComVisMD*, we are considering data of cricket playing countries and their players. We are considering the countries and players playing only One Day International (ODI) cricket matches. The data of countries and their players (as on March 2017) are taken from Wikipedia web page[3]. Similarly, all other 15 countries data are taken. In terms of relational tables, first relation is: *Country (cid, countryName, playingSince, matches, won, lost, tied, noResult, winPer, roundWinPer)*. The second relation is: *Player (Pid, Player, First, Last, Matches, Innings, NO, Runs, HS, Avg, Balls, Maidens, Runsg, Avgb, Wkt, Catches, Stumpings)*.

[1] en.wikipedia.org/wiki/OLAP_cube
[2] en.wikipedia.org

Pid	Player	First	Last	Matches	Innings	NO	Runs	HS	Avg	Balls	Maidens	Runs	Wkt	Avg
1	Syed Abid Ali	1974	1975	5	3	0	93	70	31	336	10	187	7	26.71
2	Bishan Singh Bedi	1974	1979	10	7	2	31	13	6.2	590	17	340	7	48.57
3	Farokh Engineer	1974	1975	5	4	1	114	54*	38	0	0	0	0	0
4	Sunil Gavaskar	1974	1987	108	102	14	3092	103*	35.13	20	0	25	1	25
5	Madan Lal	1974	1987	67	35	14	401	53*	19.09	3164	44	2137	73	29.27
6	Sudhir Naik	1974	1974	2	2	0	38	20	19	0	0	0	0	0
7	Brijesh Patel	1974	1979	10	9	1	243	82	30.37	0	0	0	0	0
8	Eknath Solkar	1974	1976	7	6	0	27	13	4.5	252	4	169	4	42.25
9	Srinivas Venkataraghavan	1974	1983	15	9	4	54	26*	10.8	868	7	542	5	108.4
10	Gundappa Viswanath	1974	1982	25	23	1	439	75	19.95	0	0	0	0	0
11	Ajit Wadekar	1974	1974	2	2	0	73	67	36.5	0	0	0	0	0
12	Gopal Bose	1974	1974	1	1	0	13	13	13	66	2	39	1	39
13	Ashok Mankad	1974	1974	1	1	0	44	44	44	35	0	47	1	47
14	Mohinder Amarnath	1975	1989	85	75	12	1924	102*	30.53	2730	17	1971	46	42.84
15	Anshuman Gaekwad	1975	1987	15	14	1	269	78*	20.69	48	0	39	1	39
16	Karsan Ghavri	1975	1981	19	16	6	114	20	11.4	1033	12	708	15	47.2
17	Syed Kirmani	1976	1986	49	31	13	373	48*	20.72	0	0	0	0	0
18	Parthasarathy Sharma	1976	1976	2	2	0	20	14	10	0	0	0	0	0
19	Dilip Vengsarkar	1976	1991	129	120	19	3508	105	34.73	6	0	4	0	0
20	B. S. Chandrasekhar	1976	1976	1	1	1	11	11*	0	56	0	36	3	12
21	Pochiah Krishnamurthy	1976	1976	1	1	0	6	6	6	0	0	0	0	0
22	Sudhakar Rao	1976	1976	1	1	0	4	4	4	0	0	0	0	0
23	Surinder Amarnath	1978	1978	3	3	0	100	62	33.33	0	0	0	0	0

Table 1: Indian players playing ODI matches (partial display)

Here, we consider 16 countries that play One Day Internationals (ODIs) affiliated to International Cricket Board: Afghanistan, Australia, Bangladesh, Canada, England, India, Ireland, Kenya, New Zealand, Pakistan, Scotland, South Africa, Sri Lanka, United Arab Emirates, West Indies and Zimbabwe.

Table 1 shows part of the table of Indian players. It contains 17 columns (fields / attributes) and 216 objects or records. In Table 1 we can just see one-seventh of records or objects on part of screen. In order to view all records or objects we propose compact visualization. Our compact approach through *ComVisMD* is explained in the next section.

4. ComVisMD

In our system, *ComVisMD*, taking the inspiration of region and coloring of geographical data of choropleth map, we compactly present the multidimensional data. We designed a special mechanism, where we bring all the records on the part of the display screen and perform the task of analysis. Initially we define the terms used for compact view and then we discuss data scale construction (section 4.2).

4.1. Definitions

With reference to Figure 2, we define the different terms used for the compact visualization.

Map: Our map is a graphical display to represent all the records (objects) of a database table. All *n* number of records of a table *t* with reference to selected dimensions can be displayed on the map.

Row: A row on the map in horizontal direction displays *n* number of records from the table based on classification with respect to values of an attribute *a*. For example, the second row of map represents 26 players who started playing ODIs in one of the years between 2010 and 2014. The attribute *a* is *playingSince*.

Square: A square in a row represents a complete record *rec* of a database table, displaying a summary computed according to the given attributes, *b* and *c*. For example, the attribute *b* is Runs, the number of runs scored by the player. The attribute *c* is matches, the number of ODI matches played by a player. In each square are displayed,

- a color: it is an indicator based on attributes *b* and *c* to visualize a record *rec* according to a 5-value color scale (see section 4.2). It indicates significance of an attribute. For example, *batting average* based on the attributes, *Runs* and *matches*.
- a drawn circle: its size is proportional to the value of an attribute *d*. It represents magnitude of an attribute. For example *HS*, highest score from all played innings of a player.
- a number: the value of an attribute *c*, for example matches, the number of matches played.

[3]For example, https://en.wikipedia.org/wiki/List_of_India_ODI_cricketers
and https://en.wikipedia.org/wiki/List_of_One_Day_International_cricket_records

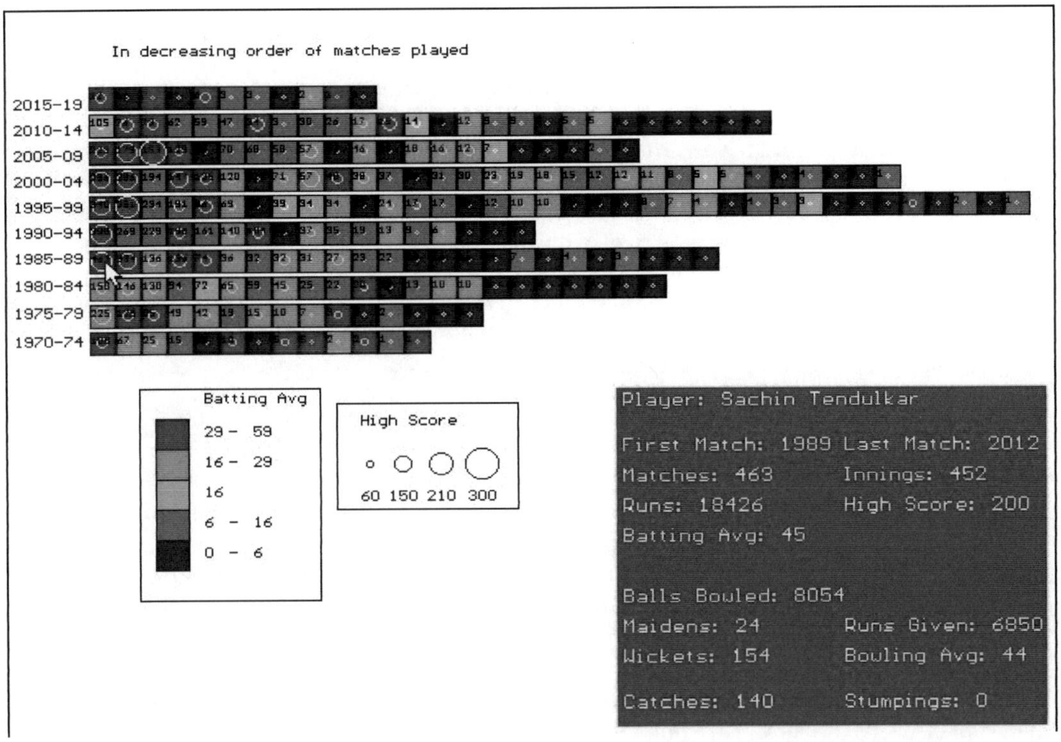

Figure 2: Indian Cricketers

For example, in the first square of the seventh row represents the player named Sachin Tendulkar, one of the players from India who has started to play in the year 1989, his batting average is in best slot (red), he has good high score (large circle but not the largest), the number of matches played are 463.

Figure 2 illustrates compact visualization of data from Indian players table. The 216 records of the *Player* table are first classified on a selected attribute *playingSince* to represent all of them on the map. These squares are distributed in 10 classes (rows), each class is a group of 5 years. If a player started playing ODI in one of these years, the square representing that player (object / entity) is displayed in the respective row.

Blue and light blue color squares signify least important object/s, red and light red color squares signify highly important record/s whereas gray color squares signify average object/s. For example, if the dynamically calculated scale is: 29-59, 16-29, 16, 6-16, 0-6 then 16 represents break point, average or middle value. The upper two ranges are excellent values with respect to the batting average whereas the lower two ranges are least interested values. In this way the objects or records of table are analyzed for significance with respect to an average value.

The size of circle indicates magnitude of the highest score scored by the player. The largest circle indicates largest high score among players of a country where as the smallest circle indicates lowest high score.

4.2. Dynamic scale construction

A scale is required to classify each record of the table and assign a color. We use divergent color schemes [5] to represent the break point. The break point is the mean value of the data values. With reference to this break point the multidimensional data records of the table can be first classified into three groups: average records, above average records and below average records. Again, based on break point of group of records above and below average are further grouped into two groups resulting in total 5 classes of records.

First, the break point is calculated by finding the average (mean) of n values of the table column e.g. *Avg*. Let us call it *bp*. Second, minimum and maximum values of the column are evaluated. Let us call them *min* and *max* respectively. Third, the average of columns values between *min* and *bp* is calculated. Let us call it *bpmin*. Fourth, the average between the columns values *bp* and *max* is calculated. Let us call it *bpmax*. Now, finally, five ranges of values for the classification are: *bpmax* to *max*; *bp* to *bpmax*; *bp*; *bpmin* to bp; *min* to *bpmin*.

The multidimensional data records can be classified on the basis of the dynamic scale constructed. Now using this dynamic scale, the records are clustered into 5 groups, each group represented by means of a chosen color. For example, upper two ranges by red and light red color; lower two ranges by light blue and blue color; the break point value records represented by gray color.

4.3. Discussion

In Figure 2, in total, 216 records / objects of *Player* table are displayed in the form 216 squares. The left most labels of each row show the period of years like 1970-1974, 1975-1979, and so on up to 2015-2019. This refers to the classification of the players from a country. The players started playing (represented by the field *playingSince* of the *Player* table) ODI between 1970 and 1974 (inclusive of both the years) is classified under the group, 1970-74. Similarly, all other players are classified and grouped under one of these 10 classes. In the top row represented by class 2015-19, 11 players are grouped, means 11 players started playing ODIs in one of the years of this range. Similarly, 26, 21, 31, 36, 17, 24, 22, 15 and 13 players are grouped under 2010-14, 2005-09, 2000-04, 1995-99, 1990-94, 1985-89, 1980-84, 1975-79 and 1970-74 respectively.

The decreasing order of displaying the players (objects) in respective class is based on the number shown on the square. This number represents the number of ODI matches played by the player. For example, in the first row, the numbers on squares starting from the first are: 12, 11, 7, 6, 6, 3, 3, 3, 2, 1 and 1. They represent the matches played by Manish Pandey, Jasprit Bumrah, Hardik Pandya, Barinder Sran, K L Rahul, Rishi Dhawan, Gurkeerat Singh, Yazuvendra Chahal, Karun Nair, Faiz Fazal and Jayant Yadav respectively.

The legends Batting Average (represented by the field, *Avg*) and the High Score (represented by *HS*) of the players are shown in Figure 2. In the first legend, the blue colored square represents batting average between 0 and 6. The light blue colored square represents the batting average between 6 and 16. The gray colored square represents the batting average (mean value) 16, light red colored square represents batting average between 16 and 29, and the last dark red colored square represents the batting average between 29 and 59. In the second legend, the circle (displayed on each square) represents highest score of the player. If the size of the circle is small, then the highest score is low and increase in the size of the circle indicates more higher highest score of a player. The largest circle represents the largest highest score of a player or players.

With reference to the Figure 2, 4 players in top row first, third, fifth and tenth (Manish Pandey, Hardik Pandya, K L Rahul and Faiz Fazal respectively) are represented by red color squares. Similarly, 4, 4, 6, 4, 3, 4, 1, 3 and 6 number of squares are in red color on second, third, fourth, fifth, sixth, seventh, eighth, ninth and tenth rows respectively. It indicates that out of 216 players, these 39 players have the batting average between 23 and 59. Analysis is these are better players than the others. Similarly, 51 players have the batting average between 16 and 29, represented by light red color. 8 players have batting average 16 equal to the mean value (break point) represented by gray color. 55 players have the batting average between 6 and 16, represented by light blue color. 63 players have batting average between 0 and 6 represented by blue color, those are least significant objects (players). Observer of the map can focus on only the best players looking at red colored squares very easily among all 216 objects.

With reference to again the Figure 2, the squares having small circles indicate that these players have low highest scores. Similarly, the squares having big circles clearly indicate larger higher scores, and analysis is these players are performing well.

When the user clicks left mouse button keeping mouse pointer on a square, the object details are displayed in a rectangle below the map for final view of selected player, with same background color of the player-square. On click of the right button of the mouse, the user goes back to the first map (cross reference), *Country* map, basically which is the initial point of the system *ComVisMD* for compact visualization.

In this way, the user of the compact visualization system *ComVisMD* can analyse the map to produce different meanings. We can pick few interesting players among a large number of players and display further details of only those interesting players to know more about them one-by-one.

5. Related work

In [4] a set of concepts are organized into a multidimensional cube. The cube focuses on a given subset which is defined by a query. An attribute used in a query is a dimension. This approach of dimensions is inspiration for the dimensions of our system, *ComVisMD*. In the system VisDB [6] users specify queries and an interface permits to use the visualization. Each pixel of the screen is used to visualize results. In *ComVisMD*, the queries are not specified by the users and to visualize multidimensional data, objects are represented on a map as colored squares. In Visage [7], bar charts are used to display attributes and objects are dragged for visualization. We bring all objects simultaneously and through dynamic scaling squares (representing objects) are colored for analysis. The authors in [3] were inspired by word-cloud, we are inspired by choropleth map. They claim that largest molecule structure catches the eyes of the observer. In *ComVisMD*, the 5-colored objects are caught by the observer for the analysis.

6. Conclusion

ComVisMD, compact visualization system for multidimensional data, focuses on compact and multidimensional data visualization. It uses dynamic scaling for coloring squares representing objects. The system displays all objects of the table, moving across tables. Through *ComVisMD* we have contributed 4-dimension visualization and object comparison without query. In the experiment on cricket players' data the compacting factor was about 4.

References

[1] Leonowicz A 2003 Research on two-variable choropleth maps as a method for portraying geographical relationships *Proceedings of the 21st International Cartographic Conference Durban-SA* 2290–2296.

[2] Andrienko G, Andrienko N and Savinov A 2001 Choropleth maps: classification revisited *ResearchGate.*

[3] Ertl P and Rohde B 2012 The molecule cloud - compact visualization of large collections of molecules *Journal of Cheminformatics.*

[4] Ferr´e S, Allard P and Ridoux O 2012 Cubes of concepts: multi-dimensional exploration of multi-valued contexts *International conference on formal concept analysis-Springer* 112–127.

[5] Horrower M and Brewer C A 2003 Colourbrewer.org: An online tool for selecting colour schemes for maps *The Cartographic Journal* **40** 27–37.

[6] Keim D A and Kriegel H P 1994 Visdb: database exploration using multidimensional visualization *IEEE Computer Graphics Applications* 40–49.

[7] Roth S F, Lucas P, Senn J A, Gomberg C C, Burks M B, Stroffolino P J, Kolojejchick J A and Dunmire C 1996 Visage: a user interface environment for exploring information *In proceedings Information Visualization-IEEE* 3–12.

IOP Conf. Series: Materials Science and Engineering **331** (2018) 012007 doi:10.1088/1757-899X/331/1/012007

Degree of Approximation by a General C_λ- Summability Method

S Sonker, A Munjal

Department of Mathematics, National Institute of Technology Kurukshetra, Kurukshetra-136119, Haryana, India
Email: alkamunjal8@gmail.com

Abstract. In the present study, two theorems explaining the degree of approximation of signals belonging to the class Lip(α, p, w) by a more general C_λ- method (Summability method) have been formulated. Improved estimations have been observed in terms of $\lambda(n)$; where $(\lambda(n))^{-\alpha} \leq n^{-\alpha}$ for $0 < \alpha \leq 1$ as compared to previous studies presented in terms of n. These estimations of infinite matrices are very much applicable in solid state physics which further motivates for an investigation of perturbations of matrix valued functions.

1. Introduction

Let *f* be a 2π-periodic function, integrable over the interval $[-\pi, \pi]$. Let the trigonometric Fourier series associated with *f* be

$$s_n(f;x) = \frac{a_0}{2} + \sum_{k=1}^{n} (a_k coskx + b_k sinkx), \quad \forall \ n \geq 1 \ with s_0(f;x) = \frac{a_0}{2}, \tag{1}$$

denotes the $(n+1)^{th}$ partial sums, called trigonometric polynomials of degree (or order) n, of the Fourier series of f. The conjugate series of the Fourier series of f is defined by $\sum_{n=1}^{\infty}(b_n cosnx - a_n sinnx) = \sum_{n=0}^{\infty} v_n$ and its nth partial sum is defined as

$$\tilde{s}_n(f;x) = \sum_{k=1}^{n} (b_k coskx - a_k sinkx), \forall n \geq 1 \ and \tilde{s}_n(f;x) = 0, \tag{2}$$

where,

$$a_k = \frac{1}{\pi} \int_{-\pi}^{\pi} f(x) \cos kx \, dx, \ k = 0,1,2,... \quad and \ b_k = \frac{1}{\pi} \int_{-\pi}^{\pi} f(x) \sin kx \, dx \ , k = 1,2,3,... \tag{3}$$

1.1. Approximation of Functions.

The problem of approximation of a function is to select a function among a well-defined class that closely matches (approximates) a target function. The need for function approximations arises in many branches of mathematical, physical and engineering science. Approximation theory is the branch of analysis that investigates the operation of certain known signals (functions) approximated by a specific class of functions (for example, polynomials, Fourier series etc.) that often have desirable properties (inexpensive computation, continuity, integral and limit values, etc.). Depending on the structure of the domain and co-domain of the problem, several techniques for approximating the function may be applicable.

A real-valued continuous function defined on a bounded interval on the real line can be well approximated arbitrarily using the sup-norm if the order of the polynomial goes to infinity. Even functions that have a discontinuity can be well approximated arbitrarily, if another norm is used instead of the sup-norm. Taylor series expansion is used for the polynomial approximation of the function and the quality of the approximation depends on the number of terms taken.

Approximations of a real-valued continuous function *f*(*x*) by an approximation function *T*(*x*) (by Fourier series, orthogonal series) have a finite number of terms. The main aim of this study is to solve the engineering problem (a set of sparse or noisy training data points) by approximation of the continuous function using the summability theory and to find the degree of approximation with minimized error. Based on the developed theorems, some theoretical explanations have been given to estimate the degree of approximation and to minimize the error in data.

The major items in an approximation problem are the type of the approximating function applied and measure of the goodness of an approximation. This is also known as the question of choosing the *form* and *norm*. The choice of the approximating function (form) is more important than the choice of a measure of goodness, that is, a distance function or norm that measures the distance between f and f_a.

1.2. Trigonometric Fourier series approximation with the help of summability method (Cesàro sub-methods).

Let f be 2π-periodic (bounded, integrable) and $f \in L_p[0, 2\pi] = L_p$ for $p \geq 1$. Then,

$$s_n(f;x) = \frac{1}{2}a_0 + \sum_{k=1}^{n}(a_k \sin kx + b_k \cos kx) \equiv \sum_{k=1}^{n} u_k(f;x) \quad (4)$$

is the partial sum of the first $(n+1)^{th}$ terms of the Fourier series of $f \in L_p (p \geq 1)$ at a point x, then it is known that

$$s_n(f;x) - f(x) = \frac{2}{\pi}\int_0^{\pi} \emptyset_x(t) \frac{\sin(n+1/2)t}{2\sin(t/2)} dt \quad (5)$$

where
$$\emptyset_x(t) = \frac{1}{2}\{ f(x+t) + f(x-t) - 2f(t)\}.$$
(6)

A measurable 2π -periodic function w: $[0, 2\pi] \rightarrow [0, \infty]$ is said to be a weight function if the set $w^{-1}(\{0,\infty\})$has the Lebesgue measure zero and $f \in L_w^p([0,2\pi])$ is the weighted Lebesgue space of all measurable 2π-periodic functions if

$$\|f\|_{p,w} = \left(\int_0^{2\pi} |f(x)|_p w(x)dx\right)^{1/p} < \infty. \quad (7)$$

A weight function w belongs to the Muckenhoupt class $A_p(1 < p < \infty)$, if

$$\sup \left(\frac{1}{|I|}\int_I w(x)dx\right)\left(\frac{1}{|I|}\int_I [w(x)]^{-1/(p-1)} dx\right)^{p-1} < \infty \quad (8)$$

where the supremum is over all intervals I with length $|I| \leq 2\pi$.

Let $f \in L_w^p$ and $w \in A_p(1 < p < \infty)$. Then modulus of continuity of the function f is defined by

$$\Omega(f,\delta)_{p,w} = \sup_{|h|\leq\delta} |\Delta_h(f)|_{p,w}, \quad \delta > 0, \quad (9)$$

where

$$\Delta_h(f(x)) = \frac{1}{h}\int_0^h |f(x+t) - f(x)|\, dt. \quad (10)$$

For $0 < \alpha \leq 1$, the Lipschitz class $Lip(\alpha,p,w)$ is give as,

$$Lip(\alpha,p,w) = \{f \in L_w^p : \Omega(f,\delta)_{p,w} = O(\delta^{\alpha}), \delta > 0\}. \quad (11)$$

Let $T \equiv (a_{n,k})$ be a lower triangular matrix with non-negative entries. For each n, the sequence $\{a_{n,k}\}$ is either non-increasing or non-decreasing in k, $0 \le k \le n$, the matrix T is said to have monotonic rows.

A positive sequence $c = \{c_n\}$ is called almost monotonically increasing (or decreasing) if there exists a constant $K=K(c)$, depending on the sequence c only, such that for all $n \ge m$,

$$c_n \ge K\, c_m, \qquad (or\ c_n \le K\, c_m).$$

Such sequences will be denoted by $c \in$ AMDS (or $c \in$ AMIS).

If $e \in$ AMIS (or $e \in$ AMDS), then the sequence e is almost monotonically increasing (or decreasing) mean sequence, denoted by $e \in$ AMIMS (or $e \in$ AMDMS) and have the following relations,

$$\text{AMIS} \subset \text{AMIMS}$$

and

$$\text{AMDS} \subset \text{AMDMS}.$$

Let E be an infinite subset of N and consider E as strictly increasing sequence of positive integers, say $E = \{\lambda(n)\}_{n=1}^{\infty}$. The Summability method C_λ is defined as

$$(C_\lambda(x))_n = \frac{1}{\lambda(n)} \sum_{k=1}^{\lambda(n)} x_k, \tag{12}$$

where $x = (x_k)$ is a sequence of real or complex numbers. It is clear that C_λ is regular for any λ. Thus, the C_λ-method yields a subsequence of the Cesàro method C_1. C_λ is obtained by deleting a set of rows from Cesàro matrix. The basic properties of C_λ-method can be found in [1, 18].

Armitage and Maddox [1] proved inclusion and tauberian results for the C_λ- method. Osikiewicz [6] expanded the work of the Armitage and Maddox by examining further inclusion properties of the C_λ- method for bounded sequences and its relationship to statistical convergence. Liendler [7] have dropped the monotonicity on generating sequence $\{p_n\}$ and Mittal *et al.* [8] used more general matrix and found some valuable results. In the present study, C_λ- method has been used for generalizing the result of Mittal [8].

Several investigators have studied the degree of approximation of a function using different summability methods of series [17, 4]. In a recent work carried out by Nayak *et al.* [9, 10], the rate of convergence of Fourier series in the generalized Holder metric by *DCM* and second-type *DAM* has been studied. A result on the degree of approximation of functions (signals) in Besov space by N*p*-matrix has also been established by Mohanty *et al.* [5].

In 1986, Bor [2] found the relationship between two summability techniques $|C, 1|_k$ and $|\bar{N}, p_n|_k$ and in [3], he generalized a theorem based on minimal set of sufficient conditions by using the $|\bar{N}, p_n|_k$ for infinite series. In 2016, Sonker and Munjal [11] determined a generalized theorem on absolute Cesàro summability with the sufficient conditions for infinite series and in [12], they used the concept of triangle matrices for obtaining the minimal set of sufficient conditions of an infinite series to be bounded. In 2017, Sonker and Munjal [13] found the approximation of the function $f \in Lip\,(\alpha, p)$ using infinite matrices of Cesàro sub-method and in [14], they obtained boundness conditions of absolute summability φ - $|A|_k$ factors. In this way by using the advanced summability method, we can improve the quality of the filters as in [15, 16].

2. Known Results

Mittal *et al.* [8] has generalized and improved the approximation results by dropping monotonicity on the elements of the matrix rows and proved the following theorem.

***Theorem 1* [8]:** *Let $f \in Lip(\alpha, p)$ and let $T = (a_n, k)$ be an infinite regular triangular matrix.*

(i) If $p > 1, 0 < \alpha < 1, \{a_{n,0}\} \in AMS$ in k and satisfies

$$(n+1)\, max\{a_{n,0}, a_{n,r}\} = O(1), \tag{13}$$

$where\ r := \left[\frac{n}{2}\right],\ then\ \|f - \tau_n(f)\|_p = O(n^{-\alpha})\ is\ satisfied.$

$$(ii)\ If\ p > 1, \quad \alpha = 1\ and \sum_{k=0}^{n-1}(n-k)|\Delta_k a_{n,k}| = O(1), or$$

$$(iii)\ If\ p > 1, \quad \alpha = 1\ and \sum_{k=0}^{n}|\Delta_k a_{n,k}| = O(a_{n,0}), \qquad or$$

$$(iv)\ If\ p = 1, \quad 0 < \alpha < 1\ and \sum_{k=0}^{n}\left|\Delta_k a_{n,k}\right| = O\left(a_{n,0}\right),$$

$and\ also\ (n+1)a_{n,0} = O(1), hold\ then\ \|f - \tau_n(f)\|_p = O(n^{-\alpha})\ is\ satisfied.$

3. Main Results

Many researchers have studied the error estimation $E_n(f)$ through trigonometric-Fourier approximation for the situations in which the summability matrix dropped its monotonicity. In 2017, Sonker and Munjal [13] found the approximation of the function f ϵ Lip (α, p) using infinite matrices of Cesàro sub-method. In the present work, some theorems describing the degree of approximation of signals belonging to the class *Lip(α, p, w)* have been explained.

Theorem 2: *Let* $f \in Lip(\alpha, p, w)$, *and* $T = (a_{\lambda(n),k})$ *be an infinite regular triangular matrix s.t.* $a_{n,k} \geq 0$ *and* $\sum_{k=1}^{n} a_{n,k} = 1, \forall\, n, k$. *If, for* p>1, $0 < \alpha < 1$,
(*i*) $\{a_{\lambda(n),k}\} \in AMDMS\ in\ k\ and$
(*ii*) $(a_{\lambda(n),0}) = O((\lambda(n)+1)^{-1})$ *satisfies,*

$$then\ \left\|f - \tau_n^{\lambda}(f)\right\|_{p,w} = O((\lambda(n))^{-\alpha}).$$

Theorem 3: $Let\ f \in Lip(\alpha, p, w), and\ let\ T = (a_{\lambda(n)}, k)\ be\ an\ infinite\ regular\ triangular\ matrix\ s.t.\ a_{n,k} \geq 0\ and\ \sum_{k=1}^{n} a_{n,k} = 1, \forall\, n, k.$ *If, for* p>1, $0 < \alpha < 1$,
(*i*) $\{a_{\lambda(n),k}\} \in AMIMS\ in\ k\ and$
(*ii*) $(a_{\lambda(n),0}) = O((\lambda(n)+1)^{-1})$ *satisfies,*

$$then\ \left\|f - \tau_n^{\lambda}(f)\right\|_{p,w} = O((\lambda(n))^{-\alpha}).$$

In this Approximation, the sharpness of the estimation is improved by using the AMIMS (AMDMS) and more general summability method and with the help of trigonometric polynomials.

4. Lemmas

The following lemmas are used in the proof of theorem 3 and 4.

Lemma 1: $If\ f \in Lip(\alpha, p)\ for\ 0 < \alpha \leq 1\ and\ p > 1, then$

$$\|f - s_n(f)\|_{p,w} = O(n^{-\alpha}).$$

Lemma 2: $Let\ matrix\ have\ AMS\ rows\ and\ satisfy$

$$(\lambda(n)+1)max\{a_{\lambda(n),0}, a_{\lambda(n),r}\} = O(1).$$

$Then, for\ 0 < \alpha < 1,$

$$\sum_{k=0}^{\lambda(n)} a_{\lambda(n),k}(k+1)^{-\alpha} = O((\lambda(n)+1)^{-\alpha}).$$

Proof: Suppose that the rows of the matrix are AMDS. Then there exists a K>0 such that

$$\sum_{k=0}^{\lambda(n)} a_{\lambda(n),k}(k+1)^{-\alpha} \leq \sum_{k=0}^{\lambda(n)} K a_{\lambda(n),0}(k+1)^{-\alpha}$$
$$= K a_{\lambda(n),0} \sum_{k=0}^{\lambda(n)} (k+1)^{-\alpha}$$
$$= O\left(a_{\lambda(n),0}(\lambda(n)+1)^{1-\alpha}\right)$$
$$= O(((\lambda(n)+1)^{1-\alpha}).$$

A similar argument applies if the rows of the matrix are AMIMS.

5. Proof of Main Theorem

If $\boldsymbol{p > 1, 0 < \alpha < 1}$.

$$\tau_n^\lambda(f) - f = \sum_{k=0}^{\lambda(n)} a_{\lambda(n),k}\, s_k^\lambda(f) - t_n^\lambda f + (t_n^\lambda - 1)f$$
$$= \sum_{k=0}^{\lambda(n)} a_{\lambda(n),k}\, (s_k^\lambda(f) - f) + (t_n^\lambda - 1)f.$$

Using Lemmas 1 and 2,

$$\left\|\tau_n^\lambda(f) - f\right\|_{p,w} \leq \sum_{k=0}^{\lambda(n)} a_{\lambda(n),k} \left\|(s_k^\lambda(f) - f)\right\|_{p,w} + \left|(t_n^\lambda - 1)\right| \|f\|_{p,w}$$
$$= \sum_{k=0}^{\lambda(n)} a_{\lambda(n),k}\, O((k+1)^{-\alpha}) + O\left(\left(\lambda(n)\right)^{-\alpha}\right)$$
$$= O((\lambda(n)+1)^{-\alpha}) + O((\lambda(n))^{-\alpha})$$
$$= O((\lambda(n))^{-\alpha}).$$

Thus, $\left\|\tau_n^\lambda(f;x) - f(x)\right\|_{p,w} = O((\lambda(n))^{-\alpha}).$

This completes the proof of theorem.

***Remark*:** If $\lambda(n) = n$, the methods $N_n^\lambda(f;x)$ and $R_n^\lambda(f;x)$ give us classical known Nörlund and Riesz means. For $\tau_n^\lambda(f;x)$, it coincides with Cesàro arithmetic mean.

6. Conclusion

In the present research work, the approximation degree of the function for generalization of the summability methods using the concept of AMIMS (AMDMS) has been established. The work is a motivation for the researchers, interested in studies of function with the help of trigonometric polynomials. By reducing the restrictions of the filter, the functions of the filter (like Automatic compensation, permanently unit power factor, overcome of unbalancing situation, etc.) have been improved. The result presented here has a direct application in electrical engineering, e.g. the electrical load signal can be represented as a summation of sinusoids with different frequency. Using the outcomes from the present research work, the output of the signals can be made stable, bounded and may be used to predict the behavior of the input data and the changes in the complete process.

Acknowledgement

The authors express their sincere gratitude to the Department of Science and Technology (India) for providing financial support to the second author under INSPIRE Scheme (Innovation in Science Pursuit for Inspired Research Scheme).

References

[1] Armitage D H and Maddox I J 1989 *Analysis* **9** (1-2) 195-204.
[2] Bor H 1986 *Proc. Amer. Math. Soc.* **98** (1) 81-84.
[3] Bor H 1986 *Proc. Amer. Math. Soc.* **118** (1) 71-75.
[4] Nigam H K and Sharma A 2010 *Int. J. Math. & Math.Sci.* **2010** (351016) 7 pages.
[5] Mohanty H, Das G and Ray B K 2011 *J. Orissa Math. Soc.* **30** (2) 13–34.
[6] Osikiewicz J A 2000 *Analysis* **20** (1) 35-43.
[7] Leindler L 2005 *Mathematical Analysis and Application* **302** (1) 129-136.
[8] Mittal M L, Rhoades B E, Sonker S and Singh U 2011 *Appl.Math.Comp.* **217** (9) 4483-4489.
[9] Nayak L, Das G and Ray B K 2014 *J. Math. Anal. Appl.* **420** (1) 563-575.
[10] Nayak L, Das G and Ray B K 2014 *Int. J. Anal.* **2014** 171675.
[11] Sonker S and Munjal A 2016 *Int. J. Math. Anal.* **10** (23) 1129-1136.
[12] Sonker S and Munjal A 2016 *Nonlinear Studies* **23** (4) 533-542.
[13] Sonker S and Munjal A 2017 *Nonlinear Studies* **24** (1) 113-125.
[14] Sonker S and Munjal A 2017 *Proceedings ICAST-2017 Type A* **67**, 208-210.
[15] Sonker S and Munjal A 2017 *Journal of Inequalities and Applications* **2017** (168) 1-7.
[16] Sonker S and Munjal A 2017 *Int.J. Engg. Tech.* **9** (3S) 457-462.
[17] Lal S and Nigam H K 2001 *Tamkang Journal of Mathematics* **32** 143–149.
[18] Deger U, Dagadur I and Kucukaslan M 2012 *Proc. Jangjeon Math. Soc.* **15** (2) 203-213.
[19] Degree of Approximation by a General $\boldsymbol{C_\lambda}$- Summability Method

Design and Fabrication of Multi Quantum well based GaN/InGaN Blue LED

K Meel[1], P. Mahala[2] and S Singh[3]

[1]Physics Department, BKBIET, CEERI road, Pilani-333031, Rajasthan, India
[2]Birla Institute of Technology and Science (BITS), Pilani-333031, Rajasthan, India
[3]Flexible and Non-Silicon Electronics Group, CSIR-CEERI, Pilani-333031, Rajasthan, India

krishna.meel@bkbiet.ac.in

Abstract. This paper presents the optimization of the multi-quantum well based Light Emitting Diode (LED) structure. We investigate the electrical and optical properties of the device on several factors like well width, barrier width, the number of quantum wells and then optimize the structure. The device is optimized for a well width and barrier width of 3nm and 6nm respectively, consisting of five quantum wells. Simulations were carried out using Silvaco ATLAS TCAD simulation program (Silvaco International, USA). The optimized structure was grown by MOCVD and fabricated. The I-V characteristic was also measured.

1. Introduction

Light Emitting Diodes have acquired a lot of attention towards Solid-State lighting due to their continuous improvement in efficiency and longer life time when compared to other lighting systems including Compact Fluorescent Lamp (CFL) bulbs, filament bulbs etc. [1]. LEDs are capable of producing light in the visible region ranging from violet to red. Technology has been developed for LEDs of red, violet, green, blue etc but for Solid-State lighting it is also important to concentrate on white lights due to their high-energy efficiency and reliable lifetime. The progress in phosphor-converted white LEDs has primarily been enabled by advances in the InGaN-based LEDs as excitation sources reported by several researchers [2, 3]. People have also reported Growth and Fabrication of GaN/InGaN Violet Light Emitting Diode on Patterned Sapphire Substrate [4]. People have also compared performance analysis of InGaN/GaN multi-quantum-well light-emitting diodes with p- and n-type step-doped barriers [5]. Technological developments have been made and are under process for producing white light using LEDs by several methods like 1) color mixing technology, 2) Phosphor coating on Ultra Violet LEDs and 3) Yellow Phosphor coating on Blue LEDs [6]. In this paper we concentrate on the third method of producing white light using Blue LED. Recently Singh et al. have compared Performances of p-side down vertical InGaN/GaN blue light-emitting diodes with chip size [7].

2. Device Structure

The structure has been considered to have epitaxial layers of Gallium Nitride and Indium Gallium Nitride material to be grown using Metal Organic Chemical Vapour Deposition on C-plane Sapphire substrate of thickness 300nm.

The basic structure consists of a buffer layer of GaN onto which a 2µm n-GaN layer is deposited followed by active region consisting of unintentionally doped GaN/InGaN layers and a p-GaN layer of thickness 0.2µm. In the

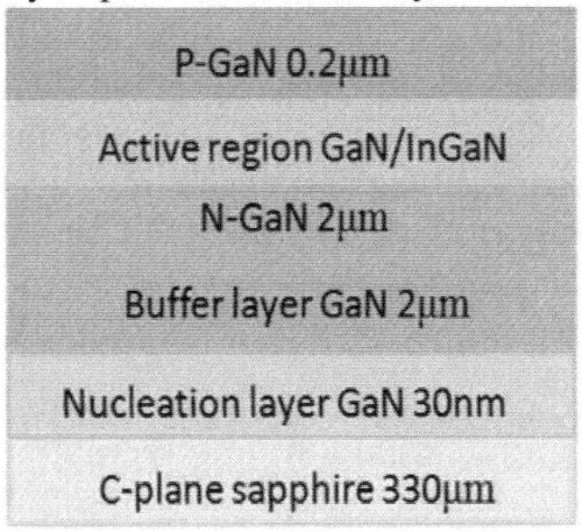

Figure 1. Device basic structure

figure 1 shows the schematic diagram of the basic structure.The doping concentrations are assumed to be $1x10^{19}cm^{-3}$and $5x10^{17}$ cm^{-3}in n-GaN and p-GaN respectively.

3. Structure Optimization

The active region consists of Multi-Quantum Well (MQW) structure in order to avoid and obtain an optimum carrier overflow and efficiency droop caused using a single quantum well structure or a double heterostructure. Hence there is a need for the optimization of multiple quantum well structure. The structure was optimized in a systematic manner starting from a number of quantum wells to well-width and barrier-width optimization. The optical and electrical properties of InGaN/GaN MQW LED are investigated numerically with Silvaco ATLAS TCAD simulation program (Silvaco International, USA) by extracting the integrated radiative recombination rate (radiative efficiency) and total integrated recombination rate [8]. Both optical and electrical characteristics of the device like Internal Quantum Efficiency (IQE) and turn-on voltage are examined carefully to obtain optimum results.

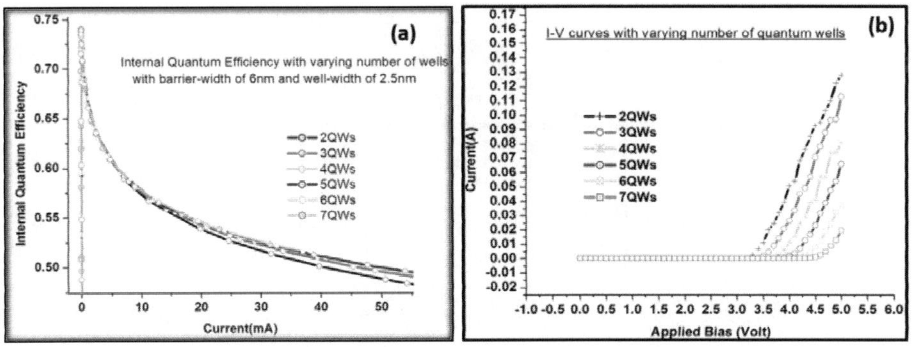

Figure 2. (a) Shows the variation of IQE with number of wells and (b) Shows the variation of I-V characteristics with number of wells.

From the above-shown results in figure 2, it is evident that five quantum wells are optimized. If we see the graph of IQE versus current density as we increase or decrease the number of quantum wells from five, we notice a decrease in efficiency with increasing current density, the reason may be attributed to carrier overflow or less number of carriers being captured resulting in radiative recombination. If the number of quantum wells is reduced the total number of carriers being captured in the well region is reduced and hence the radiative efficiency is reduced. This implies that one should have a high number of quantum wells to capture more and more carriers for recombination. But one can observe that as we increase the number of quantum wells above five efficiencies is again reduced, this may be attributed to the increase in nonradiative recombination rate caused due to the lower mobility of holes as compared to electrons which are not able to reach the well present on the n-GaN side. This results in a lower concentration of holes as compared to electron concentration as we move from p-GaN to n-GaN. Based on the same parameters well-width and barrier-width are also optimized.

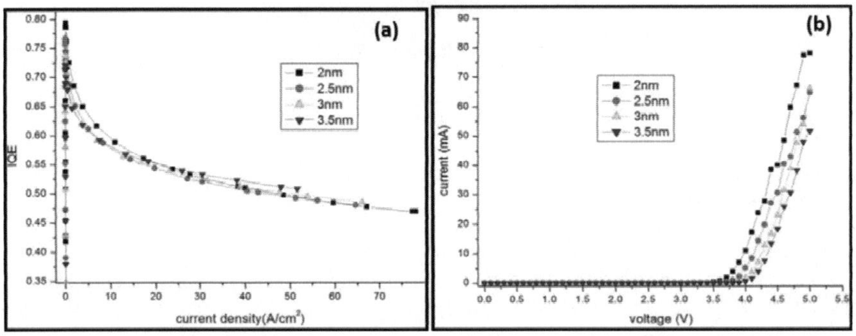

Figure 3. (a) Shows the variation of IQE with varying well-width. (b) Shows the variation in I-V curve with varying well-widths.

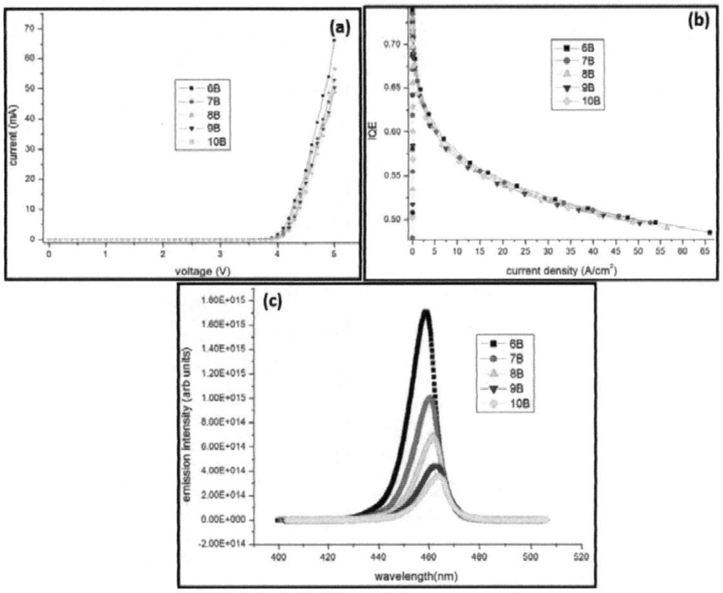

Figure 4. (a) Shows variation in I-V curve with varying barrier width. (b) Shows variation in IQE with barrier width. (c) Shows the effect on wavelength emission intensity with varying barrier width.

From the above IQE and I-V curves shown in figure 3, it is evident that a well-width of 3nm can be taken as optimized based on its low forward voltage and less droop in IQE as compared with less or higher well-width values. Also for the wider quantum width, the quantum confined stark effect becomes strong, the hole wave functions and electron wave functions are separated more, resulting in less electron and hole recombination. Therefore, the efficiency reduces with larger quantum well width. Also if the well thickness is very thin than carrier overflow increases that again results in low efficiency. Hence the well-width has been optimized to 3nm.

Similarly, as we fix the well width to 3nm and vary the barrier width we notice from figure 4 that with the increase in barrier width the emission intensity decreases and a slight increase in series resistance as obtained from I-V curve. When the thickness of barrier is thin enough, the wave functions of electron and hole would penetrate the adjacent well layer, resulting in inter-well transition. If the inter-well transition of the MQWs occurred, the Photo Luminous (PL) peak might be red-shifted because the band structure of MQW would be inclined due to the large piezoelectric field. If the barrier width is thinner enough, tunnelling current through the LED structure begins to dominate. Reduction of IQE due to tunnelling leakage of carriers from the quantum well to defect states in barriers.

Now if the width of barriers is increased, the overflowing carriers probably tend to diffuse into the barriers rather than going into the quantum wells. This results in a decrease in the intensity. Also as the thickness of the GaN barrier layer is increased, the abruptness of the interface between InGaN and GaN layers deteriorated, probably due to the generation of defects induced by the strain accumulation in the MQWs. Accordingly, the intensity of the line-width of the PL taken from the MQWs is reduced and broadened respectively with an increase of barrier thickness.

Therefore, the final structure has been optimized as five quantum wells with 3nm and 6nm well and barrier width respectively. Figure 5 shows the 1D schematic of the final optimized structure.

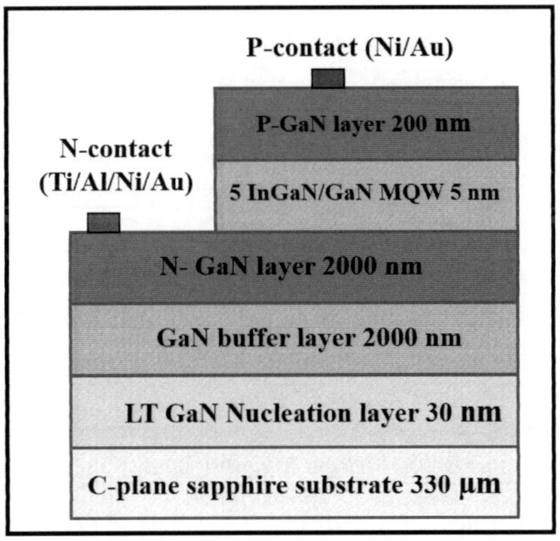

Figure 5. 1D optimized structure of blue LED

4. Fabrication

The above optimized epitaxial layers were grown by using MOCVD system. Then LED of multi finger structure with chip size (500μmx500μm) was fabricated. I-V characteristics of the chip were measured by using Keithley 4200 SCS measurement system. A uniform emission of blue light from the LED chip was observed. In figure 6 shows the measured I-V characteristics and blue light emission from a multi finger LED chip. Current-voltage characteristics of figure 6(a) shows a turn-on voltage of 2.8 V.

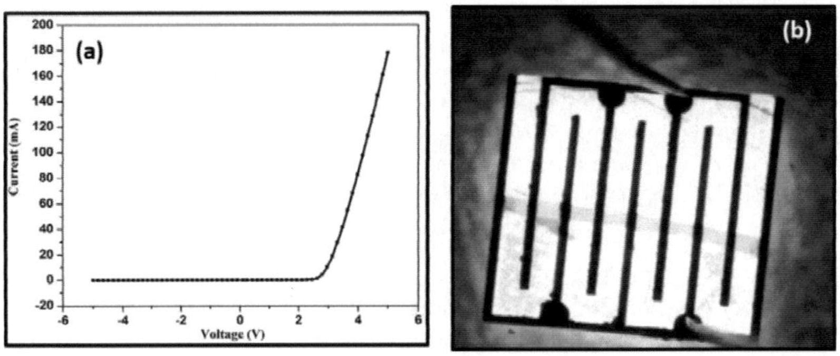

Figure 6. (a)shows the I-V curve measured for the planar LED chip (500 μm x 500 μm) (b)shows uniform blue light emission under biased conditions.

5. Conclusion

In conclusion, MQW GaN/InGaN based LED structure has been simulated by varying various parameters like number of quantum wells, well-width and barrier-width. The result shows that the structure is optimized with five quantum wells, well widths with 3nm and barrier width with 6 nm respectively. The optimized structure was then grown by MOCVD. The chip level fabrication was successfully done and its I-V characteristic was measured.

Acknowledgements

Authors are thankful to the Director, BKBIET, Pilani and Director, CSIR-CEERI, Pilani for their encouragement and support. Authors express their gratitude towards all the colleagues of CEERI who helped in experimental work.

References

[1] Held G, 2008 Introduction to light emitting diode technology and applications. *New York: CRC press, Auerbach publications.*
[2] Li X-H, Zhu P, Liu G, Zhang J, Song R, Ee Y-K, Kumnorkaew P, Gilchrist J F, and Tansu N 2013 *Journal of Display Technology* **9(5)**, 324-332.
[3] Zhu P, Liu G, Zhang J, and Tansu N 2013 *J. of Disp. Tech.* **9(5),** 317-323.
[4] Adhikari S, Patra S K, Lunia A, Kumar S, Parjapat P, Kushwaha B, Kumar P, Singh S, Chauhan A, Singh K, Pal S, Dhanavantri C 2014 *Journal of Applied Mathematics and Physics* **2**, 1113-1117
[5] Singh S, Utpalla P, Pal S, and Dhanavantri C 2016 *Journal of Computer Electronics* **15**, 1040–1045.
[6] Schubert E F 2006 Light-Emitting Diode *Cambridge University Press,* Cambridge, U.K.
[7] Singh S, Kumar S, Pal S, and Dhanavantri C 2017 *Optical Laser Technology* **95**, 165-171.
[8] Silvaco ATLAS TCAD user manual [www.silvaco.com].

IOP Conf. Series: Materials Science and Engineering **331** (2018) 012009 doi:10.1088/1757-899X/331/1/012009

Design and Implementation of Viterbi Decoder Using VHDL

Akash Thakur and Manju K Chattopadhyay

VLSI Design and Embedded System Lab, School of Electronics,
Devi Ahilya University, Indore, India
E-mail: akashth28@gmail.com

Abstract. A digital design conversion of Viterbi decoder for ½ rate convolutional encoder with constraint length k = 3 is presented in this paper. The design is coded with the help of VHDL, simulated and synthesized using XILINX ISE 14.7. Synthesis results show a maximum frequency of operation for the design is 100.725 MHz. The requirement of memory is less as compared to conventional method.

***Index Terms*—VD, VHDL, RTL, BMU, PMU, ACSU, SMU.**

1. Introduction

Encoders and decoders play an important role in the field of communication and data storage industries. There is a need of reliable system with least probability of error and high speed of operation. Error correcting coding [1] has algorithms for expressing a sequence of numbers, so that if any error due to noise arises, it can be detected and corrected (within certain limitations). These algorithms come into operation by converting them into digital design. The digital conversion will work at certain frequency. Recent developments have contributed towards achieving high reliability required by today's high-speed digital systems. The use of coding for error control has become an integral part in the design of modern communication and digital storage systems.

A typical transmission or storage system may be represented by the block diagram shown in figure1. Efficient error control code in encoder and decode can help in gaining the following advantages:

a. Power gain by coding in power/bit for cellular systems in digital communication.
b. Increase in storage capacity (no. of bits per sq. inch increases)

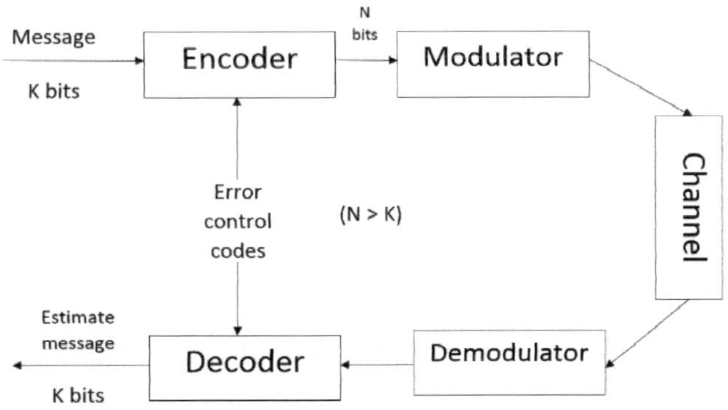

Figure 1. Simple Communication Model

Even though the concept of Viterbi algorithm is very old, it is still very much researched for its optimization. Jinjin He *et al.* [2] presented paper on low power design of Viterbi decoder without degrading the decoding speed. K. S Arunlal and S. A Hariprasad [3] shows the reduction in computational time and hardware requirement. The crucial point among the digital design conversion is the operating speed and the power reduction, which is presented by many researchers with variation in operating frequency and reduction in digital complexity [4-9]. D Vaithiyanathana *et al.* [10] proposed a modified pipelined architecture for ACS of Viterbi decoder and shows optimization in frequency of operation, up to 165 MHz with reduced area. The design proposed by us reduce the memory requirement for the trace back of the trellis.

2. Architecture

Viterbi Algorithm is an efficient way to implement the M L Decoding, for both hard and soft decision. It is easy to implement for small constraint length for convolutional encoders which is based on the trellis method. A J Viterbi proposed Viterbi algorithm in 1967 for convolutional codes. The basic architecture of the Viterbi Decoder (VD) based on Viterbi algorithm is shown in figure 2.

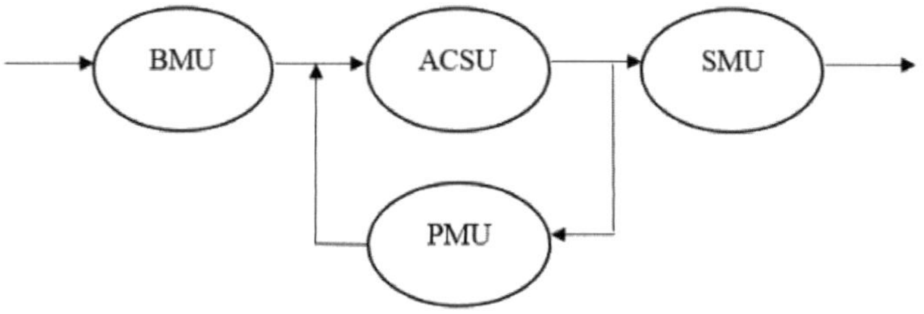

Figure 2. Block Diagram of Viterbi Algorithm

The main components of the VD are Branch Metric Unit (BMU), Path Metric Unit (PMU), Add-Compare and Select Unit (ACSU) and Survivor Management Unit (SMU). Viterbi Decoder decodes the convolutional encoded data on the basis of trellis. An intermediate state of the trellis for a Viterbi decoder of ½ rate convolutional encoder with constraint length k=3 is shown in figure 3.

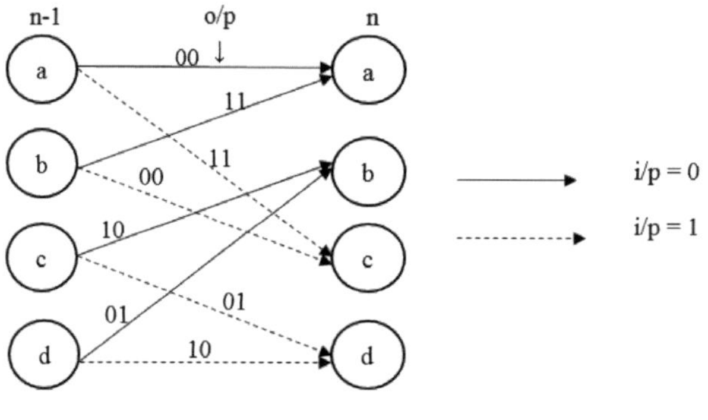

Figure 3. Intermediate trellis state of the decoder

A Branch Metric Unit-BMU calculates the branch metrics; An Add-compare-Select Unit -ACUS recursively accumulates the branch metrics. Path metrics (PM), compares the incoming path metrics, and makes a decision to select the most likely state transitions for each state of the trellis and generates the corresponding decision bits. Viterbi algorithm is a special case of Bellman Ford shortest distance algorithm.

3. Hardware Description

The proposed digital design of the Viterbi decoder for ½ rate convolutional encoder of constraint length of k=3 is shown in the figure 4. In this design, instead of storing the path metric for each stage, the decision is taken directly. Following are the components used in the decoder:

3.1 Branch Metric Unit (BMU)
Branch Metric (BM) calculation is the distance between received and estimated output code word. It is also called the Hamming distance, for e.g. if the received code word is 10 then the hamming distance with 00, 01, 10 and 11 will be 01, 10, 00, 01 respectively. This function is done by the BMU.

3.2 Add Compare & Select Unit (ACSU)
Path Metric (PM) calculation is done by ACSU in iterative manner. It is repeated for every encoder state. Previous path metric will be added to the corresponding branch metric, and the output will be sent to the comparator and select unit. Two paths ending in a given state are compared and the greater metric is dropped, whereas the smaller one is selected for further calculation, known as survivor path.

3.3 Path Metric Unit (PMU)
Path Metric Unit takes input from the output of the ACSU and store it for the next clock pulse. It gives input to ACSU for calculation of next PM. For state a, the path metric for time instant n will:

$$PM_{(a,n)} = \min \{ PM_{(a,n-1)}+BM_{(00)}, PM_{(b,n-1)}+BM_{(11)} \} \text{ .... (1)}$$

3.4 RAM
Random Access Memory is used in the design, to store the decisions made by the survivor unit.

3.5 Survivor Unit (SU)
Survivor unit takes the path metrics of all states and select the least of the path metrics. A decision will be made by selecting the least of the path metric of all states for particular time instant and it will be stored in RAM.

The design is coded in VHDL, figure 5 shows the block view and figure 6 shows the RTL view generated after successful synthesis of the Viterbi decoder. Simulation results are taken for input code – 11, 10, 10, 00, 01, 11.

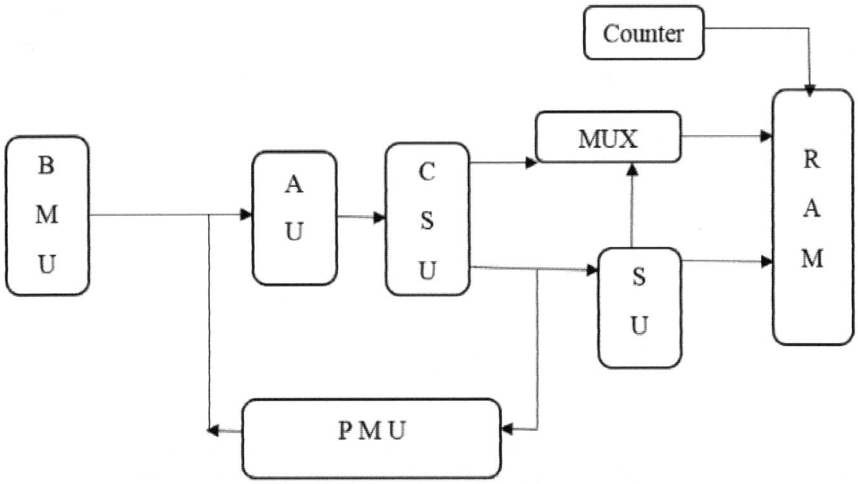

Figure 4. Proposed digital design of Viterbi Decoder for K=3

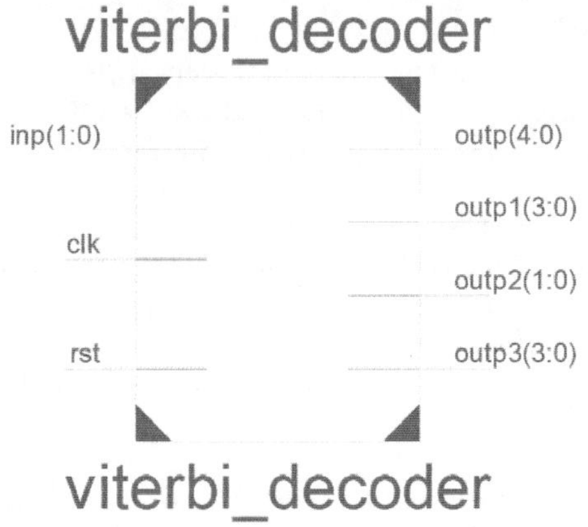

Figure 5. Block View of Viterbi decoder

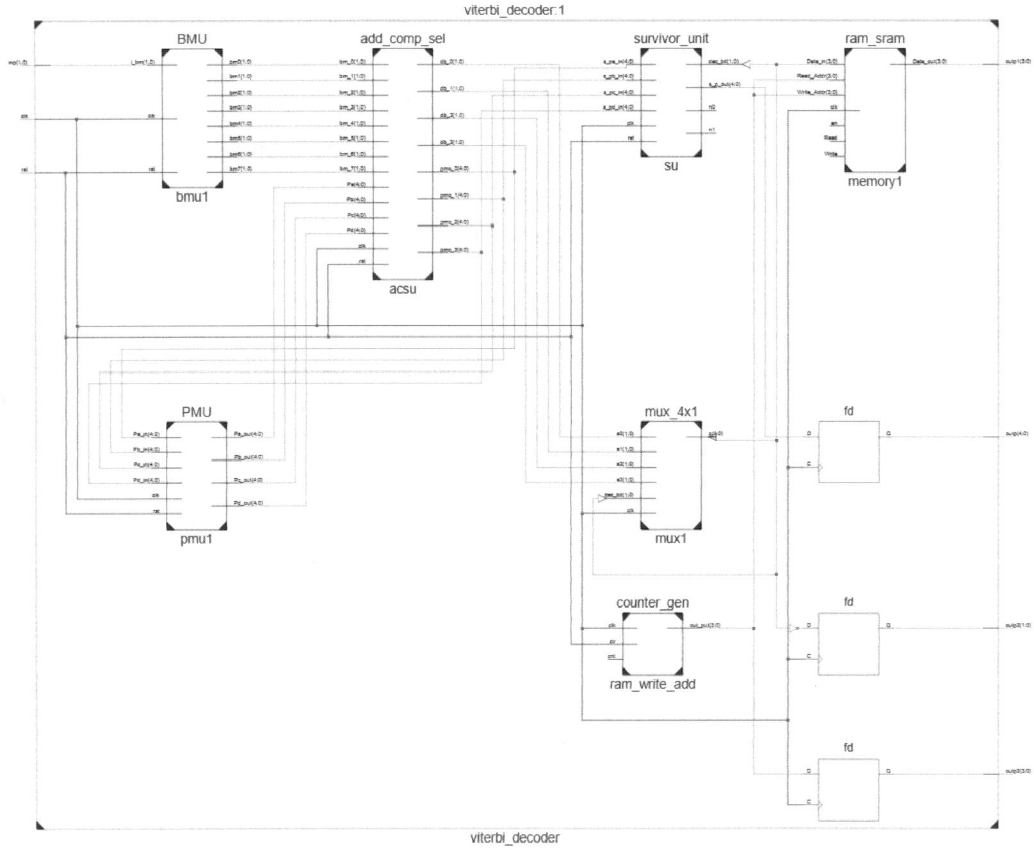

Figure 6. RTL View of Viterbi decoder

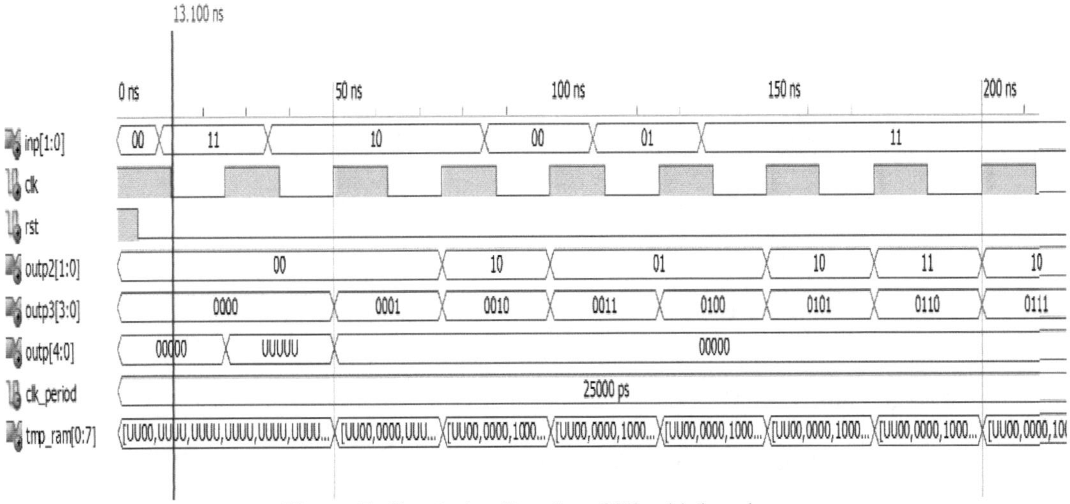

Figure 7. Simulation Results of Viterbi decoder

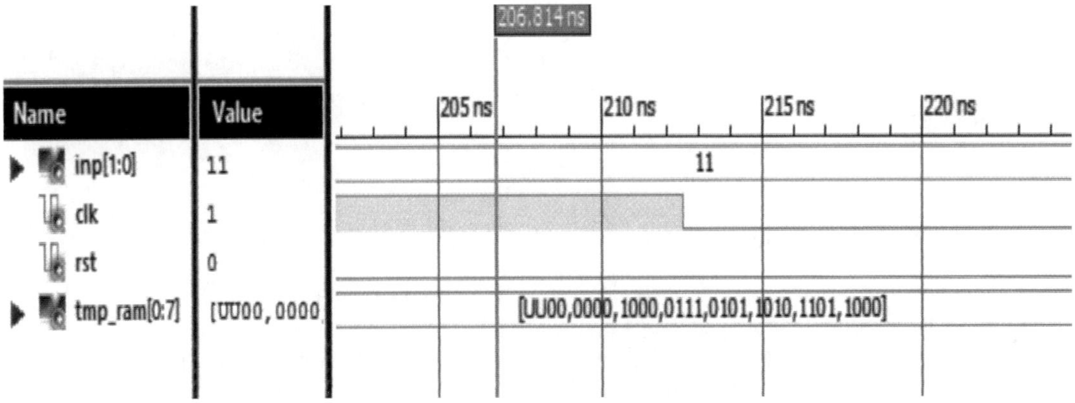

Figure 8. Data stored in RAM after 7 clock pulses

```
=========================================================================
HDL Synthesis Report

Macro Statistics
# ROMs                                                 : 1
 4x8-bit ROM                                           : 1
# Counters                                             : 1
 4-bit up counter                                      : 1
# Registers                                            : 32
 1-bit register                                        : 13
 2-bit register                                        : 11
 4-bit register                                        : 2
 5-bit register                                        : 6
# Comparators                                          : 15
 5-bit comparator equal                                : 4
 5-bit comparator greater                              : 7
 5-bit comparator less                                 : 4
# Multiplexers                                         : 1
 2-bit 4-to-1 multiplexer                              : 1
# Tristates                                            : 1
 4-bit tristate buffer                                 : 1
# Xors                                                 : 40
 1-bit xor3                                            : 40

=========================================================================
```

Figure 9. HDL synthesis report of the Viterbi decoder

```
Timing Summary:
---------------
Speed Grade: -4

   Minimum period: 9.928ns (Maximum Frequency: 100.725MHz)
   Minimum input arrival time before clock: 8.936ns
   Maximum output required time after clock: 4.283ns
   Maximum combinational path delay: 2.731ns
```

Figure 10. Timing summary of the Viterbi decoder

4. Conclusion & Future Work

The proposed digital design of the Viterbi decoder for ½ rate convolutional encoder of constraint length k = 3 is successfully synthesized in VHDL and simulated using XILINX ISE 14.7. RTL View of the complete design is shown in figure 6, whereas figure 7 and figure 8 shows the simulation results. Instead of storing the path metric of all four states for a single time instant, a decision is made by the survivor unit and reduce the memory requirement for a complete trellis length. The HDL synthesis report and the timing summary is shown in figure 9 and figure 10.

Future work will be implementing the design and evaluation of speed performance and power consumption of the decoder on different FPGA boards.

5. References

[1] Lin S & Costello D J Jr 2011 *Error Control Coding* (Prentice Hall Inc) 2e

[2] He J, Liu H, Wang Z, Huang X and Zhang K 2012 High-Speed Low-Power Viterbi Decoder Design for TCM Decoders, *IEEE Trans. on Very Large Scale Integration (VLSI) Sys.* vol 20 no 4 pp 755-9

[3] Arunlal K S and Hariprasad S A 2012 An Efficient Viterbi Decoder, *Inter. Jour. of Comp. Sci., Engg. and App. (IJCSEA),* vol 2 no 1 pp 95-110

[4] Pujara H and Prajapti P 2013 RTL Implementation of Viterbi Decoder using VHDL, *IOSR Jour. of VLSI and Sig. Pro. (IOSR-JVSP)* vol 2 issue 1 pp 65-71

[5] Reddy A S and Krishna P R 2013 High Speed Viterbi Decoder Design With A Rate Of 1/2 Convolution Code For Tcm Systems, *Inter. Jour. of Techno. Enhan. and Emerging Engg. Research* vol 1 issue 4 pp 126-30

[6] Soreng B and Kumar S 2013 Efficient implementation of Convolution Encoder and Viterbi Decoder, *Inter. Confer. on Circuits, Power and Computing Techno. (ICCPCT)* pp 1270-3

[7] Mahendran R and Sathish T M K 2013 Low Power Adaptive Viterbi Decoder Design For Trellis Coded Modulation, *Inter. Jour. of Innovative Research in Elec., Electron., Inst. and Control Engg.* vol 1 issue 3 pp 123-8

[8] Singh P and Vishvakarma S K 2013 FPGA Implementation of 413.121 MHz and 11.34 mW High Speed Low Power Viterbi Decoder, *International Journal of Modeling and Optimization* vol 3 no 1 pp 15-9

[9] Wayal K, Gore K, Waikule S and Wagaj S.C. 2015 Convolution Encoding and Viterbi Decoding Based on FPGA using VHDL, *Inter. Jour. of Advanced Tech. in Engg. and Science* vol 03 issue 01 pp 622-6

[10] Vaithiyanathana D, Nargisb J, and Seshasayanana R 2015 High performance ACS for Viterbi decoder using pipeline T-Algorithm, *Alexandria Engineering Journal* vol 54 issue 3 pp 447–55

Design of Efficient Mirror Adder in Quantum- Dot Cellular Automata

Prashant Kumar Mishra[1], Manju K. Chattopadhyay[2]

VLSI Design and Embedded Systems Laboratory, School of Electronics,
Devi Ahilya University, Indore, India
E-mail: prashantind.18@gmail.com

Abstract. Lower power consumption is an essential demand for portable multimedia system using digital signal processing algorithms and architectures. Quantum dot cellular automata (QCA) is a rising nano technology for the development of high performance ultra-dense low power digital circuits. QCA based several efficient binary and decimal arithmetic circuits are implemented, however important improvements are still possible. This paper demonstrate Mirror Adder circuit design in QCA. We present comparative study of mirror adder cells designed using conventional CMOS technique and mirror adder cells designed using quantum-dot cellular automata. QCA based mirror adders are better in terms of area by order of three.

Index Terms— Nanotechnology, quantum-dot cellular automata (QCA), Mirror Adder, Full Adder

1. Introduction

Digital Signal Processing system are the essential requirement for various portable multimedia devices. The adder circuit is the critical components of the Arithmetic Logic Unit (ALU), [1]. Demand for increasing mobile electronic devices require the efficient low power VLSI circuits. Power consumption also increases, with increase in chip density of VLSI system. With high power dissipation, packaging and cooling cost also goes up [1-3]. For IC designers, Lower power consumption and minimum area requirements is one of important design constraints.

Current transistor-based semiconductor devices are becoming resistant to scaling. For transistor circuit, the power dissipation due to leakage current is a big problem [4-7]. A possible alternative to these problems is Nanotechnology based Quantum-dot cellular automata (QCA) circuits [8-17]. The focus of this paper is on design of efficient Mirror adder circuit [3]. Prior to our work on mirror adder we examined some previous works on adder circuits in QCA technology. J. Cocorullo, Giuseppe proposed Design of Efficient BCD Adders in Quantum Dot Cellular Automata [4]. An Efficient Design of Full Adder in Quantum-Dot Cellular Automata (QCA) Technology is exhibited by M. Mohammadi, S. Gorgin [9]. Efficient Design of a Hybrid Adder in Quantum-Dot Cellular Automata is displayed by V. Pudi and K. Sridharan [12]. M. Gladshtein exhibited Quantum-Dot Cellular Automata Serial Decimal Adder [14]. M. Sangsefidi, M. Karimpour and M. Sarayloo proposed Efficient Design of a Coplanar Adder/Subtractor in Quantum-Dot Cellular Automata [16]. H. Cho and E. E. Swartzlander exhibited Adder and multiplier outlines in quantum-dot cell automata [17].We have shown plan of Mirror Adder in QCA and contrasted these outcomes and CMOS technology. To the best of our knowledge, there is no earlier work that has analyzed design of Mirror Adder in QCA. Improvement in terms of area of the order of three was obtained in our QCA based designs.

2. Background

QCA gives an altogether different computation platform than conventional CMOS, one in which polarization, instead of current, contains the digital information and one in which the cells themselves, rather than interconnecting wires, transmit this information all through the circuit [6, 8, 14-15]. This section begins with a concise diagram of QCA cells and the wires that are made out of QCA cells. This

IOP Conf. Series: Materials Science and Engineering **331** (2018) 012010 doi:10.1088/1757-899X/331/1/012010

is followed by a depiction of the QCA 3-input majority gate and the different gate structures that can be framed by reconfiguring this fundamental gate [8, 15-17].

2.1. QCA Basics

As shown in figure 1, these cells can be considered as square rectangles containing a quantum dot [2, 5, 11, 18]. In each corner two additional electrons, present inside each cell, can tunnel from dot to dot inside a cell.They are unable to go beyond the cell boundary to neighboring cells. The two polarizations framed by these electrons are utilized to represent a logic value of 1 and a logic value of 0 as shown on the left and on the right of figure 1. [4, 8-9, 19-20].

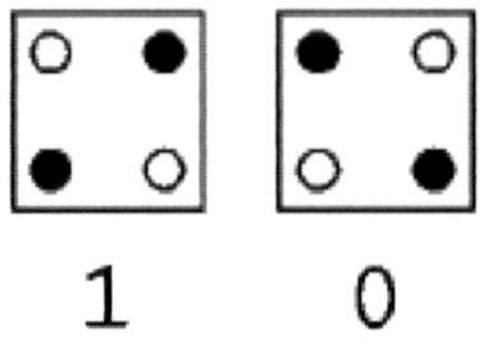

Figure 1. QCA Cells.

The QCA cells contain the interconnecting wires .An example of a QCA wire is shown in figure 2. In this example, a value of 1 is transmitted along the wire. Just a slight polarization in a cell is required to completely polarize its neighbor [7, 10, 13, 20-21]. The direction for the flow of information through a gate or a wire is controlled by a four phase clocking system which raises and brings down barrier between the cell [1, 6, 12-15].

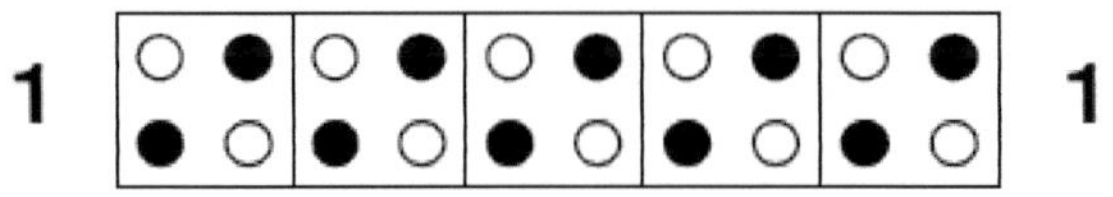

Figure 2. QCA Wire

2.2. QCA Logic Gates

The basic logic gate for QCA is the 3-input majority gate as shown in figure 3, that is made out of five cells [6-9, 12, 16-18]. Three of these, representing the inputs to the cell, are named a, b, and c. The center cell is the "device cell" that performs the calculation. The rest of the cell, marked out, gives the output [4-6, 8, 11, 21]. The circuit shown in figure 3 performs the Boolean function.

$$Out = ab + bc + ac.$$

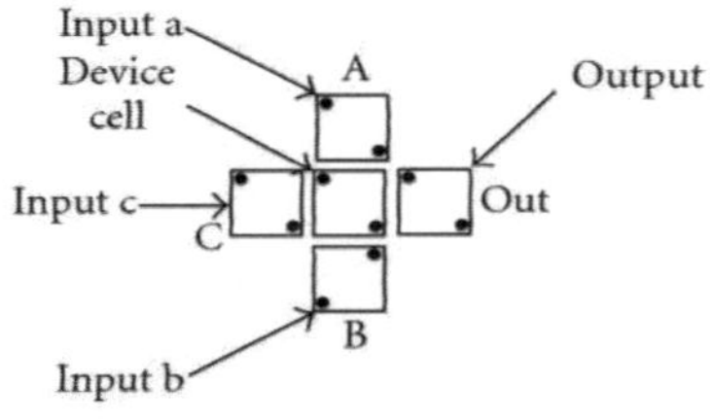

Figure 3. Majority Gate

Fixing one of the majority gate inputs to "0" or "1", the majority gate simply turns to a logic AND or OR gate[9-14].

2.3. Clocking
Unlike CMOS circuits in which electrical current transfers the data, QCA circuits use Coulomb repulsion and different clock zones to transfer the data flow. Clocking in QCA make electrons either tunnel between the wells or stay in their current situations [2, 4-8, 13-15]. In this way, data can flow in a specific direction or forced to stay in their positions. To control the direction of data flow In QCA technology, the cells of the circuits are divided into four segments, called clock zones [6-9, 14-17, 19]. The phase of each clocking zone is shifted by 90 degrees comparing to its previous zone. The phases of QCA clocking zones are called switch, hold, release and relax. figure 4 shows the propagation of input signal to output using four phases [1, 6-8, 15-18].

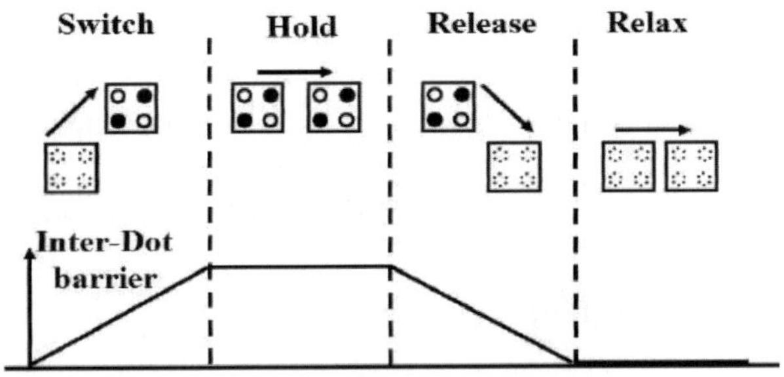

Figure 4. Four Phase Clocking for QCA

3. Adders Circuits

In this section, we discuss for designing full adder Circuits by QCA. Since the mirror adder is one of the broadly utilized economical implementations of the full adder in CMOS technology, we use it as our basis for proposing full adder circuits using QCA [3].

3.1. Generalization of Different Adders
Table 1 gives details of different types of full adder circuits eg. Kogge-Stone Adder, Carry Look Ahead Adder, Brent Kung Adders, Ladner-Fisher Adder, Hybrid Adder, Mirror Adder and their advantages.

IOP Conf. Series: Materials Science and Engineering **331** (2018) 012010 doi:10.1088/1757-899X/331/1/012010

Table 1. Types of Adders Circuits

Full Adder Types	Advantages
Kogge-Stone Adder[12]	1. The Kogge-Stone adder has Low depth and High node count (implies more area). 2. Minimal fan-out of 1 at each node (implies faster performance)
Carry Look Ahead Adder[12]	1. The carry lookahead adder(CLA) has a regular structure. 2. It achieves high speed. 3. These adders avoid feedback signals that are used in regular CMOS. 4. By the nature of QCA cells the carry look ahead adder is pipelined. Due to the pipeline diagram all sum signals are available at the same clock period.
Brent Kung Adders[12]	1. Maximum logic depth in PP adders (implies longer calculation time). 2. Minimum number of nodes (implies minimum area)
Ladner-Fisher Adder[12]	1. Low depth. 2. High fan-out nodes
Hybrid Adder[12]	1. The hybrid adder requires a substantially lower number of majority gates for different adder sizes. 2. Hybrid adder has advantages in terms of delay
Mirror Adder [22]	1. The NMOS and PMOS chains are completely symmetrical. Maximum of two series transistors in the carry-generation gate. 2. Only the transistors in the (propagate) carry chain have to be optimized for speed. All transistors in the sum stage can be minimal size. 3. Carry signals are critical transistors connected to Ci are placed closest to the output.

3.2. Conventional mirror adder

Table 2. Truth Table of full adder circuit

Inputs			Outputs	
A	B	Cin	Sum	Cout
0	0	0	0	0
0	0	1	1	0
0	1	0	1	0
0	1	1	0	1
1	0	0	1	0
1	0	1	0	1
1	1	0	0	1
1	1	1	1	1

IOP Conf. Series: Materials Science and Engineering **331** (2018) 012010 doi:10.1088/1757-899X/331/1/012010

Figure 5 shows the transistor level schematic and figure 6 shows the layout [3] of a conventional Mirror Adder (using CMOS Technology). It consists of a total of 24 transistors.

$$Sum = ABCin + A^{-} B^{-} Cin + A^{-} BC^{-}in + AB^{-} C^{-}in,$$
$$Cout = AB + A^{-} BCin + AB^{-} Cin.$$

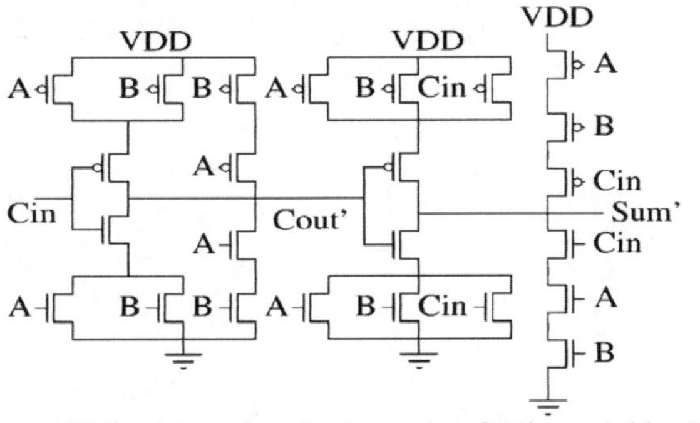

Figure. 5 Transistor level schematic of Mirror Adder [3].

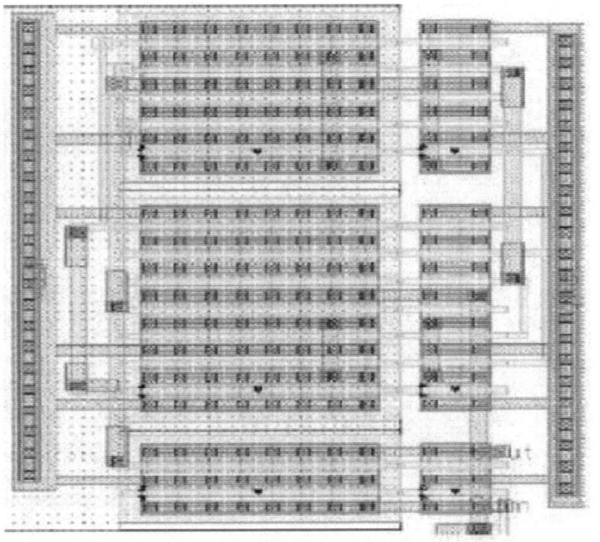

Figure 6. Layout of Mirror Adder using CMOS Technology -90 nm [3].

4. Mirror adder using QCA

In this segment, we show layout got from QCA Designer and the simulation results for Mirror Adder.

4.1. Design Rules

Cells for our design are assumed to have a dimension of 18*18 nm while the quantum dots have a diameter of 5 nm. Cells are set on a grid with a cell center to center separation of 20 nm.

4.2. Layout Diagrams and Simulation Results

Figure 7 shows the QCA Designer layout of Mirror Adder while figure 8 gives the corresponding simulation results.

IOP Conf. Series: Materials Science and Engineering **331** (2018) 012010 doi:10.1088/1757-899X/331/1/012010

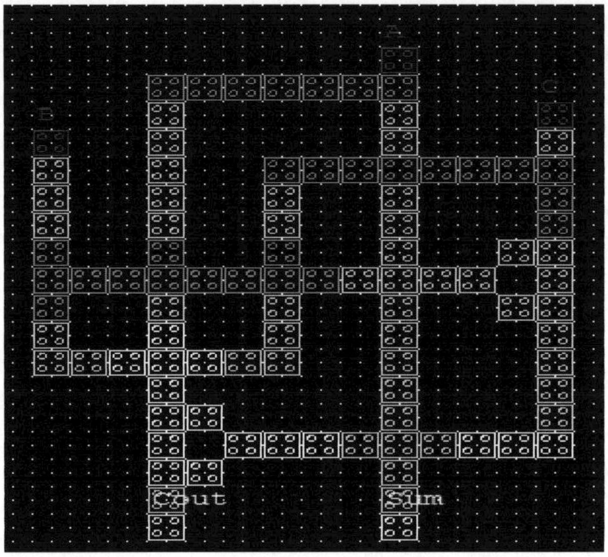

Figure 7. Layout of Mirror Adder using QCA

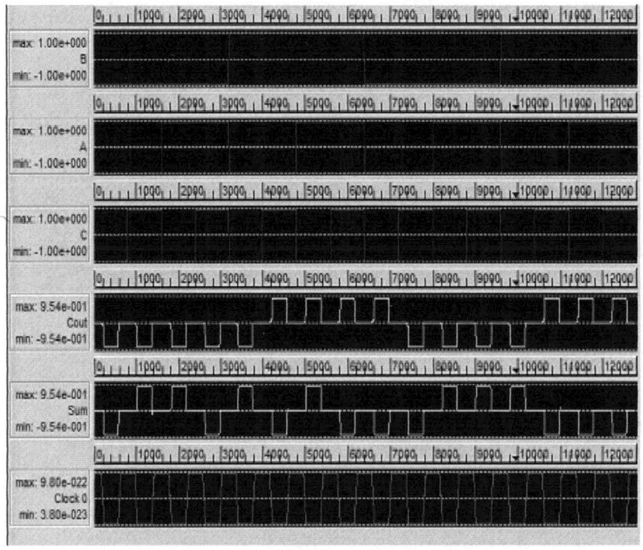

Figure 8. Simulation diagram for Mirror adder

Table 3. Layout Area of Approximate Mirror Adder using 90-nm CMOS Technology vs. Approximate Mirror Adder using QCA Cells.

MA Cell	Area	Cells
Conventional	40.66 um² (Area using 90-nm CMOS Technology[3])	24(CMOS Transistor)
Using QCA	36488 nm²(Using QCA)	98(QCA Cells)

4.3. Discussion of Results

With respect to table 3, we observe that the no. of cells and area, for QCA Mirror Adder is less as compared to conventional Mirror Adder. Our comparison results are based on Mirror based on 90-nm CMOS technology and Mirror Adder with QCA technology.

5. Conclusion

Quantum dot cellular automata is a future nanotechnology architecture for computing, best alternative for CMOS technology. QCA offers a new method of computation and information transformation. We have examined a several aspect of QCA method with their working methodology. These designes give better outcomes when contrasted with the CMOS technology as far as small area, fast speed, and low power dissipation.

In this paper, we have presented efficient Mirror Adder Design using QCA. The designs are based on new concepts concerning majority logic gate. Comparisons with conventional Mirror Adder and QCA Mirror Adder are presented. Detailed simulation results are also given.

6. References

[1] Lent C.S. , Tougaw P.D., Porod W. and Bernestein G.H. 1993 Quantum Cellular Automata (Nanotechnology), vol. 4, no. 1, pp. 49–57.

[2] Lent C.S. and Tougaw P.D 1997 A device structure for Computing with Quantum Dots *proceedings of the IEEE, vol.*85, no.4, pp.541-557.

[3] Gupta V., Mohapatra D., Raghunathan A. and okay. Roy 2013 Low-energy virtual signal Processing using Approximate Adders in *IEEE Transactions on Computer-Aided layout of incorporated Circuits and systems, vol.* 32, no. 1, pp. 124-137.

[4] Cocorullo J., Giuseppe and et al. 2016 design of efficient BCD Adders in Quantum Dot cellular Automata *IEEE Transactions on Circuits and structures II: specific Briefs*.

[5] Nath R., Sen B. and Sikdar B. 2017 top-rated Synthesis of QCA logic Circuit getting rid of wire-c Lent C.S.,Isaksen B. and Lieberman M.L. 2003 Molecular Quantum-Dot cellular Automata J. Am. Chem. Soc., vol. 125, pp. 1056-1063

[6] Campos, Caio Araujo T. and et al. 2016 A regular, Scalable, and efficient Clocking Scheme for QCA *IEEE Transactions on pc-Aided design of included Circuits and systems, vol* , 35, no.3, pp. 513-517.

[7] Toth G. 2000 Correlation and Coherence in Quantum-Dot mobile Automata *Ph.D. Thesis, university of Notre Dame*, pp. 56-63.

[8] Mohammadi M., Mohammadi M. and Gorgin S. 2016 An green design of complete Adder in Quantum-Dot cell Automata (QCA) technology *Microelectronics magazine, vol.* 50, pp. 35–forty three.

[9] Cocorullo G., Corsonello P., Frustaci F. and Perri S. 2017 design of green BCD Adders in Quantum-Dot cellular Automata in *IEEE Transactions on Circuits and structures II: express Briefs, vol.*64, no. 5, pp. 575-579.

[10] "The global technology avenue map for semiconductors (ITRS), Semi-conductor affiliation,2015":*http://www.itrs.net.*

[11] Pudi V. and Sridharan ok. 2011 green design of a Hybrid Adder in Quantum-Dot cellular Automata in IEEE *Transactions on Very huge Scale Integration (VLSI) structures, vol.* 19, no. 9, pp. 1535-1548.

[12] Perri S., Corsonello P. and Cocorullo G. 2014 layout of efficient Binary Comparators in Quantum-Dot cellular Automata in *IEEE Transactions on Nanotechnology, vol.* 13, no. 2, pp. 192-202.

[13] Gladshtein M. 2011 Quantum-Dot mobile Automata Serial Decimal Adder in *IEEE Transactions on Nanotechnology, vol.* 10, no. 6, pp. 1377-1382.

[14] QCAdesigner. (2005). model 2.0.three. [Online]. to be had: *http://www.qcadesigner.ca/*

[15] Sangsefidi M., Karimpour M. and Sarayloo M. 2015 efficient layout of a Coplanar Adder/Subtractor in Quantum-Dot cellular Automata *IEEE european Modelling Symposium (EMS)(Madrid),* pp. 456-461.

[16] Cho H. and Swartzlander E.E. 2009 Adder and multiplier designs in quantum-dot mobile automata IEEE Trans. Comput., vol. fifty eight, no. 6, pp. 721–727.

[17] Walus okay., Dysart T., Jullien G., and Budiman R. 2004 QCADesigner: A rapid design and simulation tool for quantum-dot mobile automata *IEEE Trans. Nanotechnol., vol.* 3, no. 1, pp. 26–29.

[18] Xia Y. and Qiu k. 2009 Comparator layout based totally on quantum-dot mobile automata *J. Electron. Inf. Technol., vol.* 31, no. 6, pp. 1517–1520.

[19] Timler J. and Lent C.S. 2002 energy gain and dissipation in quantum-dot cellular automata *J. Appl. Phys., vol.* 91, pp. 823–831.

[20] Y. Lu, M. Liu, and C. Lent 2007 Molecular quantum-dot cellular automata:From molecular structure to circuit dynamics *J. Appl. Phys. vol.* 102, no.3, p. 034311.

[21] Rabaey J.M. 1996 virtual incorporated Circuits A layout attitude 2nd version (NJ: Prentice-corridor).rossings *IET Circuits, device & structures.*

IOP Conf. Series: Materials Science and Engineering **331** (2018) 012011 doi:10.1088/1757-899X/331/1/012011

Door Security using Face Detection and Raspberry Pi

Venkatesh Bhutra[1,a], Harshav Kumar[2,b], Santosh Jangid[3c] and L Solanki[4d]

[1] Student, Department of Electronics and Communication, BKBIET, Pilani, Rajasthan, India-333031.
[2] Student, Department of Computer Science & Engineering, BKBIET, Pilani, Rajasthan, India-333031.
[3] Assistant Professor, Department of Electronics and Communication, BKBIET, Pilani, Rajasthan, India-333031.
[4] Principal (Academics), BKBIET, Pilani, Rajasthan, India-333031.

[a]v.bhutra1995@gmail.com
[b]harshavk97@gmail.com
[c]jangid.santosh@gmail.com
[c]dean.ecee@gmail.com

Abstract. With the world moving towards advanced technologies, security forms a crucial part in daily life. Among the many techniques used for this purpose, Face Recognition stands as effective means of authentication and security. This paper deals with the user of principal component and security. PCA is a statistical approach used to simplify a data set. The minimum Euclidean distance found from the PCA technique is used to recognize the face. Raspberry Pi a low cost ARM based computer on a small circuit board, controls the servo motor and other sensors. The servo-motor is in turn attached to the doors of home and opens up when the face is recognized. The proposed work has been done using a self-made training database of students from B.K. Birla Institute of Engineering and Technology, Pilani, Rajasthan, India.

1. Introduction

Face recognition, which deals with recognition and authentication of a face, is a modern security feature and a field that has been studied extensively for a long time. Used in most of the real world applications, varying from surveillance to security and access, face recognition has found many techniques for itself. Principal component analysis (PCA) is one of the most extensively used and simplest procedure for face recognition. The PCA algorithm uses mathematical tools such as eigen value, eigenvectors, eigen faces and euclidean distance for recognition. As each image can be viewed as matrix of dimension M x N, this process is quiet handy and easy to implement. The image with the least euclidean distance predicts the most accurate face. As our field of interest lies only the facial part, we use Viola Jones method for face detection. This security technique finds lot of applications, home security being one of them.

2. Viola-Jones Method

Viola Jones method is a widely used method for object detection. In this proposed work, Viola Jones method is used to detect and capture the facial part. This method works by looking for feature. Each feature are represented by small set of rectangles, group of pixels contained in it. The face in this method is represented by a rectangle, which is formed by the set of smaller rectangles. The basic reason for use of this method is that the detection in this case is very fast.

3. Principal Component Analysis

Principal component analysis, a method to abridge huge high dimensional data, is a statistical approach for face recognition and image compression. The PCA uses mathematical tool such as eigenvector, eigen faces, euclidean distance, covariance matrix for its functioning. This 2-dimensional data feature extraction method, computes feature of each image in training set into their respective covariance matrix. This covariance matrix is used to determine the eigenvector and eigen value for each corresponding image. From the derived values, eigenvector corresponding to the highest eigen value is considered. This eigenvector is the principal component of that image and is projected in the face space. Similarly, eigen vector of the test image is calculated and the euclidean distance is evaluated. The euclidean distance is the average distance between the test vector and the training set. The minimum distance represents the most similar image.

4. PCA Algorithm

Principal component analysis, a mathematical tool is largely used in face recognition using different statistical approach. PCA instead of acting upon quantized image works directly on unquantized image using complexity reduction algorithm.

Let the two dimension face image be represented by I(x, y) of dimension M*N. This image can be represented by vector of reduced dimension M*N. Each high dimensional image in the training set is represented by a vector of coefficient in a free space of reduced dimension..

4.1. Training Dataset

4.1.1. If we considered the dataset of image to be {I1,I2,I3,....IN}, then the average face for the training images can be give as mean of the images, that is Average face J=1/N $\sum =_1 i$.

4.1.2. The next step is to calculate the covariance matrix. It represents the degree up to which the faces are linearly correlated w.r.t average faces.

Let D represents the difference between i^{th} *training face and average face:*

$$D= I_i - J$$

The covariance matrix C is,

$$C= D\, D^T$$

4.1.3. The covariance matrix is used to determine the eigenvector and eigen value for each image in training set. Therefore, the eigenvector and eigen value associated with each image is given by:

$$C\, V_i=\mu_i\, V_i$$

V_i *is the* i^{th} *eigenvector and* μ *is its eigen value.*

4.1.4. For each image we sort the eigen vector in descending order with respect to its eigen values. Henceforth, we keep only 'k' eigen vector starting from top of the descending list

4.1.5. Eigen faces are ghostly facial image represented by U_i and is given by

$$U_i=D\ V_i$$

4.1.6. The training face image is now projected into the Eigen face space by

$$W_i=U_i\ (I_i - J) = U_i^T D$$

Where, $W_i= \{W_1, W_2, W_3, \ldots W_N\}$ represented feature weights for training images. These weights are represented by vector P_i.

4.2. Sample Test

The Process flow adopted for testing is a mentioned:

4.2.1. Obtain the test image from camera and detect the face using Viola Jones method.

4.2.2. Project the test image in the face space and calculate the feature vector and weight. The weights are given as

$$W_{test} = U_i^T\ (\ I_{test} - J)$$

Here, $W_{test} = \{W_1, W_2, W_3, \ldots W_N\}$ feature represent corresponding to value of i, i.e. the i_{th} eigen faces. These weights be represented by vector P_{test}.

4.2.3. Calculate euclidean distance between the test feature vector and all training feature vector of the training set. The Euclidean distance, average distance is given by:

$$Y_i^2= (||P_{test}-P_i||^2)$$

Where, $i=\{1,2,3,\ldots K\}$. This distance gives the similarity between the test image and training image.

4.2.4. Find the minimum euclidean distance. The one with minimum euclidean distance describes the most similar face.

5. Hardware Interfacing

The face recognition algorithm interacts with the different home appliance such as camera and servo motor using the Raspberry Pi. The Raspberry Pi, a series of credit card sized single board computer, is a low cost ARM based computer. Its GPIO ports facilitates the interaction between the software and hardware.

A specially configured Linux operating system from MATLAB is used in this case for communication between the two parties, the hardware and code. The camera linked with the Raspberry Pi capture the image. The MATLAB using Viola Jones method detect the face. The detected face is matched with the training facial dataset using MATLAB code. If the face is recognized MATLAB communicates with the Raspberry Pi. Which in turn instructs the servo motor to open the doors. If the face is not recognized, MATLAB instructs the Raspberry Pi to turn on the alarm.

6. Diagrammatic Representation Flowchart

6.1. Diagrammatic Representation

IOP Conf. Series: Materials Science and Engineering **331** (2018) 012011 doi:10.1088/1757-899X/331/1/012011

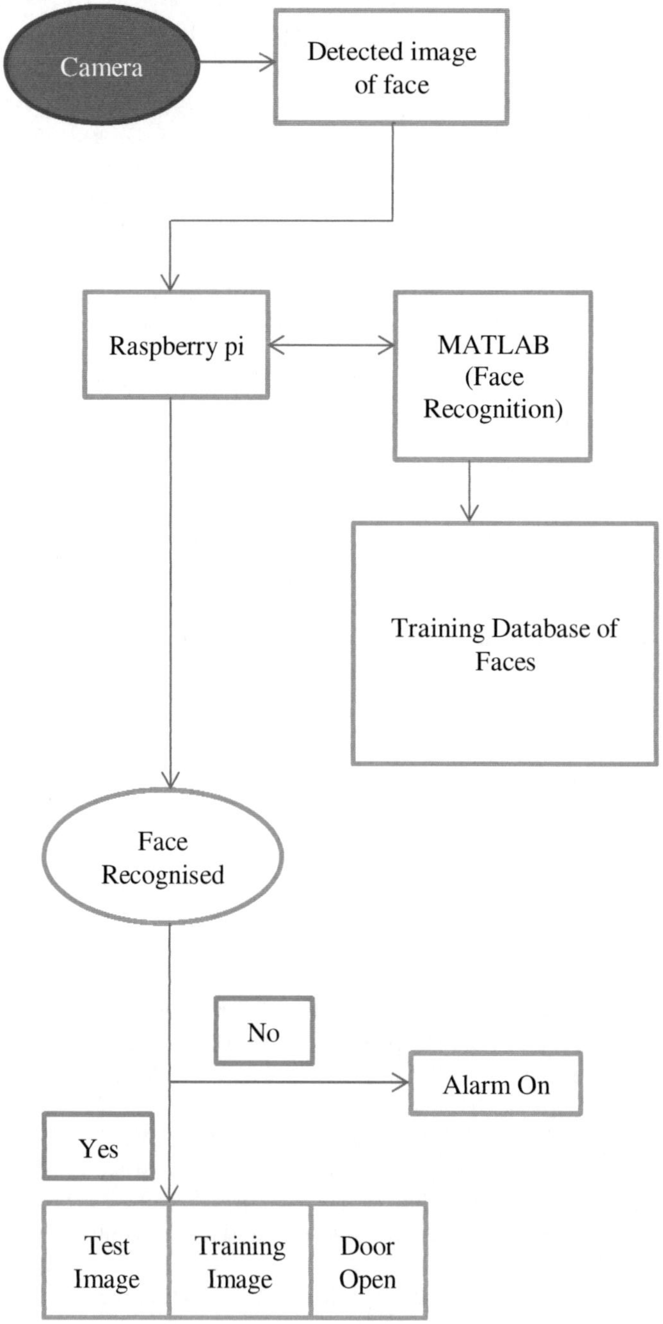

Figure 1: Representation of complete work flow taking into consideration the software and hardware part.

IOP Conf. Series: Materials Science and Engineering **331** (2018) 012011 doi:10.1088/1757-899X/331/1/012011

6.2. Data Flow

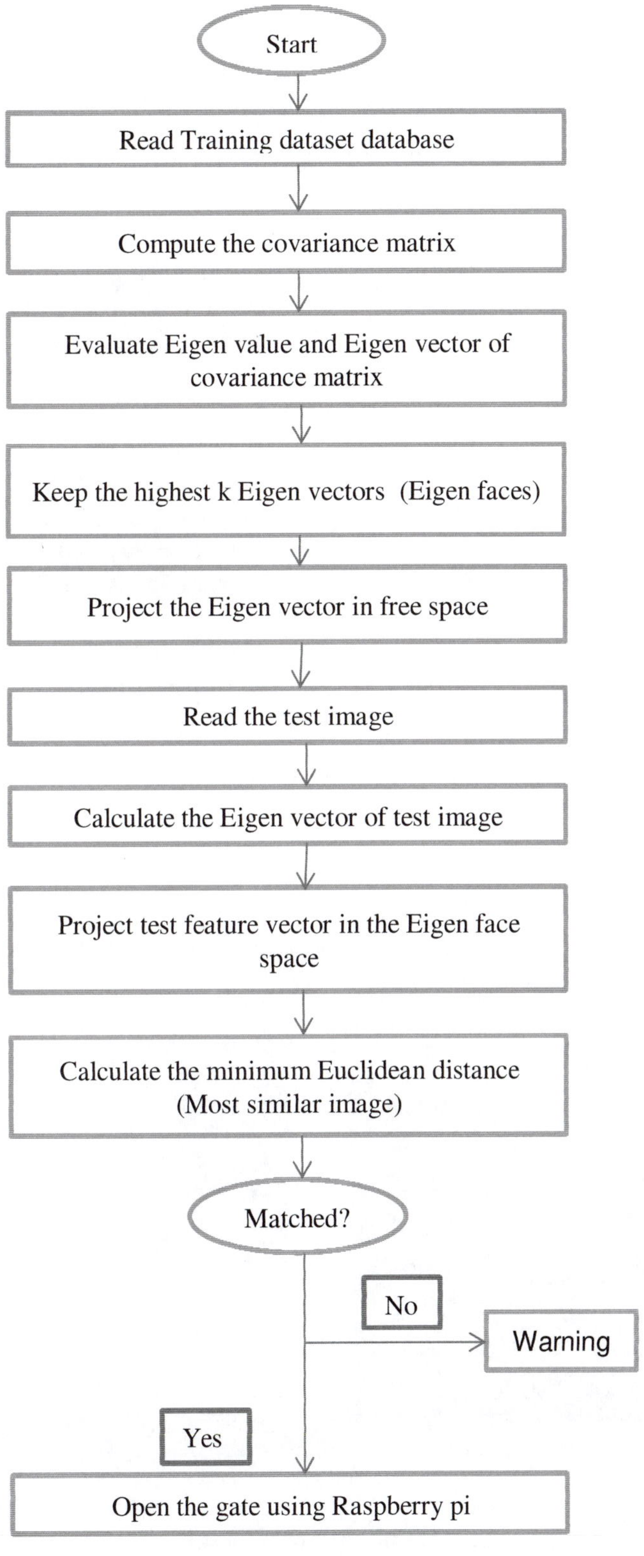

Figure 2: Process flow for the proposed work.

7. Experimental Result

The complete proposed model works efficiently with the self-trained face database at different as well as similar facial expression. It is also found to work successfully with multiple faces.

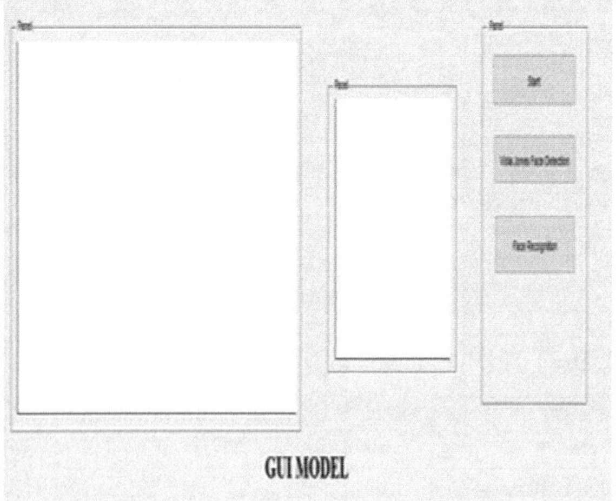

Figure 3: GUI Model of the project.

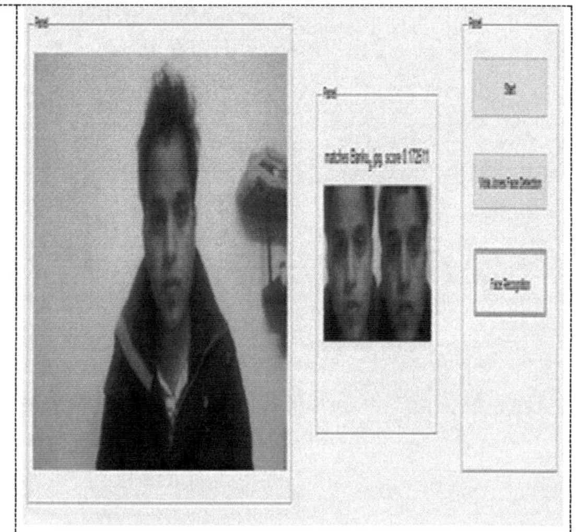

Figure 4 : Experimental result for single face.

In case when multiple faces are in consideration, the face which is most similar with that in the training database is recognized.

Figure 5 : Experimental result for multiple face.

IOP Conf. Series: Materials Science and Engineering **331** (2018) 012011 doi:10.1088/1757-899X/331/1/012011

8. Conclusion

In this research, principal component analysis plays the major role. It is considered as the best approach as it not only decrease computational complexity but also support compact data distribution. Though the PCA efficiency is effected by variation of light intensities, increasing the number of training images with variation can neutralize the fluctuation. Care must be taken in increasing the number of training images as the neutralization effect is possible only up to a limit. The proposed work works well with different image format but variation in image size limits the work. The interaction between the PCA algorithm and the hardware is quiet commendable. Thus, the principal component approach provides a solution that fits best in the proposed work

9. References

[1] A tutorial on Principal Component Analysis, Lindsay I Smith, February 26, 2002.
[2] Analysis and Recognition in Image and Videos, Face Recognition using corselet transform, Project Report, Rami COHEN, Technion - Israel Institute of Technology, arXiv:1107.2781v1 [cs.CV] 14 July 2011.
[3] I T Joliffe; Principal component analysis 1986 .
[4] R C Gonzalas, P A Wintz; Digital processing 1987.
[5] K.Karhunen;Uber Linear methoden in der wabrscheinlichkeit's Rechnug;1946 M.M.Loeve;probability theory 1955.
[6] W Zhao, R Chellappa, P J Phillips, Face recognition: A literature survey, "ACM Computing surveys (CSUR)", December 2003.
[7] Paul viola, Michaelj. Jhon, Robust Real time face detection.
[8] D.chen, and H Jiu-qiang "An FPGA-based face recognition using combined 5/3 DWT with PCA method". Journal of communication and computer, ISSN 848-7709, USA, \volume 6, no.10 (Serial No.59), oct.2009.
[9] H Moon, and P Phillps, "Computational and performance aspects of PCA based face-recognition algorithm", perception, 2001, volume 30, pp303-321.
[10] M Sharkas, "Application of DCT blocks with principal component analysis for face recognition" proceedings of the 5th WSEAS intconf. on single, speech and image processing, Corfu, Greece, (pp107-111), 2005
[11] M Turk, A. Pentland: Face recognition using eigen faces, conference on computer vision and pattern recognition , 3-6 June 1991,Maui, HI,USI, pp.586591.
[12] PRINCIPAL COMPONENT ANALYSIS: A BEGINNER'S GUIDE - I. Introduction and application By IAN T. JOLLIFFE Institute of Mathematics, University of Kent, Canterbury
[13] Door Lock System through Face Recognition Using MATLAB, Ayushi Gupta*1, Ekta Sharma2, Neha Sachan3 and Neha Tiwari4 *1, 2, 3, 4 Kanpur institute of technology, Kanpur, India.
[14] MATLAB Programming for engineers, Stephen J. Chapman, BAE SYSTEM, Australia, CENGAGE Learning, third edition.
[15] Digital image processing, S Jayaraman,S Esakkirajan, T Veerakumar,McGraw Hill Education(India)private limited.
[16] Applied Numerical Analysis using MATLAB, Second edition, Laurene V.Fausett, Texas A&M University-commerce.

IOP Conf. Series: Materials Science and Engineering **331** (2018) 012012 doi:10.1088/1757-899X/331/1/012012

Excitation of Higher Order Modes of Cylindrical Dielectric Resonator Antenna using Dual-slot feed

A K Ojha[1] and A.V. Praveen Kumar[2]

[1,2] Department of Electrical and Electronics Engineering,
Birla Institute of Technology & Science (BITS), Pilani, Rajasthan, India
Email: ojha.anuj11@gmailcom[1], praveen.kumar@pilani.bits-pilani.ac.in[2]

Abstract. Excitation of the higher order modes (HOM) of a cylindrical dielectric resonator antenna(DRA) of high relative permittivity, using dual feed scheme is investigated. The feed scheme uses a pair of narrow slots and is chosen on the basis of the field distribution of the desired DRA modes. Numerical studies using ANSYS HFSS show that the dual-feed excited a combination of two HOMs, which are identified as $HEM_{21\delta}$ and $TM_{01\delta}$. The mixed-up nature is further verified through studying the radiation pattern of the DRA which shows azimuthal asymmetry and low gain. It is suggested that if one of the HOM is suppressed, better antenna performance can be achieved.

1. Introduction

Possibility of using a cylindrical dielectric resonator (DR) as an antenna was first reported by the S. A. Long et.al. in 1983[1]. Later in 1984, D. Kajfez et.al. presented the detailed modal analysis of cylindrical DRs, and emphasised on the near-field distribution of various DRA modes for the purpose of exciting / suppressing them [2]. Even though $TE_{01\delta}$ mode is identified as the fundamental mode of a cylindrical DRA, most of the DRA studies were focused on the $HEM_{11\delta}$ and $TM_{01\delta}$ modes because of the easiness in exciting them and the usefulness of the radiation patterns they provided [3]. Investigations on the excitation of $HEM_{12\delta}$ mode was first reported by D. Guha et.al [4-5] which showed gain improvement compared to the lower order $HEM_{11\delta}$ mode. Other works focused on exciting new HOMs [6-8]. In the present study, HOMs of a cylindrical DRA are excited using a simple dual-slot feed and mode analysis has been carried out for the purpose of improving the mode purity. ANSYS HFSS is used for numerical analysis.

2. DRA Configuration

Figure 1 shows the dual-slot fed cylindrical dielectric resonator (DR). The feed consists of two slots made on the ground plane of an FR4 substrate with two opposing microstrips running on the other side. The cylindrical DR is placed on the slots and the microstrips are excited with in-phase currents in order to excite higher order modes. Having in-phase feed currents also suppresses the fundamental broadside mode ($HEM_{11\delta}$). The FR4 substrate is of dielectric constant ε_r = 4, loss tangent $\tan\delta = 0.02$ and substrate thickness 1.6 mm. The DR used has diameter $2a$ = 19.43 mm, height d = 7.3 mm and loss tangent $\tan\delta = 0.002$. On the basis of the preliminary studies conducted on a single slot fed DRA, a slot of 7 mm × 2 mm size and a substrate of 90 mm × 90 mm size are chosen. The microstrip has a width of 3.2 mm for 50 Ω impedance.

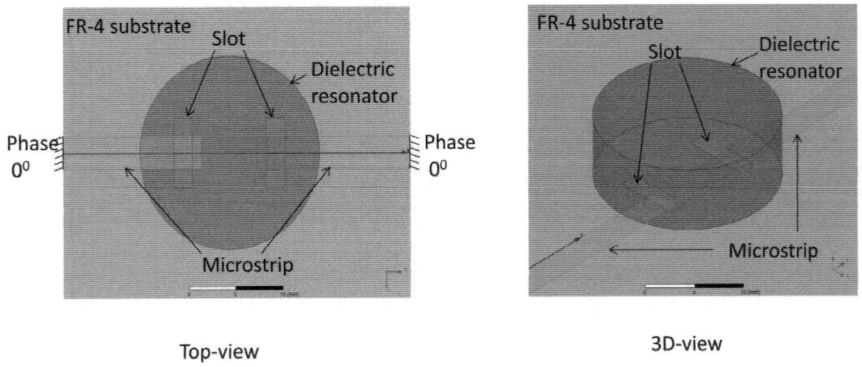

Figure 1. Schematic of the dual-slot fed cylindrical DRA

3. Principle of Operation

Any coupling scheme should be designed with the proper understanding of the near-field pattern of the particular mode. When two opposing microstrips are carrying in-phase currents, their magnetic fields will be directing opposite to each other as shown in Figure 2 (top). The magnetic fields will then couple through the narrow slots to the DR at two diametrically opposite positions. Ideally this should excite the $HEM_{21\delta}$ mode, but there is the possibility of exciting $TM_{01\delta}$ mode also as shown in Figure 2(bottom). Also note that the magnetic field is maximizing at about the middle of the radial distance for both modes, hence both modes are likely to be excited as their eigen frequencies are close by [2].

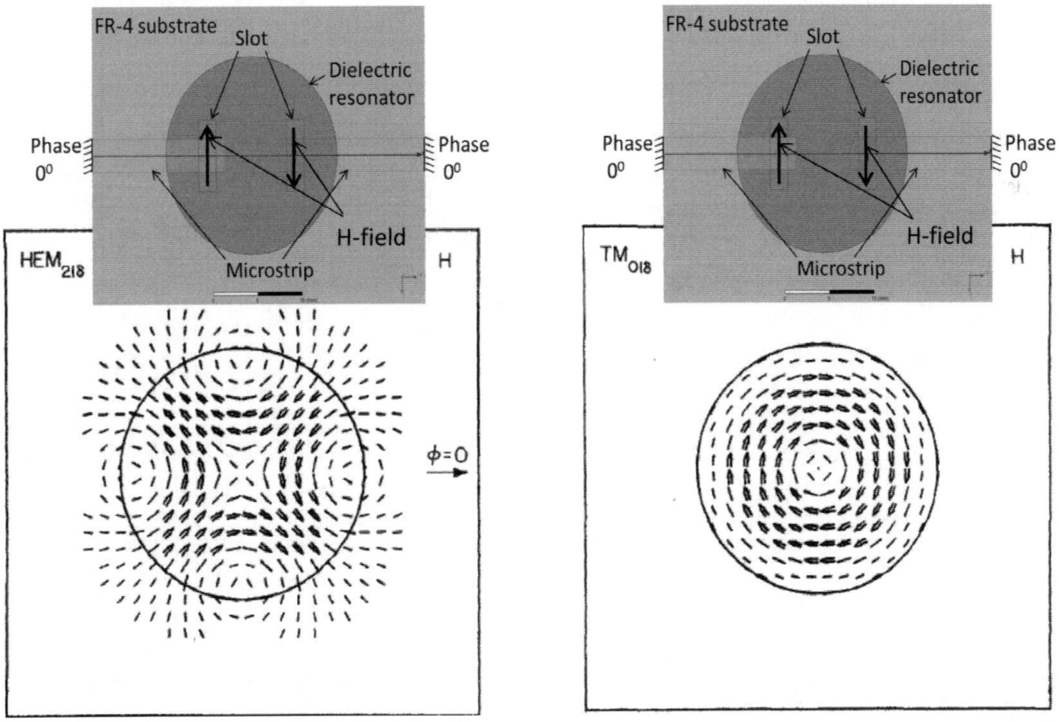

Figure 2. Magnetic field plots for the $HEM_{21\delta}$ and $TM_{01\delta}$ [2] with respect to the dual-slot feed

Thus the dual-slot feed is expected to excite a hybrid of the two above HOMs. To predict the resultant mode distribution hence the radiation pattern, an illustration is made as shown in Figure 3. Assuming the modes are excited in phase, the magnetic fields of the modes add constructively on the left and right while destructively on the top and bottom as indicated. In any case, the mode has a monopole-like behaviour.

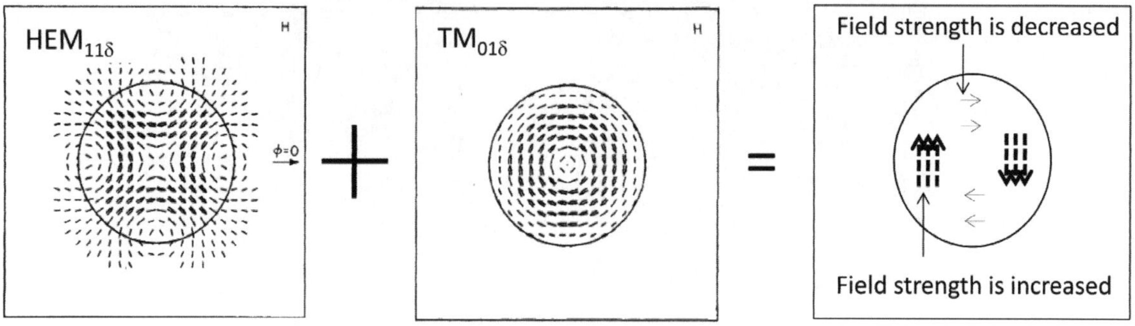

Figure 3. Illustration of forming a hybrid of $HEM_{21\delta}$ and $TM_{01\delta}$ modes

In the practical environment, a single microstrip feed will divide equally between the two slots, but is not modelled here. Instead, two separate feeds with in-phase source excitations have been modelled. Table I shows the DRA characteristics for various slot positions. Being a dual-feed, the overall input reflection coefficient is computed as $|\Gamma_{in}| = |S_{11}| + |S_{12}|$ [9]. Trend shows that as the slots move away from the centre of the DR, the impedance matching is improved. The reflection coefficient is shown for a slot position of 5 mm in Figure 4. Maximum bandwidth of 3.54 % is found for the slot position of 3 mm, maximum gain of 3.78 dB is achieved for the slot position of 5 mm and minimum cross-polarization of -13.8 dB for the slot position of 6 mm .

Table 1. Comparison of the DRA characteristics for different slot /feed position

Slot position from the center of the DR (mm)	**Microstrip position from the port**	f_0 **(GHz)**	$|\Gamma_{in}|$ **(dB)**	**% Band width**	**Max. E-plane gain (dB)**	**Max. H-plane gain (dB)**	**Max. E-plane cx-pol gain (dB)**	**Max. H-plane cx-pol gain (dB)**
6	40	5.455	-35	1.83	3.29	3.29	-13.8	-13.8
5	42	5.572	-22	3.23	3.78	3.78	-11.4	-11.4
4	43	5.563	-32	4.21	3.56	3.56	-10.0	-10.0
3	44	5.365	-20	3.54	3.78	3.78	-9.1	-9.1

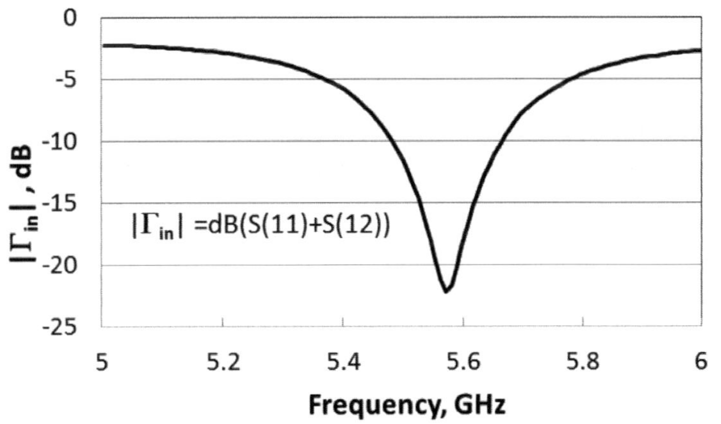

Figure 4. Reflection coefficient at the slot position = 5 mm and feed position = 42 mm

Radiation patterns for various plane cuts ϕ = 0, 90 degrees and 45, 135 degrees are shown in Figure 5 and 6 respectively. There is about 3-4 dB difference in the co-polar gains of the 0 and 90 degree planes, which is the effect of the mode mix-up. Now the near-field distribution on the DR at the frequency of 5.572 GHz is plotted in Figure 7. From the figure it is clear that the magnetic field strength is increased on the left and right sides whereas it is decreased on the top and bottom sides. This result is in agreement with the qualitative view presented above (Figure 3). Future works are aimed at suppressing one of the modes so that single mode operation with improved symmetry and gain is possible. External metallic perturbation [10] is one possible technique for unwanted mode suppression.

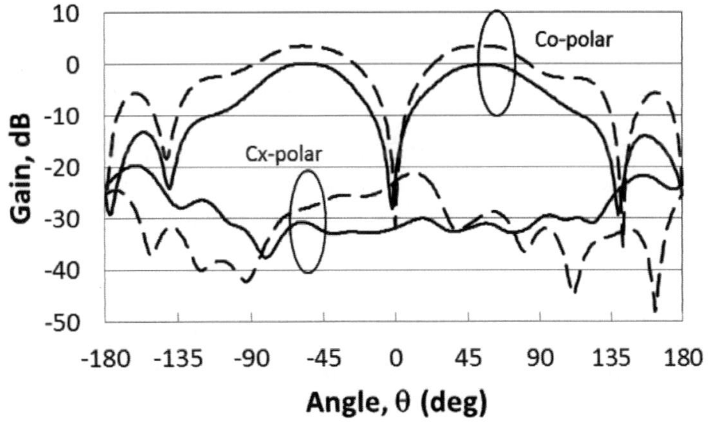

Figure 5. Radiation Pattern for planes $\phi=0^0$, 90^0 for slot position= 5 mm and feed position = 42mm

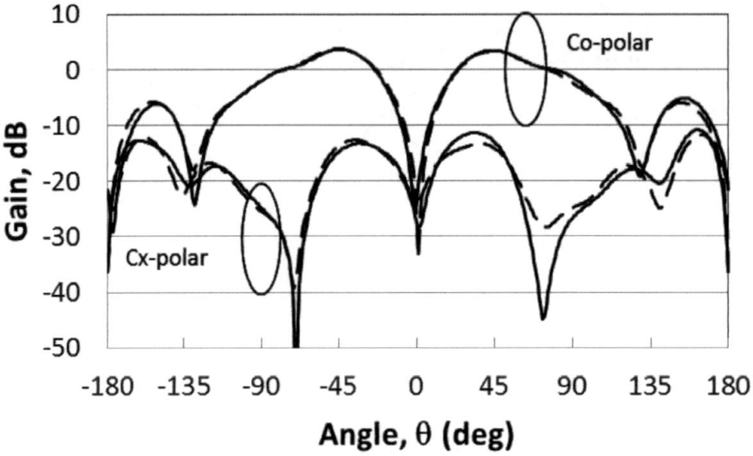

Figure 6. Radiation Pattern for planes $\phi=45^0$, 135^0 at slot position= 5mm and feed position = 42mm

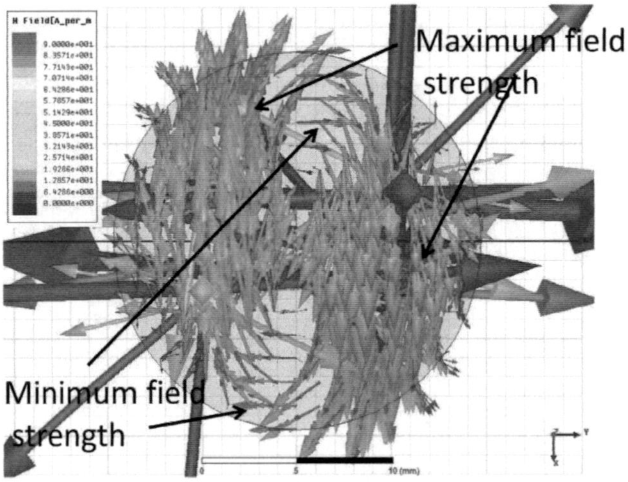

Figure 7. Top view of the near-field distribution at 5.572 GHz for slot position = 5 mm and feed position = 42 mm

4. Conclusion

The paper described the excitation of the higher order mode (HOM) of a cylindrical DRA using dual-slot feed with in-phase currents. From the simple modal analysis, the resulting field distribution is predicted as a hybrid between $HEM_{21\delta}$ and $TM_{01\delta}$ modes. Simulations in HFSS confirmed the predictions. Pattern asymmetry and low gain were observed for the combined mode. Improvement in performance can be expected if one of the modes is suppressed which is possible using field perturbations.

Acknowledgement

This work was supported by the Department of Science and Technology (DST), India through the FIST program under Grant Ref: SR/FST/ETI-346/2013.

References

[1] Long S A, McAllister M W, and Shen L C 1983 The resonant cylindrical cavity antenna *IEEE Trans. Antennas Propag*ation **5** 406–412.
[2] Kajfez D, Glisson A W and James J 1984 Computed Modal Field Distribution for Isolated Dielectric Resonators *IEEE Transaction Microwave Theory Technology* **32** 1609-1616.
[3] Hady L K, Kajfez D, and Kishk A A 2009 Triple Mode Use of a Single Dielectric Resonator *IEEE Transactions on Antennas and Propagation* **57** 1328- 1335.
[4] Guha D, Banerjee A, Kumar C and Antar Y M M 2012 Higher order mode excitation for high gain broadside radiation from cylindrical dielectric resonator antennas," *IEEE Transaction Antennas Propagation*, **60** 1 71–77.
[5] Guha D, Banerjee A, Kumar C and Antar Y M M 2014 New Technique to Excite Higher-Order Radiating Mode in a Cylindrical Dielectric Resonator Antenna *IEEE Antennas and Wireless Propagation Letters* **13.**
[6] Zhong L, Hong J S and Zhou H C 2016 A Novel Pattern-Reconfigurable Cylindrical Dielectric Resonator Antenna With Enhanced Gain IEEE Antennas And Wireless Propagation Letters **15** 1253-1256.
[7] Mrnka M and Radia Z 2016 Enhanced- Gain Dielectric Resonator Antenna Based on the combination of Higher-Order Modes *IEEE Antenna and Wireless Propagation Letters* **15** 710-713.
[8] Sharma A et.al 2016 Novel Feeding Mechanism to stimulate Triple Radiating Modes in Cylindrical Dielectric Resonator antenna *IEEE Access* **4** 9987-9992.
[9] Li B. and Leung K W 2008 On the Differentially Fed Rectangular Dielectric Resonator Antenna *IEEE Transactions On Antennas And Propagation* **56**.
[10] Kumar A V P 2016 A Probe fed Cylindrical Dielectric Resonator Antenna with Improved Polarization Purity *International Conference Proceedings of IEEE Indian Antenna Week Thiagarajar College of Engineering, Madurai.*

IOP Conf. Series: Materials Science and Engineering **331** (2018) 012013 doi:10.1088/1757-899X/331/1/012013

Facial Emotions Recognition using Gabor Transform and Facial Animation Parameters with Neural Networks

Aditya Harit; Col J. C. Joshi; K. K. Gupta

Department of Electrical and Electronics Engineering, Birla Institute of Technology & Science, Pilani

kgupta@pilani.bits-pilani.ac.in

Abstract. The paper proposed an automatic facial emotion recognition algorithm which comprises of two main components: feature extraction and expression recognition. The algorithm uses a Gabor filter bank on fiducial points to find the facial expression features. The resulting magnitudes of Gabor transforms, along with 14 chosen FAPs (Facial Animation Parameters), compose the feature space. There are two stages: the training phase and the recognition phase. Firstly, for the present 6 different emotions, the system classifies all training expressions in 6 different classes (one for each emotion) in the training stage. In the recognition phase, it recognizes the emotion by applying the Gabor bank to a face image, then finds the fiducial points, and then feeds it to the trained neural architecture.

1. Introduction

Facial emotion recognition is a highly challenging area in machine vision due to the variations in facial gestures, illuminations, poses, background, etc. Different cultures in different parts of the world do not express same emotion in exactly the same way. Even human emotions are compounded by the fact that humans being often experience simultaneous emotions. For example, a person being upset with something may also show anger or fear. This makes it difficult for an automated recognition system to grasp the emotion accurately. Researchers have identified six basic emotions that all humans show [15]: anger, fear, disgust, happiness, surprise and sadness. Apart from these, there are other expressions which are "voluntary" expressions, which differ from culture to culture. Those are hard to distinguish, and not the focus of this paper.

Emotion recognition can take two main approaches: geometrical based and general techniques [1-2]. Geometrical techniques rely on ratios and measurements between various facial aspects such as eyes, nose, lips, etc. General techniques treat the whole face as a single object. In this paper, we describe an approach which uses both Gabor transform and Facial Action Coding System (FACS), and then uses supervised learning to train a neural network to recognize emotions.

2. Proposed Algorithm

The feature extraction is the backbone of the whole system. If good features are not extracted, the training phase will invariably fail. The feature extraction in Facial Expression Recognition (FER) problems is of two types: aggregated template matching structures and geometric feature based [5]. In aggregated setups, a feature vector is obtained after processing the image as a whole. No attention is paid to the individual components of a face. In geometric based setups, a dimensionality reducing technique such as PCA or ICA and MPANN are judiciously used to achieve a low dimensional

IOP Conf. Series: Materials Science and Engineering **331** (2018) 012013 doi:10.1088/1757-899X/331/1/012013

portrayal. In geometric feature dependent structures, primary face and fiducial points are first found in face images. The separations between fiducial points and the normalized representation of the primary face components are calculated to form a feature vector. These holistic approaches are usually computationally costlier than template based, but are less dependent upon changes in illumination, scale, size, orientation of subjects' head, and location of the face in an image. The work in this paper is a mixed approach. We first find a set of fiducial points, and then use them to extract FAPS [14] and then extract a set of Gabor wavelet coefficients at each point through the operation of convolution. After that, we feed the feature vector thus obtained in a two layered MPANN for training purposes, and the network trained is used for testing purposes later on.

3. Implementation

There are many algorithms for extracting fiducial points. The one we used to be developed by Kazemi and Sullivan [4]. In this approach, the facial image is first normalized by moving it to a coordinate system based on a current prediction of the points. Afterwards, the fiducial points are found to estimate an update vector for the shape parameters. This procedure is repeated many times until the error rate in the estimated and current points become negligible. With this, a total of 68 points are extracted.

Two operations are performed on these points: First, a Gabor filter bank of 6 orientations and 3 spatial frequencies is created. Then, Gabor transform is applied at each of the extracted points for all the specific filters present in the filter bank. This is done by taking convolution at a point (a, b) in the image. Such a Gabor transform applied at a point results in a complex Gabor coefficient of that point, containing both real and imaginary points. Since the phase of the Gabor varies greatly, while the magnitude remains relatively constant, we take the magnitude as the single feature of a point. When the whole filter bank is applied, we get 18 such magnitudes for each fiducial point in the image. If we take a maximum of 68 such fiducial points in the image, we get a vector of 18x68 = 1224 elements for each image.

Twenty-one of the extracted fiducial points are chosen for extracting relevant facial ratios. These are: 5 points for each eye-ball (left, right, top, bottom and center of the eye), 4 for the mouth, and 2 for each eyebrow, and 3 for the jaw Next, 14 relevant ratios are chosen using these points, as listed in the Table 1.

Table 1. Iterations and accumulated error.

Serial number	Description
1	Left outer eyeball to left outer eyebrow
2	Right outer eyeball to right outer eyebrow
3	Left inner eyeball to left inner eyebrow
4	Right inner eyeball to right inner eyebrow
5	Height of the left eyeball
6	Height of the right eyeball
7	Width of the left eyeball
8	Width of the right eyeball
9	Distance between center of two eyeballs
10	Distance between the eyebrows and the forehead.
11	Middle of the mouth to jaw
12	Width of the mouth
13	Middle of the upper lip to middle of the lower lip

IOP Conf. Series: Materials Science and Engineering **331** (2018) 012013 doi:10.1088/1757-899X/331/1/012013

14	Lower lip to jaw

4. Training the Neural Network

Once the features are extracted then neural network is design to train. A two-layered architecture was used. The input vector for each image was a vector of dimensionality 1238. Out of these, 1224 were from the Gabor bank, while the last 14 were from the face ratio. The features from the Gabor bank are indicative of a holistic approach, while the ones from the facial ratios indicate a geometric approach. For the hidden layer, a hyper-tangent function was used.

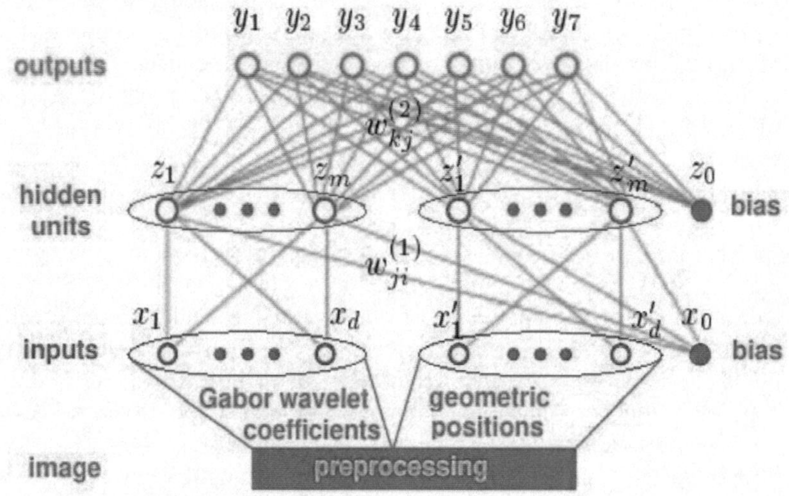

Fig 1. Architecture of neural network

As shown, the input was broken into two parts, which are fed to separate hidden layers. Note that there are no connections in the first layer among geometric and Gabor coefficient parts, since they are two separate pieces of information (holistic and geometric). The second layer then makes a decision depending upon the low dimensional set of features in the hidden nodes. Each output node is bound to a different emotion, so our system contains six output units (7 in the case where a neutral emotion is also present). When provided a test image, each output unit expresses the probability of correct emotion, which is related to that output unit. When training the neural network, RPROP was used, as it provides faster response than simple backpropagation.

5. Testing the proposed Model

The model was tested on two datasets: The Yale Face Database A, provided by Computer Science and Engineering U.C. San Diego. It contains The 165 grayscale images in GIF format of 15 individuals. There are 11 images per subject, one per different facial expression or configuration: center--light, w/glasses, happy, left--light, w/no glasses, normal, right--light, sad, sleepy, surprised, and wink. Since the task is for gesture recognition so the images are removed which had no categorization from this dataset. This left 90 images in the dataset.

The Japanese Female Facial Expression (JAFFE) Database, provided by Michael Lyons, Miyuki Kamachi, and Jiro Gyoba of Psychology Department in Kyushu University. This database contains 213 images of 7 facial expressions (6 basic facial expressions + 1 neutral) posed by 10 Japanese female models. Each image has been rated on 6 emotion adjectives by 60 Japanese subjects.

The testing of results was done in various ways. A general methodology was adopted for all 3 ways: First partition the data into two sets, one for training, and another for testing randomly. Then train the two-layer perceptron on the training data, and tests its performance on the testing data. This is done multiple times, and the final network is the average of all the resulting networks. The training was usually stopped after 250 epochs.

When tested on Yale database, the algorithm gave a result of 67% accuracy (33% error) in emotion recognition rate on the test data. This was when the number neuron in the hidden layer was 7. As the number of neurons increased, over-fitting was observed, increasing sharply after it rose above 10 units, giving an accuracy of less than 25% for 11 units. This is not surprising, as the number of images used was very low and the Yale database has greater ethnic diversity than the JAFFE database.

When tested on the JAFFE database however, the algorithm gave an accuracy of 81%, with similar trends in overfitting as for Yale database [Fig 3]. This is a good result; as conventional databases typically have around 500 or more images in them. The similar ethnicity of people in JAFFE databases also probably contributed to this effect.

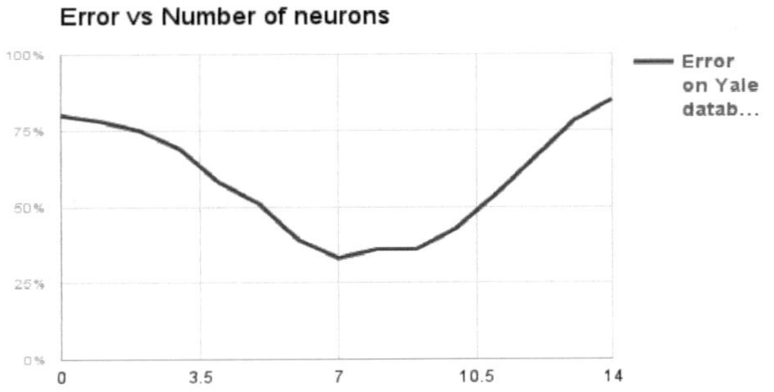

Figure 2. Error on Yale database

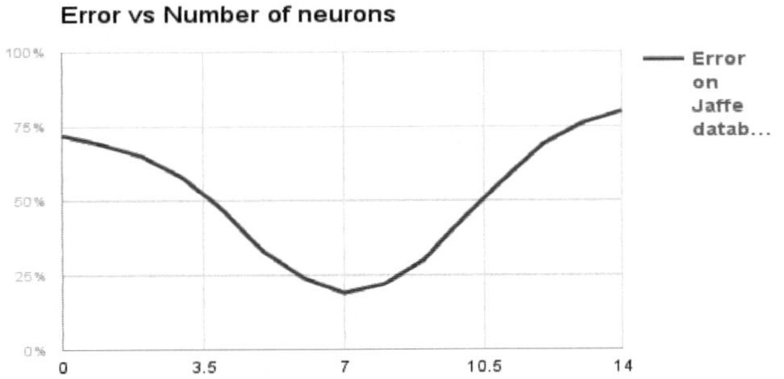

Figure 3. Error on Jaffe database

6. Conclusions and Future work

In this paper, facial emotion recognition is investigated. A hybrid model based on both geometric and holistic approaches is developed, which performed admirably considering the low number of images used. The accuracy could have been improved if automatic fiducial point marker improves. The ethnicity of subjects in YALE database caused problems, as the system could not effectively learn from the a relatively small sample of pictures. This is clearly shown as the JAFFE database yields a result of about 20% more accuracy. A single emotion was easily detected with a very high accuracy, of over 90%. This shows that although the system is not very adept at recognising different emotions, it can easily serve as a second layer of a system where only an approximate guess of emotion is made. Less than 5 units in the hidden layer do not give a satisfactory result, as the dimensionality of the input is not able to be conveyed by such a small number of nodes. Increasing the number of hidden nodes is not always beneficial, and leads to overfitting More work is currently being done on this, integrating new approaches.

References

[1] M. Suma, N. Sugie, and K. Fujimora. A preliminary note on pattern recognition of human emotional expression. In Proceedings of the 4th International Joint Conference on Pattern Recognition, pages 408–410, 1978.
[2] K. Mase. Recognition of facial expression from optical flow. IEICE Transactions E, 74(10):3473–3483, 1991.
[3] Z. Zhang et al. Comparison Between Geometry Based and Gabor Wavelets Based Facial Expression Recognition Using Multi-Layer Perceptron.
[4] Vahid Kazemi and Josephine Sullivan KTH. One Millisecond Face Alignment with an Ensemble of Regression Trees. Royal Institute of Technology Computer Vision and Active Perception Lab.
[5] T. Hastie, R. Tibshirani, and J. H. Friedman. The elements of statistical learning: data mining, inference, and prediction. New York: Springer--Verlag, 2001
[6] Christian Igel and Michael Hüsken. Empirical Evaluation of the Improved Rprop Learning Algorithm. Neurocomputing 50:105-123, 2003.
[7] Lee, Tai S. Image Representation Using 2D Gabor wavelets. IEEE Transactions on Pattern Analysis and Machine Intelligence, October 1996.
[8] M. Slavkovic et al. Face Recognition using Gabor filters, PCA, and Neural Networks 2013 20th International Conference on Systems, Signals and Image Processing (IWSSIP)
[9] C. Bishop. Neural Networks for Pattern Recognition. Clarendon Press, Oxford, 1995.
[10] PJ Werbos. Backpropagation through time: what it does and how to do it. Proceedings of the IEEE, 1990.
[11] JR Movellan. Tutorial on Gabor filters. Open Source Document, 2002 - cs.olemiss.edu
[12] L Shen, L Bai. A review on Gabor wavelets for face recognition, Pattern analysis and applications, 2006.
[13] A. Hyvarinen, J. Hurri, P.O. Hoyer. Natural Image Statistics A probabilistic approach to early computational vision. Springer, February, 2009.
[14] A. Raouzaiou, S. Ioannou, K. Karpouzi,N. Tsapatsoulis, S. Kollias, R. Cowie. An Intelligent Scheme for Facial Gesture Recognition. Springer, 2003.
[15] P. Ekman and W. Friesen. The Facial Action Coding System. Consulting Psychologists Press, San Francisco, CA, 1975

IOP Conf. Series: Materials Science and Engineering **331** (2018) 012014 doi:10.1088/1757-899X/331/1/012014

Generation of high power sub millimeter radiation using free electron laser

J Panwar, S C Sharma, P Malik, M Yadav and R Sharma

[1]Department of Applied Physics, Delhi Technological University (DTU), Bawana Road, Delhi- 110042, India

E-mail: panwar.jyots2010@gmail.com

Abstract. We have developed an analytical formalism to study the emission of high power radiation lying in the sub millimetre range. A relativistic electron beam (REB) is velocity modulated by the pondermotive force exerted by the laser beams. After passing through the drift space, the beam gets density modulated which further interacts with the strong field wiggler and acquires a transverse velocity that couples with the modulated density of the beam in the presence of ion channel which contribute to the non-linear current density which further leads to the emission of the radiation. The output radiation can be modified by changing the wiggler parameters and the energy of the electron beam. The power of the output radiation is found to increase with the modulation. The obtained radiation can be employed for various applications.

1. Introduction

The emission of high power terahertz (sub millimetre) radiation has attained a great interest in the field of research due to its potential applications including spectroscopy, remote sensing, biomedical, communication and security imaging [1-4]. THz waves are preferred over microwaves for broadband mobile communications and local area network due to their high frequency and more bandwidth [5]. A modulated electron beam is an efficient source for the generation of radiation using free electron laser as a high frequency device [6-7]. Xiang *et al* [8] have studied experimentally the generation of terahertz radiation using electron beam, modulated by the down conversion of the frequency of optical lasers and observed that the radiation can be altered by the by varying the lasers frequency and the electron beam energy. Bhasin and Sharma [9] have employed an energy modulated electron beam in a dielectric loaded waveguide in the presence of external magnetic field. They observed that the gain and efficiency of the radiation increases with the modulation and the growth rate decreases monotonically with the radiation frequency. Ion channel acts as guiding medium for the electron beam. The emission of terahertz radiation using helical wiggler in the presence of ion channel has been studied by Hasanbeigi *et al* [10]. They concluded that the power of the output radiation increases with the ion density, energy and radius of the electron beam.

In this paper, we have analysed the power of the output terahertz radiation, emitted by the interaction of a density modulated relativistic electron beam with an electromagnetic wave wiggler in the presence of an ion channel.

2. Physical Model

We consider two laser beams which employ a pondermotive force on interacting with the REB which leads to velocity modulation of the beam. When this beam further travels through the drift space, the current modulation converts into density modulation. This mechanism of premodulation is similar to that of klystron [11]. The modulated beam density (n_{mb}) [12] is obtained as

$$n_{mb} = -2n_0 J_1\left(\frac{\omega_b \rho \xi}{v_b}\right)\cos\left(\omega_b \tau_3 - \phi\right), \tag{1}$$

where n_0 is the initial beam density, v_b is the initial velocity of the beam, ω_b is the frequency difference of two lasers in terahertz range, ρ is the length of drift space, $\xi = \frac{k_b \rho}{2\gamma_0^4}\frac{c^2}{v_b^2}\frac{eA_I}{m\omega_I c}\frac{eA_{II}^*}{m\omega_{II} c}\frac{\sin(\varphi)}{\varphi}$, where $\varphi = \frac{(\omega_b - k_b v_b)\rho}{2v_b}$ is the phase angle of the modulation, k_b is the beam wave vector, A_I, A_{II} and ω_I, ω_{II} are the amplitudes of electric field and frequencies of two laser beams, respectively.

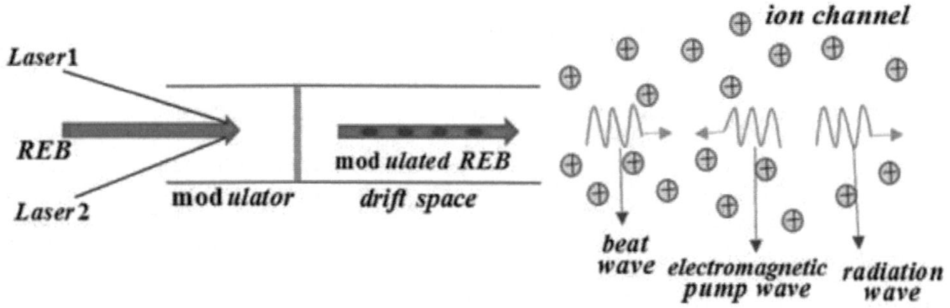

Figure 1: Interaction of laser modulated REB with the electromagnetic pump wave in the presence of ion channel.

Further we consider the propagation of modulated REB in the field of the electromagnetic wave and the uniform density ion channel. The modulated REB acquires normalized oscillatory velocity $\left(\overrightarrow{\beta_{bw}}\right)$ due to the pump wave and the ion channel field.

$$\overrightarrow{\beta_{bw}} = \frac{\Omega_0(\beta_b + \beta_0)^2}{\omega_+^2 - (\beta_b + \beta_0)^2}\cos(k_0 z + \omega_0 t), \tag{2}$$

where $\Omega_0 = \frac{eB_0}{\gamma_0 mc^2 k_0}$ is the normalized frequency of wiggler, B_0 is the wiggler magnetic field, $\left(\omega_0, k_0\right)$ is the frequency and wave vector of the wiggler wave respectively, β_b is the normalized beam velocity $\omega_+ = \left(\frac{2\pi e^2 n_+}{\gamma_0 m_+ k_0^2 c^2}\right)^{1/2}$ is the normalized frequency of ion channel, n_+ is the ion density, m_+ is the mass of ions.

This oscillatory velocity couples with the modulated beam density to give non-linear current density. The magnetic field of the terahertz wave calculated using vector potential helps in evaluating the average power ($4\pi r^2 S_{av}$, $S_{av} = \hat{r}\frac{c}{2\mu_0}\left|\overline{B}\right|^2$ is the average poynting vector) of the radiation.

IOP Conf. Series: Materials Science and Engineering **331** (2018) 012014 doi:10.1088/1757-899X/331/1/012014

$$\text{Average power} = \frac{\pi c \mu_0 \omega_1^2 n_b^2 e^2 \Omega_0^2 l_b^2 r_0^2 (\beta_b + \beta_0)^4}{2[\omega_+^2 - (\beta_b + \beta_0)^2]^2} J_1^2\left(\frac{\omega_b \rho \xi}{v_b}\right), \quad (3)$$

where l_b is the electron bunch length, r_0 is the size of electron beam, ω_1 is radiation frequency.

3. Results and Discussion

Typical parameters used for calculation are: $v_b = 0.95c$, $\gamma_0 = 3.2$, $n_b = 4\times10^{14} cm^{-3}$, $\rho = 500\mu m$, $l_b = 0.3cm$, $\frac{eA_I}{m\omega_I c} = 0.05$, $\frac{eA_{II}}{m\omega_{II} c} = 0.05$.

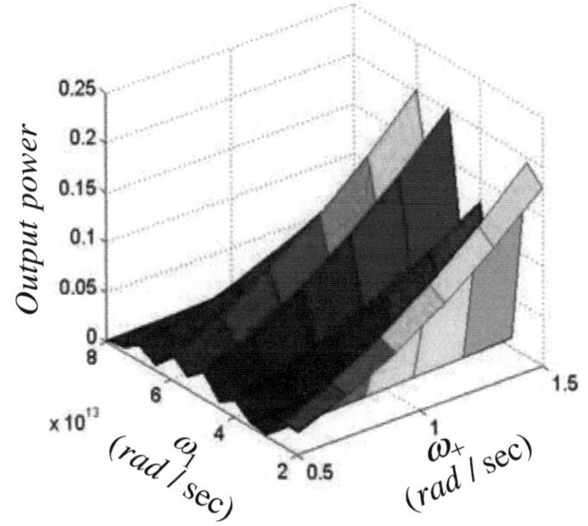

Figure 2: The figure shows the variation of output average power ($4\pi r^2 \vec{S}_{av}$) with radiation frequency (ω_1) and ion channel frequency (ω_+).

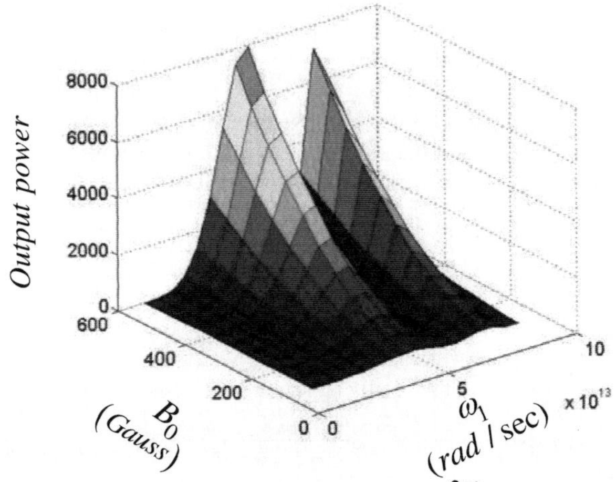

Figure 3: The figure shows the variation of output power ($4\pi r^2 \vec{S}_{av}$) with radiation frequency (ω_1) and pump field amplitude (B_0).

From figure 2, it can be seen that the average output power of the radiation for a fixed value of ion density, increases with the radiation frequency upto a certain maximum value and then decreases and further repeats the pattern. This pattern is followed by the Bessel's function which appears due to the modulation of the electron beam density. The power also increases with the frequency of the ion channel. From figure 3, it is observed that the output power increases with the amplitude of magnetic field of the pump wave. It also increases with the radiation frequency and shows Bessel's function character.

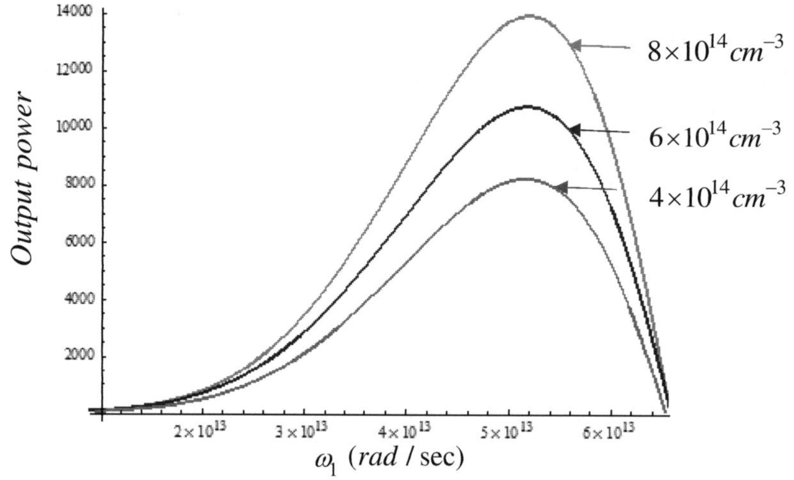

Figure 4: The figure illustrates the variation of output average power with the radiation frequency for different beam densities.

It can be inferred from figure 4 that the output average power of the radiation wave increases with the increase of density of the electron beam as more the number of electrons will be available to transfer their energy to the radiation wave. Further, it increases initially with the output frequency, attains a maximum value at a point where resonance occurs and then starts decreasing.

4. Conclusion

The terahertz radiation emitted using a density modulated beam with the effects of ion channel can be modified by varying the parameters of electron beam, amplitude of the pump field and frequency of the ion channel. Increase in ion channel density helps in enhancement of the output power. The modulation of the beam is affected by the parameters of the laser beams and the electron beam. The power scales as square of the electron beam density and the amplitude of the pump wiggler field. The tuned radiation thus obtained can be used for future practical applications.

References

[1] Malik A K, Malik H K and Kawata S 2010 Investigations on terahertz radiation generated by two superposed femtosecond laser pulses *J. Appl. Phys.* **107** 113105
[2] Chen Z Y, Li X Y and Yu W 2013 Intense terahertz emission from relativistic circularly polarized laser pulses interaction with overdense plasmas *Phys. Plasmas* **20** 103115
[3] Pickwell E and Wallace V P 2006 Biomedical applications of terahertz technology *J. Phys. D: Appl. Phys.* **39** 301
[4] Fitch M J and Osiander R 2004 Terahertz waves for communication and sensing *Johns Hopkins Apl TechnicaL Digest* **25** 348
[5] Zhu B, Chen Y, Deng K, Hu W, and Yao Z S Terahertz Science and Technology and Applications 2009 *PIERS Proceedings* 1166

[6] Freund H P, O'Shea P G and J. Neumann 2003 Simulation of prebunching in free-electron lasers *J Nucl. Instrum. Methods Phys. Res. A* **507** 400
[7] Sharma S C, Panwar J and Sharma R 2017 Modeling of terahertz radiation emission from a free-electron laser *Contrib. Plasma Phys.* **57** 167
[8] Xiang D and Stupakov G 2009 Enhanced tunable narrow band THz emission from laser modulated electron beams *Phys. Rev. ST Accel. Beams* **12** 080701
[9] Bhasin A and Sharma S C 2009 Effect of beam premodulation and guide magnetic fields on slow wave free electron laser: Nonlocal theory *IEEE J. Quantum Electronics* **45** 1129
[10] Hasanbeigi A, Mehdian H and Gomar P 2015 Enhancement of terahertz radiation power from a prebunched electron beam using helical wiggler and ion-channel guiding *Phys. Plasmas* **22** 123116
[11] Liao S Y 1990 Microwave Devices and Circuits (Prentice Hall, New Jersey)
[12] Kumar M and Tripathi V K 2012 Terahertz radiation from a laser bunched relativistic electron beam in a magnetic wiggler *Phys. Plasmas* **19** 073109

IOP Conf. Series: Materials Science and Engineering **331** (2018) 012015 doi:10.1088/1757-899X/331/1/012015

Implementation of Smart Metering based on Internet of Things

Milanpreet Kaur, Dr. Lini Mathew, Alokdeep and Ajay Kumar

Department of Electrical Engineering, NITTTR, Chandigarh

Milanpreetkaur502@gmail.com

Abstract. From the aspect of saving energy, there is a continuous modification in communication technology and information in order to satisfy all customers demand. Today customers are demanding for accurate energy measurement, timely data and for good customer services. The best solution is smart grid system with various communication technologies which can be cost effective and electrical section to have a bidirectional communication in which information about electrical energy consumption is shared between consumers as well as by utility for remote checking. This paper describes the monitoring of energy consumption with Arduino Uno board and Ethernet using IoT (Internet of Things) concept. This proposed design eliminates human inclusion in the conservation of electricity. The consumer can receive the information about consumption of energy by using IP address on their devices. The web client code is uploaded for checking the client information such as location, content, connection, and disconnection to the web server. This proposed system gives reliable and accurate information regarding electrical energy management system (EMS) through Internet of things (IoT).

1. Introduction

Energy is fundamental need of our life because of mechanical development and urbanization. Energy catastrophe is the main issue confronted by our society. One of the solutions is to control, analyze and reduce of electrical power consumption in households. The purchasers must be encouraged by providing them with a perfect arrangement- i.e. the idea of using Ethernet communication via Internet of Things (IoT). In this modern period of Internet of Things, a physical world such as machines and sensors can be interfaced through the web. The things or items can be modified into smart things by providing it exclusive identification. The sensors and devices can communicate and share information through the web. We can analyze and control the sensors and devices at whatever time from any part of the world. IoT is an essential part of the smart metering/grid. An IoT is a twenty-first-century aspect in which consumers can be associated with the web page and establish communication between consumer and supplier through sensors and actuators. The IoT is a hybrid prototype which is developing in the area of remote media transmission.

2. Literature Survey

Currently, numerous system of Automatic Meter Reading (AMR) is implemented using Bluetooth, General Packet Radio Access (GPRS) and Global System for Mobile communication (GSM) [1]. Utilizing GSM technology remote sharing of information of energy consumption is possible. But there can be chances of missing SMS thus result in decreasing the reliability and efficiency of the system. GPRS is difficult to implement for long distance two-way communication between utility and consumers.

For automated billing system, communication is being done by the use of GSM and Ad-Hoc wireless routing protocol [3]. Each home is connected through the wireless system to central/utility office to access all the readings of energy meters using radio frequency system. The utility office receives continuous updates from energy meter about the units consumed by consumer and bill to be paid through GSM messages. The main disadvantage of this system is that it needs a human workforce to read a meter. If the customer is not available at home, the operator cannot able to read the meter.

Two-way communication has been an important function of the smart energy meters. Through wireless or wired connections such as GSM/GPRS, M-bus, and ZigBee, smart energy meter communication can be realized [4]. The different communication protocols used for two-way communication between consumer and supplier have been defined in terms of data rate, frequency, coverage range and limitations, stability, cost and their limitations and is shown in Table 1.

Table1. Comparison of different communication protocols

Technology	Cost	Communication protocol	Frequency	Coverage range	Limitation
GPRS	Medium	Stable	900-1800Mhz	1-10km	Data rates low
3G	High	Stable	1.92Ghz-1.98Ghz, 2.11-2.17Ghz	1-10km	Spectrum costly
GSM	Low	Stable	900-1800Mhz	1-10km	Data rates low
WiFi-Max	Medium	Stable	2.5-3.5Ghz	10-50km	Not widespread
PLC	Low	Very Stable	1-30Mhz	1-3km	Noisy channel
SCADA	High	Stable	Up to 1.54 Mhz	Short distance	Expensive
M-Bus	High	Least Stable	2.4-4.8Mhz	1000m	expensive
ZigBee	Medium	Least Stable	2.4Ghz,868-915Mhz	30-50m	Short range

The wireless embedded based AMR system is implemented using ARM7 Microcontroller [5]. The readings of energy consumption are sent to the utility by using RF link and then to the Web server through GSM. It also provides the facility of tamper detection and recharging of energy meter.

Advance metering infrastructure (AMI) based Smart Energy Meter are connected to each other by gateway presents an implementation and execution of Automatic meter reading system [6], [7]. By using PLC lines send data to utility computer and gateway communicates with utility computer via GSM. The power line communication is implemented by using LonWork bus technology.

The overview of smart grid technologies in electrical utility based on EMS and IoT has been discussed in [8], [14]. The different communication protocols used for energy management involves the two-way communication between supplier and consumers. Fig 1 shows idea about various different protocols of communication and technologies can be used.

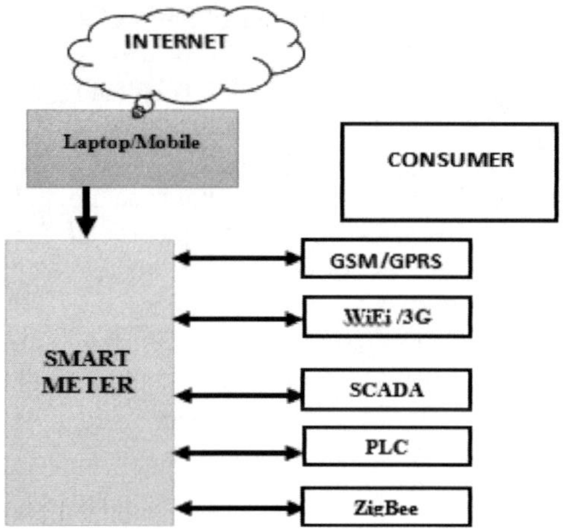

Fig. 1 Different technologies for EMS

For large and ordinary power consumers GSM based AMR has implemented called hybrid automatic meter reading [9], [11]. AMR is enabled by using ZigBee and GSM module in which ZigBee is interfaced with energy meter to collect data and GSM is used for transmitting message/SMS to the utility.

The better metering solution is developed with an energy chip and the measurement of energy through digital communication technique [12]. The information about energy is sent to the utility through ZigBee and a software is used for data management along with billing of total energy consumed.

The energy meter within home display based on GPRS has been developed [13]. The data is sent to the cloud through GPRS. LabVIEW is used for home display for the intelligent load scheduling of appliances with respect to time of use (ToU) pricing.

3. Block Diagram Description

The goal of this proposed energy metering system improves the customer demand for higher energy measurement, accuracy, and more timely data. The information of energy measurement can be used by the consumer as well as by utility. The block diagram consists of an Arduino Uno board, an energy meter, and IoT through Ethernet. This paper presents communication through Ethernet to send data to the server. The simplified block diagram of the system is shown in Fig.2.

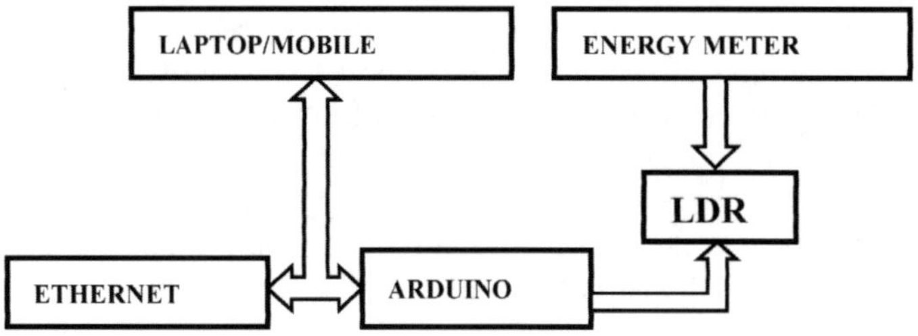

Fig. 2 Simplified block diagram of system

4. Over view of Arduino and Ethernet shield

4.1. Arduino

Arduino is an open-source gadget simple-to-use for programming and hardware. It can influence devices, similar to motor, lights and different actuators by getting inputs from different sensors. All the activity performed by Arduino is modified and programmed by the microcontroller available on board by means of Arduino programming language. The projects of Arduino can communicate with other programming applications running on a PC and different types of hardware. The Arduino Uno board is shown in Fig. 3.

Fig. 3 Arduino Uno Board

The features of Arduino Uno microcontroller board is shown in Table 2.

Table.2 Arduino Uno features

Microcontroller	ATmega 328
Operating Voltage	5V
Supply Voltage	7-12V
Maximum supply voltage	20V
Analog I/O Pins	6
Analog I/O Pins	14 (out of which 6 used as PWM output)
DC current per I/O Pin	40mA
For 3.3 V DC current	50mA
Flash Memory	32kb out of which 0.5kB used by boot loader
EEPROM	2kB
SRAM	1kB
Clock	16MHz

4.2. Ethernet shield

The Ethernet shield can be connected to Arduino Uno board through a library of Ethernet. This shield is compatible with all former version which is W5500 chip. The Ethernet shield has authorized MAC address and an internet protocol (IP) address which is fixed using the Ethernet.Begin(). A MAC address of every device is unique globally for identification. The Arduino Ethernet shield Uno board as shown in Fig. 4.

Fig. 4 Arduino Ethernet shield

The features of Arduino Ethernet shield features is shown in Table 3.

Table.3 Arduino Ethernet shield features

Operating voltage	5V
Controller of Ethernet	W5500 with internal 32K buffer
Speed	10/100Mb
Connection	Arduino on SPI port

5. Circuit Diagram & Its Working

The hardware interface circuit consists of an Arduino-Uno board, Energy meter, a Light dependent resistor (LDR), Ethernet shield, IoT such as Wi-Fi. The role of Wi-Fi is to share the energy meter information on the web page which can be achieved via IP address as well as on device also. The energy meter consists of LED which blinks continuously to indicate the unit and cost of energy consumption by customers. For experimental purpose, an LDR is placed on the LED of Energy meter. The voltage of LDR is set to receive the accurate result of energy measurement and the code is uploaded to Arduino board. The output is fed to the Arduino board at analog pin A0, 5V, and ground, and the energy is calculated. On serial monitor of Arduino board, the unit and cost of electrical energy consumed can be displayed. During two-way communication between customers and utility, if the web service fails while sending a message then it will be difficult to share or send data. The best solution to this problem is Ethernet communication. The initial cost of Ethernet is only required and communication is achieved by small bandwidth along with telephone /mobile network. The circuit diagram of smart metering based on IoT is shown in Fig.5.

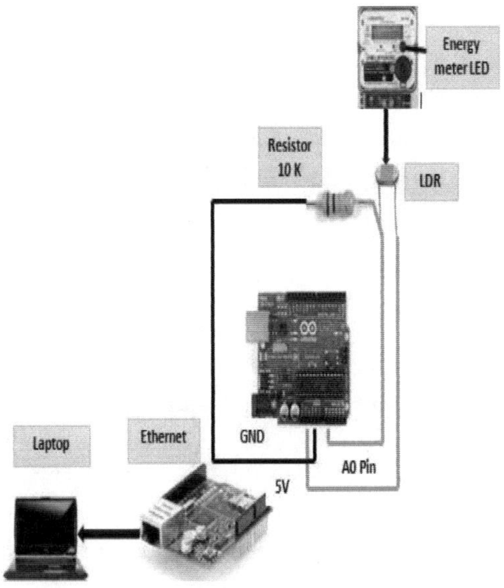

Fig. 5 Circuit diagram of system

Two-way communication between customer and utility can be done by using IoT through Ethernet shield connected to Arduino Uno board. The communication part for sending the data of energy consumption to a web page through Ethernet communication with low cost. Web server program is uploaded for the two-way communication between the consumer and utility. This Ethernet communication enables consumer and utility to share information about the energy consumption through an IP address. The utility also receives the client's information using same IP address by web client program.

By using the Ethernet communication the customer and utility receive the real status of energy consumption with less cost as compared to other communication protocols. The information can be made private by using web client program uploaded in Arduino board to get the details of the client connected and disconnected to IP address for monitoring the electrical energy consumption. When the client is connected the information about the electrical energy consumption along with client information can be seen on the serial monitor of Arduino. This proposed system is cheap as compared to other communication protocols. This server system will be able to control and manage electricity usage by consumers.

6. Flowchart of Process

From the flow chart diagram, it can be seen that the energy consumption is continuously monitored by Arduino via Ethernet. LDR sensor which is used as a button state in this project checks the unit and cost of energy consumption by the consumer through the counter code uploaded in Arduino Uno board. If the LDR increases above the set voltage 550 then counter starts to count the pulses from the LED of Energy meter. With the counting of pulses by LDR, it gives the cost, unit increases with respect to time. If the LDR voltage is less than set voltage then there is no increase in unit and cost, it shows LDR off. The output of energy consumption is uploaded in web server and web client code through Ethernet. The utility/electricity board and consumer by using IoT through Ethernet can get the real time status of energy consumption only by IP address. The flowchart diagram is shown in Fig.6.

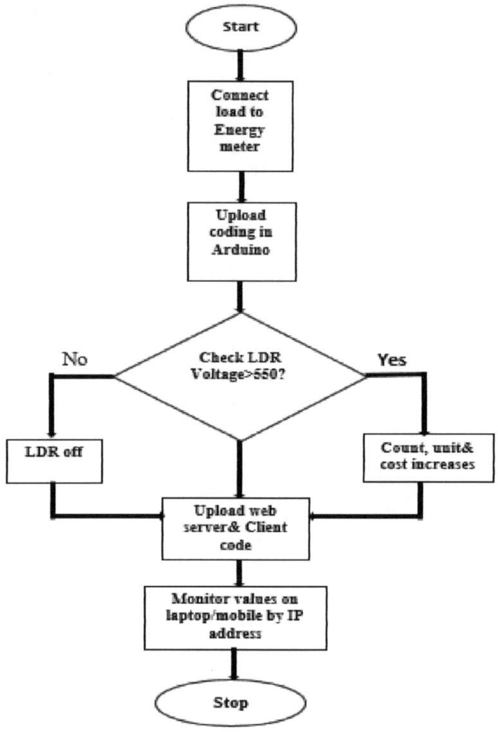

Fig.6 Shows the process of flow chart

7. Hardware Implementation

In this proposed system LDR is placed on the LED of energy meter. The counter program is uploaded in Arduino Uno board for counting the blinking of LED. The counter continuously increased with respect to blinking of LED which is detected by LDR. The LDR is used as a button state to increase the counter to give the information about unit and cost of energy consumption.

The total output of unit and cost of energy consumption by the consumer can be monitored. The implementation of proposed system hardware is shown in Fig. 7.

Fig. 7 Proposed system hardware

The data of energy consumption regarding unit and cost can be seen on Arduino serial monitor as shown in Fig.8.

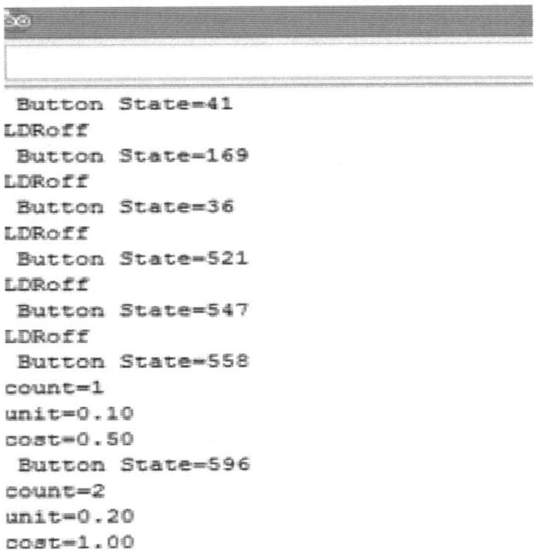

Fig.8 Output of energy consumption on Arduino serial monitor

8. Results & Discussion

The Ethernet has been used for the communication part and transmitted the information of energy consumption to a web page. The collected data of energy consumption is sent to a web page by uploading the web server code and web client code for clients connection in Arduino board connected with Ethernet shield. On Arduino serial monitor the unit and cost of energy consumed along with IP address have been displayed as shown in Fig 9.

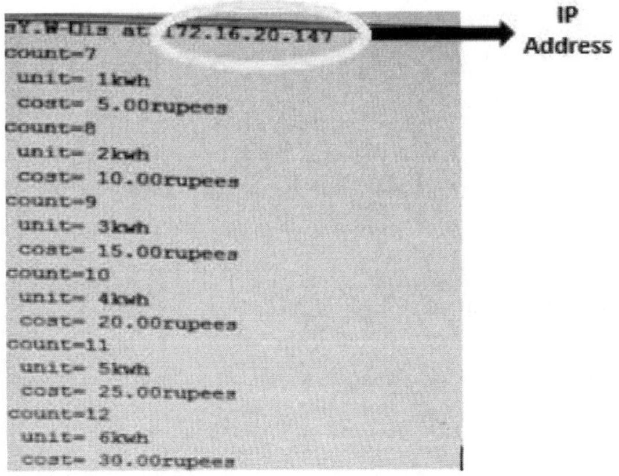

Fig. 9 Web server output on Arduino serial monitor

The consumer can receive the information about energy consumption just by filling IP address on their mobile phones at anytime and anywhere. The customer can check the information from any edge of the world by using Ethernet communication called IoT. The consumer can receive the information about energy consumption just by filling IP address on their mobile phones at anytime and anywhere. The information of electrical energy consumption on mobile through IP address is shown in Fig 10.

IOP Conf. Series: Materials Science and Engineering **331** (2018) 012015 doi:10.1088/1757-899X/331/1/012015

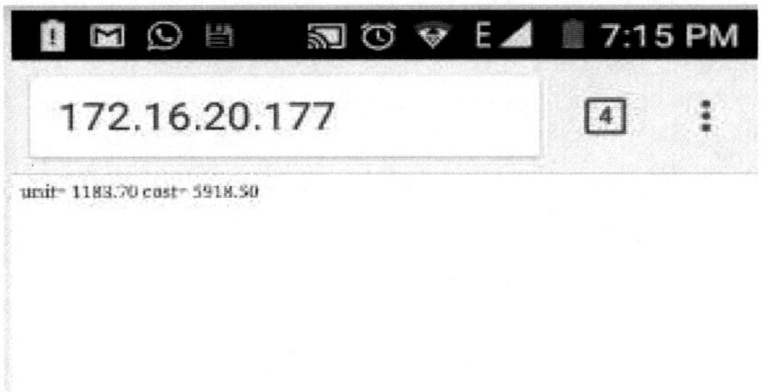

Fig. 10 Information of energy consumption on mobile

After web client program code is uploaded for the monitoring of electrical energy consumption, the client information such as location, HTTP can be seen on the Arduino serial as well as on mobile phone for information security. When the connection is successful, the information regarding unit and cost of electrical energy consumption can be seen by the consumer as well as by utility. Fig.11 and Fig.12 show the web client connection and disconnection with the web page with their location and IP address used.

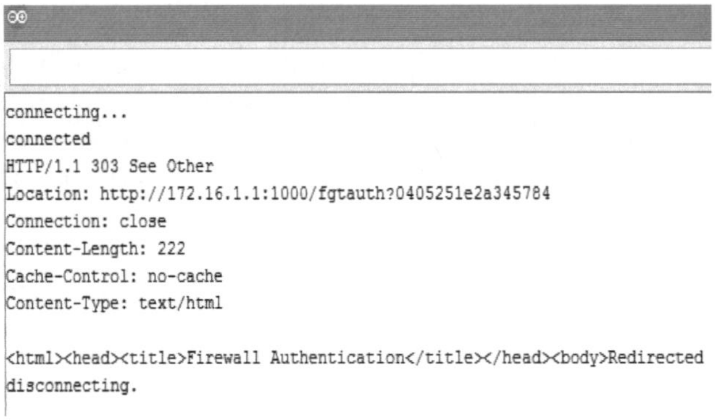

Fig 11. Shows the web client data

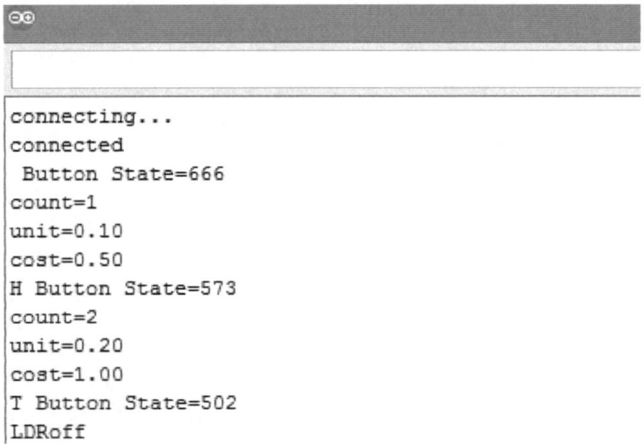

Fig 12. Shows output when client is connected

9. Conclusion & Future Scope

The goal of this project work was designed for the two-way communication of energy consumed in the home can send to the web server by using Ethernet communication (IoT). By using the Ethernet communication the customer and utility receive the real status of energy consumption with less cost as compared to other communication protocols. This proposed system is cheap as compared to other communication protocols. The collected data about energy consumption is monitored by consumer and supplier at anytime, anywhere from any part of the world.

Improvements can be achieved in the controlling and monitoring of energy online from any edge of the world.

References

[1] S. Male, P. Vethekar, K. Moore "A Smart Wireless Electronic Energy Meter Reading Using Embedded Technology", *International Journal of Engineering Research and Application*, Vol. 4, No.1, pp.145-147, January 2014.

[2] M. Wei, S.H. hang, M. Alam, "An IoT-based energy management platform for industrial purposes", *Applied Energy Elsevier Limited*, Vol. 164, pp. 607-619, Feburary 2016.

[3] A.Vijayaraj, S. Saravanan "Automated EB Billing System using GSM and Ad-Hoc Wireless Routing", *International Journal of Engineering and Technology*, Vol.2, No. 5, pp. 343-347, October 2010.

[4] M. A. Alahmad, P. G. Wheeler, A. Schwer, J. Eiden, and A. Brumbaugh, "A comparative study of three feedback devices for residential real-time energy monitoring," *IEEE Trans. Industrial Electronics*, vol. 59, pp. 2002-2013, Apr. 2012.

[5] N. Gupta, D. Shukla "Design of Embedded based automated meter reading system for real time", *IEEE Student's Conference on Electrical, Electronics and Computer Science*, pp.1-6, Bhopal, July 2016.

[6] S. Kim, E.Y.Kwon, M.Kim, J.H.Cheon, S.Ho Ju, "A Secure Smart Metering protocol over Power Line Communication", *IEEE Transactions on Power Delivery*, Vol. 26, No. 4, October 2011, pp. 2370.

[7] M. Popa, "Data Collecting from Smart Meters in an Advanced Metering Infrastructure", *IEEE 15th International Conference on Intelligent Engineering Systems (INES)*, pp. 137-142, Slovakia, 23 June 2011.

[8] S. Jain, V. Kumar, "Survey on Smart Grid Technologies – Smart metering, IoT and EMS", *International Journal of Engineering Research and Application*, pp.1-6, April 2014.

[9] A. Handoko, M. Y. Nayan, M. Awan "Hybrid Automatic Meter Reading System", *International Journal of Advanced Research in Computer Science and Software Engineering*, Volume 2, No. 7, pp.361-365, Malaysia, July 2012.

[10] K. Gill, S.H. Yang, F. Yao, and Xin Lu, "A ZigBeeBased Home Automation System," *IEEE Transactions on Consumer Electronics*, Vol. 55, No. 2, May 2009, pp. 422-430.

[11] Imran A. Zualkernan, A. R. Al-Ali, Mustafa A. Jabbar, Imad Zabalawi, and Ahmed Wasfy, "InfoPods: Zigbee-based remote information monitoring devices for smart-homes," *IEEE Transactions on Consumer Electronics*, Vol. 55, No. 3, August 2009, pp. 1221.

[12] Maity, "Intelligent Online Measurement and Management of Energy Meter Data through Advanced Wireless Network," *International Conference on Devices and Communications (ICDECOM)*, pp.1-4, Mesra, 25 Feb. 2011.

[13] L.C.Saika, H. Das, N.B Dev Choudhry, T.Malakar, "GPRS enabled smart energy meter and automation of home appliances", *IEEE Annual India Conference*, pp. 1-5, Bangalore, December 2016.

[14] H. Li, S. Gong, L. Lai, R. C. Qiu, D.Yang, "Efficient and Secure Wireless Communication for Advance Metering Infrastructure in Smart Grids," *IEEE Transactions on Smart Grid*, Vol.3, No. 3, September 2011, pp.1540.

[15] S.Wang, L.Cui, J.Que, D.H.Choi, X.Jiang "A Randomized Response Model for Privacy Preserving Smart Metering," *IEEE Transactions on Smart Grid*, Vol.3, No. 3, September 2012, pp.1317.

IOP Conf. Series: Materials Science and Engineering **331** (2018) 012016 doi:10.1088/1757-899X/331/1/012016

Improvement in reverse bias leakage current of Ni/4H-nSiC Schottky barrier diodes via MeV selective ion irradiation

Vibhor Kumar[1,2,3], A S Maan[3]

[1]Indian Institute of Technology, Delhi, India.
[2]Central Electronics Engineering Research Institute, (CSIR-CEERI), Pilani, India.
[3]Maharshi Dayanand University, Rohtak, India.

Email: vibhorsoni123@gmail.com

Abstract. Present work deals with enhancement in reverse bias leakage current of Ni/4H-nSiC schottky barrier diode (SBD) via irradiating with 200 MeV $^{107}Ag^{14+}$ ions at a fluence of 10^{13} ions/cm^2 in selective way. Compared to pristine SBD, the reverse bias I-V characteristics of selectively irradiated SBD show significant improvement in leakage current. Role of quodons have been discussed to rationalize the performance of irradiated SBD.

1. Introduction

The comfort, transportation and healthcare of modern society depend upon power devices. Therefore, significant interest has been generated in advancement of power generation, distribution and management technologies. The key ideas in development of power devices are their low switching loss, low conduction loss and high temperature tolerance [1]. Outstanding properties of SiC i.e. wide band gap, high thermal conductivity, low intrinsic carrier concentration, high breakdown electric field, high chemical inertness, etc., makes it promising material for fabrication of power devices. Miniaturization of power devices is requirement for advancement in their technology. But, proper design for reduction in electric field at the edges of miniaturized devices is a challenge yet [2]. For power devices, many edge termination techniques like field plate (FP), floating metal rings (FMR), junction termination extension (JTE), ion implantation, etc. have been reported in literature [1, 3]. Moreover, aforementioned techniques increase the size of device and hence their cost [2].

Moreover, literature reveals that swift heavy ion (SHI) irradiation can potentially modify the surface and bulk structures of the material, in a controlled way at nanoscale range [4]. Recently, it has been suggested that electronic energy loss by SHIs in crystalline medium transported as energy packets, namely, quodons [5, 6]. These quodons were found effective in interaction with lattice defects and move them to the boundary of masked (cold region) and irradiated region (hot region), so called "transition region". As a result, the atomic structure will be modified at the "transition region". This novel approach has been implemented for realization of good quality Si based p-n junction [2]. So, the purpose of this experiment is to demonstrate the implications of MeV selective ion irradiation in wide bandgap material i.e. SiC. To the best of our knowledge, this is the only report that integrates SHI irradiation with wide bandgap materials and based microelectronics. It is anticipated that this work could become a recipe for development of next generation solid state devices.

2. Experimental Details

N-type 4H-SiC wafer of <0001> orientation and 2 inch in diameter with n-type Si face of 25 μm epitaxy with doping of 5×10^{14} cm^{-3} was procured from CREE Res. Inc. USA. These wafers were cleaned as per standard chemical procedure i.e. degreasing, Radio Corporation of America (RCA), piranha followed by a short dip in dilute (2%) hydrofluoric acid. Thereafter, the wafer was baked for 4–6 hours in the oven for moisture baking [7]. Chemically cleaned molybdenum metal mask was used to (1) pattern the circular shaped SBDs of diameter 1.6 mm and (2) selectively irradiating the SBDs. This shadow mask was then fixed on the cleaned Si- face of the cleaned SiC wafer using a suitable adhesive. Nickel as schottky metal (200 nm) was deposited on the masked Si-face of the wafer using e-beam evaporation technique. Afterwards, Ohmic contact was formed on the C-face of the wafer using bilayer metallization of Ti/Au (50/150 nm). The selective ion irradiation was then performed using 200 MeV of $^{107}Ag^{14+}$ ions with fluence of 1×10^{13} ions/cm^2. A schematic diagram for experimental plan is given in figure 1.

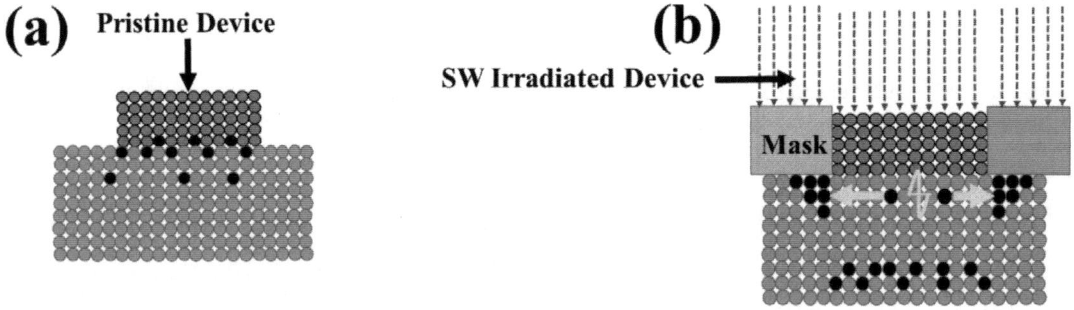

Figure 1 Schematics representation of (a) 2D atomic arrangement of chemically cleaned SiC (green balls) wafer procured from CREE Res. Inc. USA and deposited Ni as Schottky metal (red balls). Black balls represent the wafer in-buit defects in material. (b) The shadow mask was kept placed in some portions for selective way (SW) irradiation. Yellow wave represent the generated quodons which interact with defects and let them move towards edges of the SBD.

The current-voltage characteristics of these devices, before and after ion irradiation were acquired using Keithley 4200 semiconductor characterization system (SCS) .

3. Results and Discussion:

The reverse characteristics of both pristine and selectively irradiated SBDs were acquired by sweeping the DC bias from 0 to -200 V and shown in figure 2.

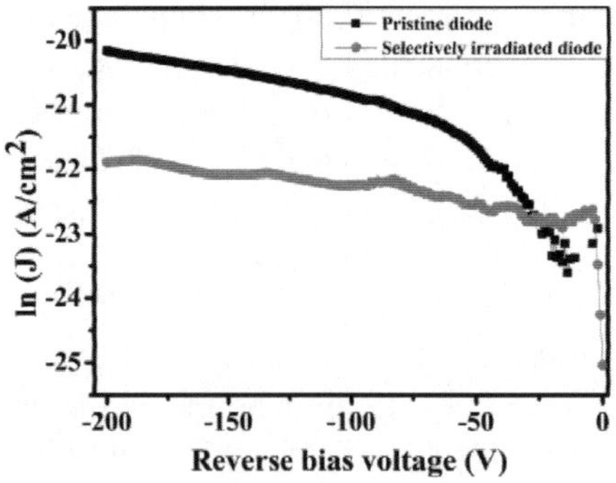

Figure 2 Reverse bias leakage current density in pristine and selectively irradiated SBDs.

It can be seen from figure 2 that selectively SHI irradiation improved the reverse bias characteristics. In case of pristine SBD, the reverse bias leakage current density is estimated to 1.77×10^{-9} A/cm^2. On the other hand, in case of selectively irradiated SBD, leakage current density is calculated to be 3.12×10^{-10} A/cm^2 which is an order of magnitude lower than pristine SBD. These findings are in contrast to reported literature, where it has generalized that SHI irradiation at higher fluence produce generation centres in material which causes more leakage current in SBDs [8-11]. Moreover, if initially the leakage current is very high in SiC based SBDs then, sometimes deactivation of dopants by SHI irradiation, may cause its reduction [10]. Further reports reveal that if initially the leakage current is very low then after few MeV (i.e. $<$ 10 MeV) ion irradiations on SiC SBDs, it will be increased. Moreover, in ion implantation induced edge termination technique, improvement in the leakage current, at -200 V was marginal [1]. In contrast to these reported results, present work shows that with selective SHI irradiation, leakage was reduced by an order of magnitude, which was already very low in pristine case. This suggest that, more studies are required to understand the fundamental physics behind SHI irradiation induced modifications in structural and electrical properties of SiC based SBDs.

The underlying phenomena in the selective area SHI irradiated SiC SBD is perceived by generation and interaction of quodons with interface defects and cause their accumulation at the boundary of masked and irradiated site i.e. at the "transition region". It has been well established that the impulsive force of SHI irradiation introduces nonlinearity in the crystalline medium. This causes transport of stored vibrational energy from the irradiation site towards transition region. Due to localization of vibrational energy at the transition region, the lattice structure gets reordered locally by changing bandgap and surface roughness [5, 6]. Based on these reported hypotheses, it is proposed in this work that SHI induced reordering of atomic structure at the boundary of the SBD (or edges of the fabricated SBD) minimize electric field there and hence causes improvement in their leakage current.

4. Conclusion

In summary, a novel approach for the realization of high quality SiC based SBD has been proposed and demonstrated. The 4H-nSiC based SBDs were fabricated and irradiated in selective way with $^{107}Ag^{+14}$ of energy 200 MeV at a fluence of 1×10^{13} ions cm^{-2}. Reverse bias performance of these diodes revealed that, as compared to the pristine, the selective way ion irradiated diode shows improvement in leakage current by an order of magnitude. These experimental findings reveal that due to localization of vibrational energy at the transition region, the lattice structure gets reordered there. This results in minimizing the local electric field and hence improves the leakage current.

Acknowledgements

One of the authors (VK) is thankful to Director CSIR-CEERI and Director IUAC for the usage of fabrication facilities, and the ion beam slots at pelletron respectively. Management staff of MDU is also thanked for overall support throughout this work.

References

[1] B. J. Baliga, SiC Power Devices, (World Scientific Publishing Co. Pte. Ltd. Singapore, 2005).
[2] Vibhor Kumar, A.S. Maan and J. Akhtar, 2013 *Journal of Physics : Conference Series* **423** 012057.
[3] L. J. Brillson, 1982 *Surface Science Reports* **2** 123.
[4] O. Ochedowski, O. Osmani, M. Schade, B. K. Bussmann, B. Ban-d'Etat, H. Lebius & M. Schleberger, 2014 *Nature communications* **5** 3913.
[5] P. Sen, J. Akhtar, and F. M. Russell, 2000 *Europhys. Lett.*, **51** 401.
[6] P. Sen, and J. Akhtar, 2000 *Pure Appl. Chem.*, **74** , 1631.
[7] V. Kumar, Shuvam Pawar, A. S. Maan, and J. Akhtar, 2015 *Journal of Vacuum Science & Technology B,* **33,** 052207.

IOP Conf. Series: Materials Science and Engineering **331** (2018) 012016 doi:10.1088/1757-899X/331/1/012016

[8] V. Raineri, F. Roccaforte, S. Libertino, A. Ruggiero, V. Massimino, L. Calcagno, 2006 *Materials Science Forum*, **527**, 1167.
[9] F. Roccaforte, S. Libertino, V. Raineri, A. Ruggiero, V. Massimino, and L. Calcagno 2006 *Journal of Applied Physics*, **99** 013515.
[10] C. Kamezawa, H. Sindoua, T. Hirao, H. Ohyama, S. Kuboyama, 2006 *Physica B*, **376,** 362.
[11] G. Izzo, G. Litrico, L. Calcagno, G. Foti, F. L. Via, 2008 *Journal of Applied Physics,* **104**, 093711.

IOP Conf. Series: Materials Science and Engineering **331** (2018) 012017 doi:10.1088/1757-899X/331/1/012017

Innovation as Road Safety Felicitator

S Sahoo [1], A Mitra [1], J Kumar [1], B Sahoo [1]

[1] Kalinga Institue of Industrial Technology, Bhubaneswar

Email: subhangee.sahoo@gmail.com

Abstract. Transportation via Roads should only be used for safely commuting from one place to another. In 2015, when 1.5 Million people, across the Globe started out on a journey, it was meant to be their last. The Global Status Report on Road Safety, 2015, reflected this data from 180 countries as road traffic deaths, worldwide. In India, more than 1.37 Lakh [4] people were victims of road accidents in 2013 alone. That number is more than the number of Indians killed in all the wars put together. With these disturbing facts in mind, we found out some key ambiguities in the Indian Road Traffic Management systems like the non-adaptive nature to fluctuating traffic, pedestrians and motor vehicles not adhering to the traffic norms strictly, to name a few. Introduction of simple systems would greatly erase the effects of this silent epidemic and our Project aims to achieve the same. It would introduce a pair of Barricade systems to cautiously separate the pedestrians and motor vehicles to minimise road mishaps to the extent possible. Exceptional situations like that of an Ambulance or any emergency vehicles will be taken care off by the use of RFID tags to monitor the movement of the Barricades. The varied traffic scenario can be guided properly by using the ADS-B (Automatic Detection System-Broadcast) for monitoring traffic density according to the time and place.

1. Introduction

Walking on roads, Indian and all over the world, have been nightmarish for pedestrians owing to the risks involved in a task as simple as crossing a lane. A global status report by WHO [7], focusing on road traffic deaths in 2009-2010, states pedestrians account for 22% of the 1.24 million deaths due to road crashes. Coming to India, quoting data recorded by the ministry of road transport and highways in India, the report further highlights that 8.7% of the deaths of 1.33 lakh people in road traffic accidents during 2010, were pedestrians- that is equal to 11,571 people. This project is a humble attempt to suggest a workable design that would enhance road safety.

2. Existing Systems and Loopholes:

Many of the roads in India have manually-operated traffic signals, with the technology of countdown timers catching up fast. But still, the number of casualties has increased by 8.7%, prompting the glaring question: where are the loopholes? The following ambiguities have been identified by the team those are addressed by this project.

2.1 *A Few Observations*

- Most of the vehicles stop on the zebra crossing (rather than before it), allowing little space for the commuters to pass.
- Some pedestrians still try to cross when the signal is Green for the motor traffic.
- The clearance time durations each for the pedestrians and the motor vehicles are not exclusive and isolated due to the spatial overlapping considerations resulting in confusion.
- Indian Traffic, where the situation is anything but stagnant, is not adaptable to fluctuating traffic density.

IOP Conf. Series: Materials Science and Engineering **331** (2018) 012017 doi:10.1088/1757-899X/331/1/012017

3. Proposed Idea As The Solution:

In busy road crossings to prevent accidents caused by the carelessness of the motor vehicle drivers and the pedestrians the proposal recommends:

- To erect a "2-rod" and one "single-rod" barricade.
- To build a barricade about 3/4 feet away from the zebra crossing, intended to prevent further movement of the vehicles when the signal is Red, allowing the pedestrians to cross the road, with their safety assured.
- The normal position of the barricade rods being vertical (allow) and automatically bend down across the road (stop) operable by the photo sensors from the traffic signal to form barricade for the motor vehicles.
- The Green light for the motor vehicles and the single –rod barricade for the pedestrians would operate simultaneously and they will be coupled with interchangeable automatic and/or manual operations. The single –rod barricade for the pedestrians will be at a horizontal level with the ground, preventing their forced movement, and the traffic will be allowed to pass. As the road traffic eases off (for manual control) or according to the timer, with the onset of yellow light, the status of the two barricades will interchange, and with the Red signal, pedestrians will be allowed to move and the vehicles will be restricted. For the last few seconds of the Red signal, the motor vehicles' side will be blocked too, to allow the last bunch of pedestrians to cross over safely. As soon as the Green light turns on, the roads will open up for the vehicles again.

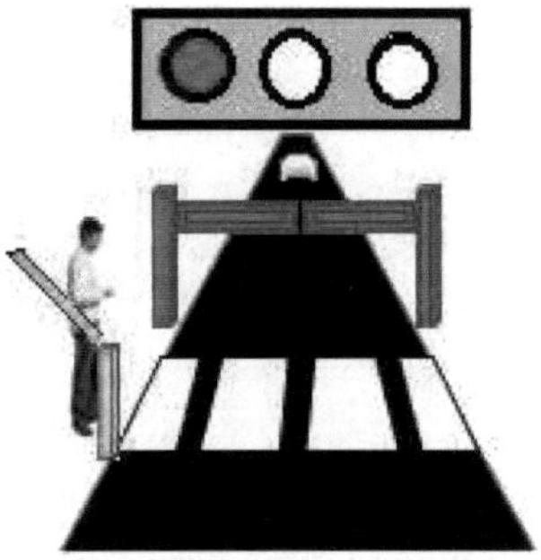

Figure 1. When the Traffic Signal is Red for motor traffic, the Barricades for the vehicles are Closed. The Barricade for the Pedestrians are Open. This allows the movement of the Pedestrians without the vehicles invading the Zebra Crossing. (Colours of the Barricades are Symbolic only).

Figure 2. The Pedestrians can cross the road hassle free when the Signal is Red for the Vehicles. (Colours of the Barricades are Symbolic only).

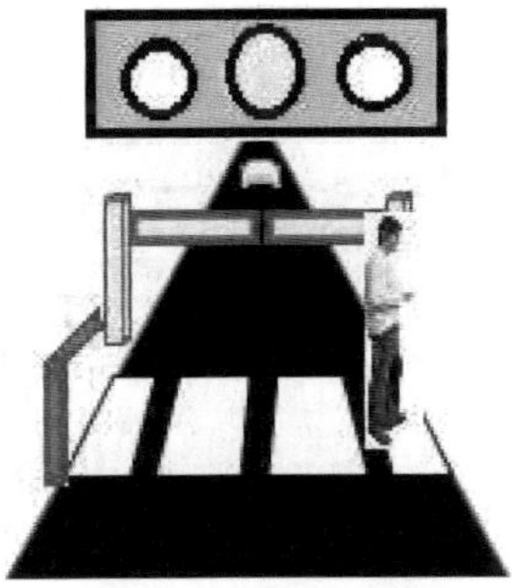

Figure 3. When the Signal turns Yellow for the motor traffic, the Barricade for the pedestrians is closed. This will prevent incoming pedestrians who tend to cross the road at the last moment. (Colours of the Barricades are Symbolic only).

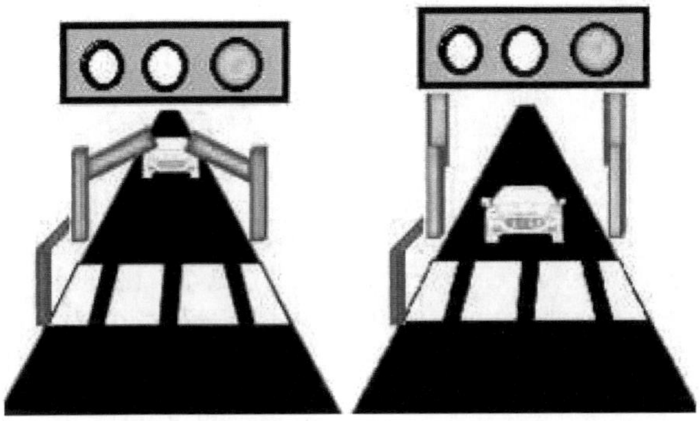

Figure 4. When the Signal is Green for the motor traffic, the barricade for the pedestrians is closed and the barricades for the vehicles are closed. (Colours of the Barricades are Symbolic only).

3.1 In Case of Emergencies/ Special Scenarios

3.1.1 *An ambulance is stuck in the traffic when the Vehicles' barricade is closed.*

3.1.2 *Any other important vehicle such as Fire Brigade and Escort vehicles have a long stretch covered by vehicles in front of them.*

3.1.3 *(SOLUTION for points 3.1.1 and 3.1.2)*: The RF module of 315 or 433 MHz ASK module [3] will have its transmitter on the emergency vehicles and the receiver on the barricades. The modulated tag code is demodulated in the reader. The electronic control unit then takes the necessary control action and opens the barricade.
When the tag from the emergency vehicles are detected, the barricades will be programmed to lift (if the current signal is Red), and as the distance between the vehicle and the system becomes zero, within two seconds, the pair of rods will drop again after 3 or 4 seconds (decided on the basis of the distance between the Tag and the Reader, as long as the emergency vehicle crosses the barrier).

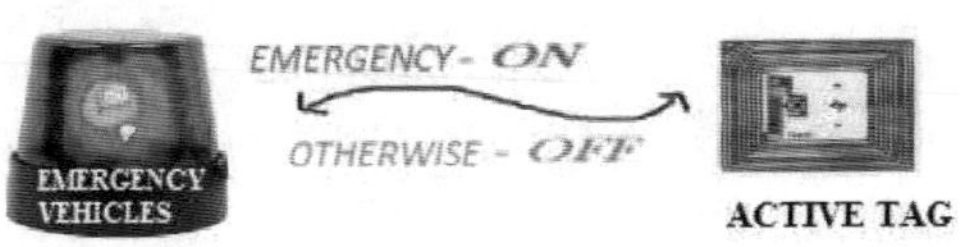

Figure 5. Emergency Vehicles exception to be detected by RFID Tags

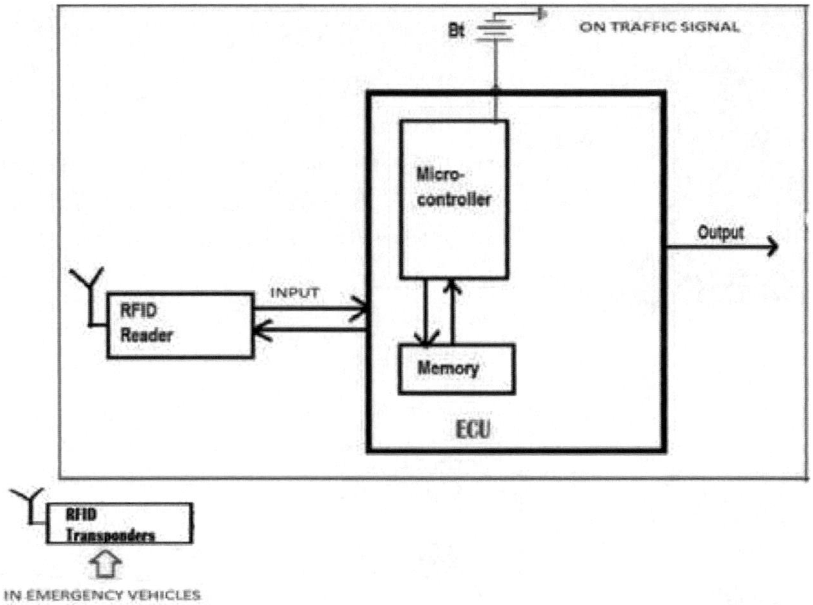

Figure 6. The RFID reader is placed in the traffic signal which detects the tag within a particular range. The RFID Tag on emergency vehicles modulates and transmits the beacon signal to RFID reader (in the traffic signal) and then to the ECU (Electronic control unit).

3.1.4 Varying traffic density
(SOLUTION):

- ADS-B (Automatic Detection System- Broadcast) to calculate the traffic density at a particular time and place and hence adjust the mechanism accordingly.

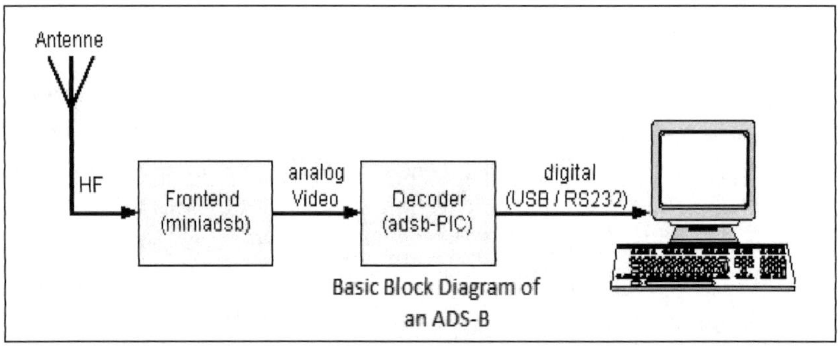

Figure 7. The ADS-B detects the Traffic Density at a given location and time.

- STEPS for the Process of Managing Real Time Traffic[5]:

STEP 1: The Antenna receives the electromagnetic wave, converts it into an electric signals and feeds this through the coaxial cable into the receiver.
STEP 2: The Receiver selects, amplifies and demodulates the signal. Its output signal is a so-called video-signal.

STEP 3: The Decoder converts the analogue video signal into digital signals and detects inside the received signal-chaos the replies from ADS-B-transponders.
STEP 4: Output of the Decoder is forwarded to the PC via USB (or RS232).
STEP 5: The Computer receives the ADS-B-data and extracts all interesting information. This is used to draw the radar picture.
STEP 6: PC-Software adsbScope (Windows): The PC-software receives ADS-B-frames from the decoder (via a virtual COM-port) and decodes the data. It will identify and calculate their positions. All the important parameters will be shown in alphanumeric form in a table and/or on a graphic display.

4. References

[1] Department of Transport, The Welsh Office, The Scottish Office and The Department of the Environment for Northern Ireland April 1995 *The Design Of Pedestrian Crossings, Local Transport Note* 2/95

[2] Automation of Railway Gate control using Microcontroller August 2015 www.electronicshub.org.

[3] Iyer S (I.I.T, Bombay) RF-ID: Technology and Application.

[4] Ray A October 2015 WHO reports 1.37 lakh road deaths in India in 2014

[5] Houston S May 2017 What's the Difference Between ADS-B Out and ADS-B In?

[6] www.ndtv.com

[7] www.who.int

IOP Conf. Series: Materials Science and Engineering **331** (2018) 012018 doi:10.1088/1757-899X/331/1/012018

Multiple Solutions for Reconfiguration to Address Partial Shading Losses in Solar Photovoltaic Arrays

Nikesh Sharma[1],Smita Pareek[2], Nitin Chaturvedi[3], Ratna Dahiya[4]

[1]Department of Electrical and Electronics Engineering, BKBIET Pilani, Rajasthan, India
[2]Department of Electronics and Communication Engineering, BKBIET Pilani, Rajasthan, India
[3]Department of Electrical and Electronics Engineering, BITS Pilani, Rajasthan, India
[4]Department of Electrical Engineering, NIT Kurukshetra, Haryana, India

Abstract-Solar photovoltaic (SPV) systems are steadily rising and considered as the best alternatives to meet the rising demand of energy. In developing countries like India, SPV's contribution being a clean energy is the most favourable. However, experiences have shown that produced power of these systems is usually affected due to day, night, seasonal variations, insolation, partial shading conditions etc. Among these parameters, partial shading causes a huge reduction in output power of PV systems. This results in lack of confidence for this technology among users. Thus, it is important and a major challenge in PV systems to minimize the effect of partial shading on their energy production. The work in this paper aims to propose solutions for reconfiguration of solar photovoltaic arrays in order to reduce partial shading losses and thus to enhance power generation.

Keywords: Array, Partial shading, Photovoltaic, Reconfiguration, Total-cross-tied interconnection.

1. Introduction

1.1 Solar photovoltaic

SPV is steadily rising as a source of renewable energy. This is because of the depletion of fossil fuels, their negative effects on environment, noticeable decrease in the cost of solar panels and other associated advantages. However, according to the reports [1-3], there is a lack of confidence for this technology among users. The output power of SPV cells depends on insolation, to day, night, seasonal variations, partial shading conditions etc, PS (partial shading) etc. [4-8]. Among the parameters listed above, PS causes major reduction in the output power. PS may be explained as when some modules in an array are receiving less insolation than others due to shading. It is a frequent phenomenon and ruins the energy yield of entire PV system. It is reported that shading causes a massive reduction in annual yields of large BIPV (building integrated PV) systems [9], [10].

To address this issue, this paper provides solutions for reconfiguration of modules in an array to reduce partial shading losses. The proposed solutions offers multiple reconfiguration possibilities in both HRPVA (half reconfigurable photovoltaic array) and FRPVA (full reconfigurable photovoltaic array) in comparison to the existing research [11], which provide only one solution in both the categories and that too using complex mathematical equations.

This paper is organized as follows: Section 2 discusses the partial shading effects and hot spot phenomenon. Section 3 reviews various techniques proposed to reduce PS losses. Section 4 discusses the mathematical analysis of reconfiguration technique on 4*4 TCT configured PV array. Section 5 presents three application case studies of partially shaded PV fields for various shading patterns followed by conclusion.

2. Partial shading effects

SPV panels are connected in series and parallel to obtain the load power requirement and thus, there is a chance that few of the panels are partially or completely shadowed by nearby clouds, trees, buildings, towers etc. [12, 13]. Partial shading causes the reduction in the insolation received by cell under shade compared to other non-shaded cells. Since in a series connected string, the string current must equally flow through cells, the result is that shaded cells operate in reverse bias region to conduct same current as that of non-shaded cells. The shaded cell will then consume power as it is operating in reverse bias region. This may lead to high bias voltage and breakdown of the shaded cell which in turn create hot-spot [14, 15].

The effect of partial shading is such severe that it need not to fall on an entire panel to deteriorate its output. Rather it is something that if blocks or cast on even small portion of the panel in a string, output of the entire string will be reduced to almost zero.

It is reported that the losses occur due to partial shading are not proportional to the PV's under shade but also depends on the shading pattern, array interconnection style [like SP (series-parallel), TCT (total cross tied), BL (bridged linked) etc.]and the location of shaded and non-shadedmodule within the array [16].

3. Partial shading loss reduction techniques

The traditional way to reduce PS effect is to connect bypass diodes. Bypass diodes are connected in anti-parallel to PV cells/modules [17-19].The incorporation of bypass diodes increases the complexity of the PV characteristic curve with multiple peaks under partial shading conditions and thus mislead the conventional MPP (maximum power point techniques) [20-23].

Another technique to reduce PS losses as reported in Refs.[24], in which a separate dc/dc power convertor with MPP controller is combined to each PV module in order to extract maximum power from each module thus enable module level MPPT. AC modules [25-27] are also proposed to reduce the PS losses by providing separate dc-to-ac conversion to each module and thus the shading of any module only affects its output power. Common drawback of all above techniques is high cost because of the requirements of separate components with each module. In addition, some require independent voltage control for each PV module and also has high power loss in switching.

Another method reported to overcome the negative effects of PS is reconfiguration of PV arrays, which is also being addressed in this work with an intention to cover the limitations of the work done in this area. Although reconfiguration technique could not gather a significant attention in past, this is because of high cost and lack of suitable hardware.

The work done in this paper is extension of reconfiguration technique proposed in reference [9] to array for TCT interconnected PV arrays in order to mitigate the negative impact of partial shading on energy yield of these arrays. The results of [9] shows that connecting the shaded modules according to the proposed technique can decrease PS losses. The aim of this paper is to check the performance of proposed reconfiguration technique by implementing them on various shade scenarios which are given by various researchers. In the performed case studies, PV arrays interconnected in TCT style, a half reconfigurable photovoltaic array and a full reconfigurable photovoltaic array (FRPVA) are considered under shading pattern given in [28-29] In HRPVA, odd numbers of columns are fixed and the even numbers of columns are reconfigurable to form a half reconfigurable photovoltaic array. In FRPVA, all columns are reconfigurable. It is work to mention that according to the proposed reconfiguration technique, the shaded modules must be connected diagonally opposite to each other (if belong to different string).

4. Mathematical analysis of reconfiguration technique on 4*4 TCT configured PV array

The dependency of the output power on the location of shaded modules is illustrated by a 4*4 TCT configured PV array as shown in Figure 1.

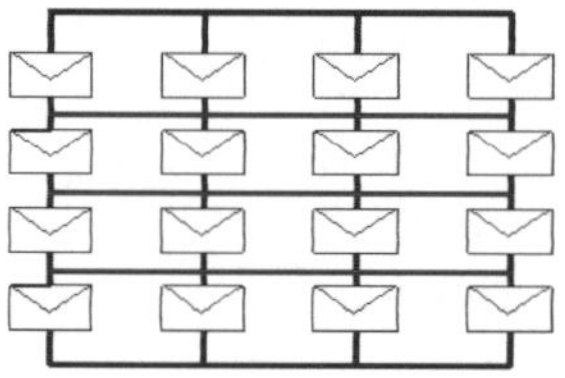

Figure 1: TCT configured 4*4 PV array

(a) Under uniform irradiance

All modules will conduct same current and it is assumed that peak current and peak voltage of all modules will be equal to I_m and V_m respectively.

In row 1, four modules are connected in parallel, thus the peak current generated by row 1 comprising of four modules will be I_{row} 1 = 4*I_m

The peak voltage of row 1 will be equal to the peak voltage of single module, as voltage in parallel remain same, thus, $V_{row}1 = V_m$

TCT configured array peak power, which has four such rows in series connected, under uniform irradiance is given by

$P_{array} = Va_{rray} * I_{array} = 4*(V_m)*(4*I_m) = 16V_m*I_m = 16 P_m$

Next, the array is considered less than three different shade scenarios i.e. horizontal shade, vertical shade and diagonal shade scenario as shown in Figure 2.

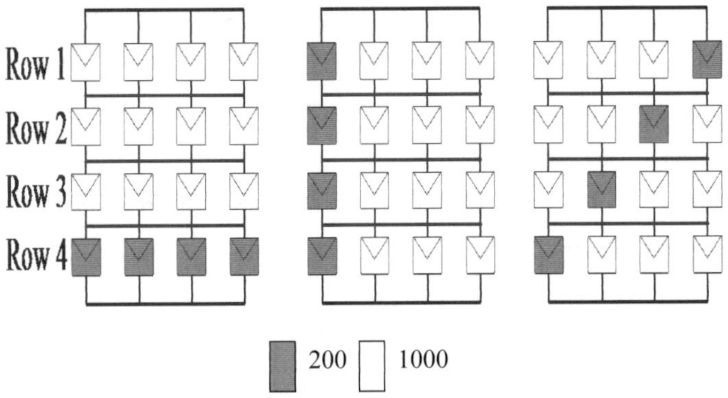

(a) Horizontal Shade Pattern (b) Vertical Shade Pattern (c) Diagonal Shade Pattern

Figure 2: TCT configured PV array under three shade patterns

(a) Horizontal shade pattern

Case 1

When array current is less than the current generated by the shaded module

Shaded modules will also contribute power and peak current generated by both shaded and non-shaded row shaded is determined by the peak current of shaded row, as all rows are in series

Shaded row, non-shaded row = 4*0.2*I_m = 0.8*I_m (as one row is comprising of four such modules and the peak current of shaded module is 0.2*I_m)

The voltage of shaded and non-shaded rows is same as that of peak voltage of single module, thus, shaded row, non-shaded row= V_m

The peak generated by the array comprising of four rows will be

P_{array}=4*(0.8*I_m)* V_m =3.2*I_m*V_m=3.2 (1)

Case 2

When string current is greater than the current generated by the shaded module (i.e. 0.2*I_m)

Shaded modules will be bypassed by bypass diode; hence shaded row will not contribute power. The peak power generated by three non-shaded rows will be given by

P_{array} = V_{array} * I_{array} = 3* (V_m)*(4*I_m) =12*V_m*I_m = 12pm (2)

According to equation (1) and equation (2), it may be concluded that the PV curve will show two peaks, one at 3.2Pm and other at 12Pm. Since peak power (pm) of single module is 197.78W, therefore two peaks will be at 632.896W (=3.2*pm=3.2*197.78) and 2373.36W (=12*pm=12*197.78). The results obtained is in close agreement with the simulated results , since the PV curve of horizontal shade pattern shows two peak with approximately the same value as obtained.

(b) Vertical and diagonal shade pattern

In these shade patterns, all rows comprises of one shaded module and three non-shaded modules, thus the peak current of each row is the given by I_{row}=I_m +I_m+I_m +0.2*I_m=3.2*I_m

The peak power generated by four such rows is given by

Prow=4*V_{row} * I_{row} =4* (V_m)*(3.2*I_m) =12.8*V_m*I_m= 12.8pm (3)

Since peak peak power (Pm) of single module is 197.78W, therefore PV curve will have only one peak at 2532.74W (=12.8*pm=12.8*197.78). The results obtained is in close agreement with the simulated results, since the PV curve of vertical and diagonal shade pattern shows only one peak with approximately the same value as obtained.

5. Application & Example [28], [29]

*5.1 Hypothetical Shading-I on 6*4 PV Array*

The PS situation as shown in Figure 3 is applied to TCT connected PV array, HRPVA and FRPVA. Some of the possible HRPVA and FRPVA solutions are shown in Figure 4 and Figure 5 respectively. Since ten modules are shaded, thus parallel connection of four shaded modules cannot be avoided.

500	500	500	1000
1000	1000	1000	1000
500	500	500	500
1000	1000	1000	1000
1000	500	500	500
1000	1000	1000	1000

Figure 3: Hypothetical shading-I irradiance levels

500	1000	500	1000
1000	500	1000	500
500	1000	500	1000
1000	500	1000	1000
1000	1000	500	500
1000	500	1000	1000

500	1000	500	1000
1000	500	1000	500
500	1000	500	1000
1000	500	1000	500
1000	1000	500	1000
1000	500	1000	1000

500	1000	500	1000
1000	500	1000	1000
500	1000	500	1000
1000	500	1000	500
1000	1000	500	1000
1000	500	1000	500

..............and many more

Figure 4: Some HRPVA solutions of hypothetical shading-I

500	1000	500	1000
1000	500	1000	500
1000	1000	500	1000
1000	500	1000	500
500	1000	500	1000
1000	500	1000	1000

500	1000	500	1000
1000	500	1000	500
1000	1000	500	1000
1000	500	1000	1000
500	1000	500	1000
1000	500	1000	500

500	1000	500	1000
1000	500	1000	500
1000	1000	500	1000
1000	500	1000	500
1000	1000	500	1000
1000	500	1000	500

................and many more

Figure 5: Some FRPVA solutions of hypothetical shading-I

For this shading, PA is equal to 2758.2 W for TCT, 3724.8 W for HRPVA and FRPVA.

$\text{Total irradiance on array} = 14 * 1000 + 10 * 500 = 19000$

IRA can be calculated as

$$\text{IRA} = \frac{\text{Total irradiance on array}}{\text{Total modules in array}} = \frac{19000}{24} = 791.67\ \text{W/m2}$$

Thus, PR of TCT can be given as

$$\text{PR} = \frac{2758.2}{791.67} * \frac{1000}{24 * 202.44} = 0.72$$

PR of HRPVA and FRPVA can be given as

$$\text{PR} = \frac{3724.8}{791.67} * \frac{1000}{24 * 202.44} = 0.97$$

As shown in Table 1, HRPVA and FRPVA have increased the generated power from 2758.2 W to 3724.8 W and percentage increase in power is 35.04 % when compared to TCT. Also the shaded modules are able to contribute power rather than dissipating. The PV characteristics of both reconfigurable arrays are smoother compared to TCT and has only one peak as shown in Figure 6, thus simplifying the task of MPPT algorithm.

Table 1: Array and modules' MPP powers

	Individual module's MPP (W)				Array 's MPP (W)	PR	Power change w.r.t. TCT (%)
TCT	96.95	96.95	96.95	196.48	2758.2	0.72	-
	96.95	96.95	96.95	96.95			
	-3.3	-3.3	-3.3	-3.3			
	96.95	96.95	96.95	96.95			
	196.48	96.95	96.95	96.95			
	96.95	96.95	96.95	96.95			
HRPVA	98.6	200.83	98.6	200.83	3724.8	0.97	35.04
(for	200.83	98.6	200.83	98.6			
solution	98.6	200.83	98.6	200.83			
a)	192.07	89.11	192.07	192.07			
	200.83	200.83	98.6	98.6			
	192.07	89.11	192.07	192.07			
FRPVA	98.6	200.83	98.6	200.83	3724.8	0.97	35.04
(for	200.83	98.6	200.83	98.6			
solution	192.07	192.07	89.11	192.07			
a)	200.83	98.6	200.83	98.6			
	98.6	200.83	98.6	200.83			
	192.07	89.11	192.07	192.07			

Shaded module location

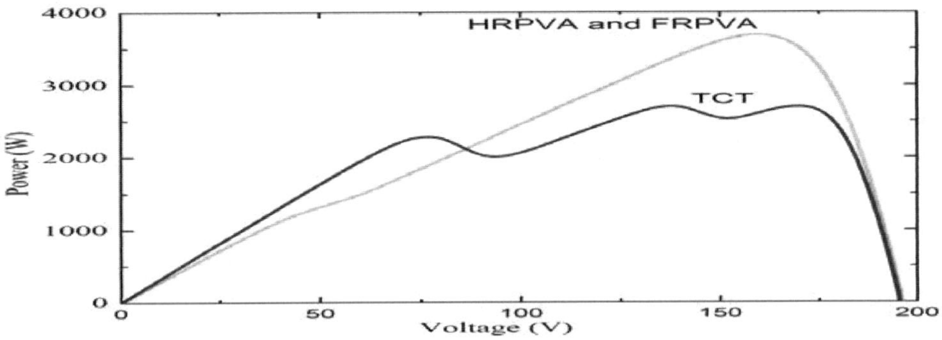

Figure 6: Arrays' P-V characteristics

*5.2 Hypothetical Shading-II on 6*4 PV Array*

The PS situation as shown in this Figure 7 is applied to TCT connected PV array, HRPVA and FRPVA. Some of the possible HRPVA and FRPVA solutions are shown in Figure 8 and Figure 9 respectively. Since eleven modules are shaded, thus parallel connection of five shaded modules cannot be avoided.

1000	1000	1000	1000
1000	1000	1000	1000
1000	1000	500	500
1000	500	500	500
500	500	500	500
500	500	1000	1000

Figure 7: Hypothetical shading-II irradiance levels

1000	500	1000	500
1000	500	1000	1000
1000	1000	500	500
1000	1000	500	500
500	1000	500	1000
500	1000	1000	500

(a)

1000	500	1000	500
1000	500	1000	1000
1000	500	500	1000
1000	1000	500	500
500	1000	500	1000
500	1000	1000	500

(b)

Figure 8: Some HRPVA solutions of hypothetical shading-II

500	1000	500	100
1000	500	500	1000
1000	1000	500	1000
1000	500	1000	500
500	1000	500	1000
1000	500	1000	500

(a)

500	1000	500	1000
1000	1000	500	1000
500	500	1000	1000
1000	1000	500	500
500	1000	500	1000
500	1000	1000	500

(b)

Figure 9: Some FRPVA solutions of hypothetical shading-II

For this shading, PA is equal to 2801.8 W for TCT, 3666.4W for HRPVA and FRPVA; G is equal to 1000W/m^2.

$\text{Total irradiance on array} = 13 * 1000 + 11 * 500 = 18500$

IRA can be calculated as

$$\text{IRA} = \frac{\text{Total irradiance on array}}{\text{Total modules in array}} = \frac{18500}{24} = 770.83\ \text{W/m2}$$

$\text{Pdc} = \text{Number of modules in array} * \text{MPP power of single module} = 24 * 202.44\text{W}$

Thus, PR of TCT can be given as

$$\text{PR} = \frac{2801.8}{770.83} * \frac{1000}{24 * 202.44} = 0.748$$

PR of HRPVA and FRPVA can be given as

$$\text{PR} = \frac{3666.4}{770.83} * \frac{1000}{24 * 202.44} = 0.98$$

Table 2: Array and modules' MPP powers

	Individual module's MPP (W)				Array 's MPP (W)	PR	Power change w.r.t. TCT (%)
TCT	152.11	152.11	152.11	152.11	2801.8	0.748	-
	152.11	152.11	152.11	152.11			
	193.87	193.87	90.56	90.56			
	186.81	92.54	87.61	92.54			
	-3.3	-3.3	-3.3	-3.3			
	90.56	90.56	193.87	193.87			
HRPVA (for solution a)	202.14	98.81	202.14	98.81	3666.4	0.98	30.86
	190.2	87.61	190.2	202.14			
	202.14	202.14	98.81	98.81			
	202.14	202.14	98.81	98.81			
	98.81	202.14	98.81	202.14			
	98.81	202.14	202.14	98.81			
FRPVA (for solution a)	98.81	202.14	98.81	202.14	3666.4	0.98	30.86
	202.14	98.81	98.81	202.14			
	190.2	190.2	87.61	190.2			
	202.14	98.81	202.14	98.81			
	98.81	202.14	98.81	202.14			
	202.14	98.81	202.14	98.81			

Shaded module location

As shown in Table 2, the HRPVA and FRPVA both has increased the generated power from 2801.8 W to 3666.4 W and percentage increase in power is 30.86 % when compared to TCT. The PV characteristics of both reconfigurable arrays are smoother as compared to TCT configuration as shown in Figure 10, thus simplifying the task of MPPT algorithm. Also the shaded modules are delivering more power compared to TCT configuration in HRPVA and FRPVA.

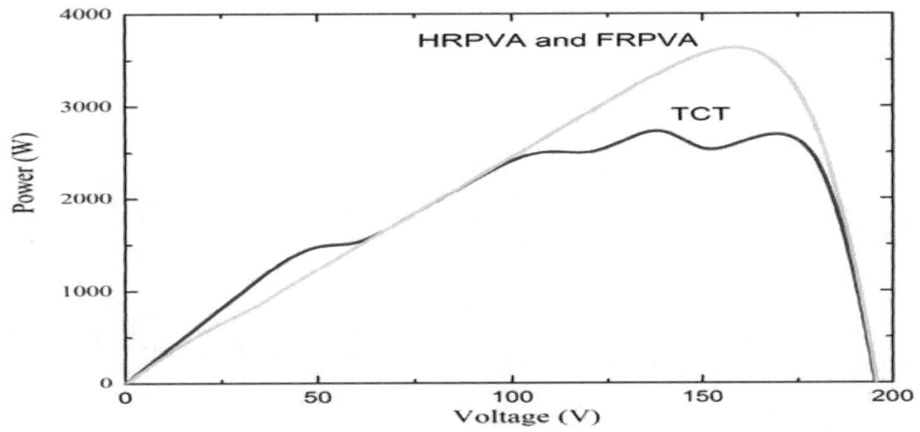

Figure 10: Array's P-V characteristic

*5.3 Double-row shading on 4*3 PV array:*

The PS situation (as shown in Figure 11), is now considered on 4*3 PV array, where numbers in the boxes indicate modules' irradiance levels in W/m^2).

1000	1000	1000
1000	1000	1000
500	500	500
500	500	500

Figure 11: Double-row shading on 4*3 PV array

5.4 Half Reconfigurable Photovoltaic Array:

By applying the proposed rules for shaded modules interconnections, there is one possible HRPVA solution (Figure 12), which yields more MPP power compared to TCT interconnections and also reported in [29]. The locations of shaded modules can be (1*2) and (2*2).

1000	500	1000
1000	500	1000
500	1000	500
500	1000	500

Figure 12: Possible HRPVA solution of 4*3 PV array

5.5 Full Reconfigurable Photovoltaic Array:

In FRPVA, all columns may be reconfigured and thus many FRPVA solutions are possible. Some of the possible FRPVA solutions are given in Figure 13.

1000	500	1000
500	1000	500
500	500	1000
1000	500	1000

(a)

500	500	1000
500	1000	500
500	1000	1000
1000	500	1000

(b)

1000	500	1000
500	1000	500
500	1000	1000
1000	500	500

(c)

Figure 13: Some of the possible FRPVA solutions

The array and module's MPP power are as shown in Table 3 together with the array's overall performance ratio (PR). For single row shading, PA is equal to 1308.5 W for TCT, 1694.6 W for HRPVA and FRPVA.

Total irradiance on array

$$= (\text{Number of non shaded modules} * 1000) + (\text{Number of shaded modules} * 500) = 6 * 1000 + 6 * 500 = 9000$$

IRA can be calculated as

$$\text{IRA} = \frac{\text{Total irradiance on array}}{\text{Total modules in array}} = \frac{9000}{12} = 750\text{W/m2}$$

$$\text{Pdc} = \text{Number of modules in array} * \text{MPP power of single module} = 12 * 202.44\text{W}$$

Thus, PR of TCT can be given as

$$PR = \frac{1308.5}{750} * \frac{1000}{12 * 202.44} = 0.718$$

PR of HRPVA and FRPVA can be given as

$$PR = \frac{1694.6}{750} * \frac{1000}{12 * 202.44} = 0.93$$

Table 3: Array and modules' MPP powers

	Individual module's MPP (W)			Array 's MPP (W)	PR	Power change w.r.t. TCT (%)
TCT	123.52	123.52	123.52	1308.5	0.718	-
	123.52	123.52	123.52			
	94.71	94.71	94.71			
	94.71	94.71	94.71			
HRPVA	183.44	82.25	183.45	1694.6	0.93	29.5
(for solution a)	183.44	82.25	183.45			
	98.71	201.24	98.71			
	98.71	201.24	98.71			
FRPVA	183.44	85.63	183.44	1694.6	0.93	29.5
(for solution a)	98.71	201.24	98.71			
	98.71	98.71	201.24			
	183.45	82.25	183.45			

● Shaded module location

It may be seen that the HRPVA and FRPVA has increased the generated power by 29.5 % when compared to TCT. Further, the performance ratio has also increased from 0.718 to 0.93 from TCT configuration to HRPVA and FRPVA configurations. These increments are because these configurations prevent bypass diodes from turning ON, which would short the 500 W/m^2 modules. The PV characteristics of TCT, HRPVA and FRPVA configurations are shown in Figure 14. The characteristics of reconfigurable arrays are smoother as compared to TCT configuration, thus simplifying the task of MPPT algorithm. The array's MPP power obtained for this partial shading situation has increased from 1308.5 for TCT to 1694.6 W for both HRPVA and FRPVA.

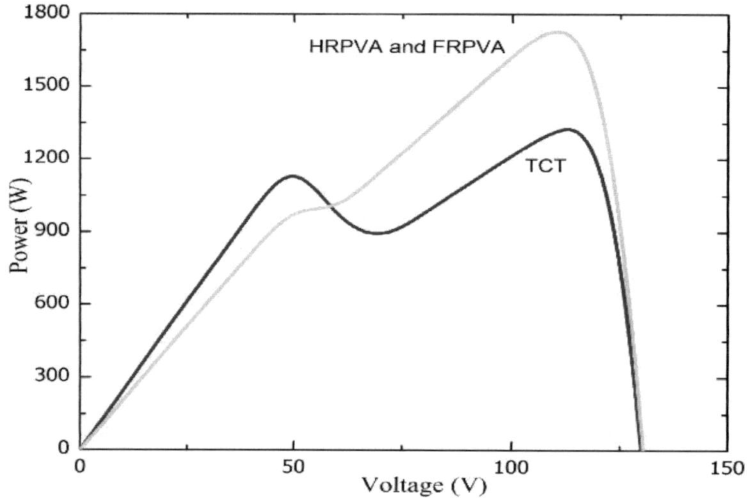

Figure 14: Array's P-V characteristics

6. Conclusion

In this paper multiple solutions for reconfigurations of PV array are given. As verified by three case studies, the proposed solutions are capable to reduce partial shading losses when compared to TCT configurations. The reconfiguration technique also results in a smoother P-V array characteristic with less number of local maxima, thus simplifying the task of MPPT. In addition, the proposed reconfiguration technique is not restricted to the case studies performed but are also valid for any partial shading PV array of any size and consisting of any number of shaded modules. Thus these partial shading losses reduction reconfiguration solutions may easily be implemented for the design of large photovoltaic structures without tedious mathematical formulation (If it is possible to connect shaded modules diagonally) to improve energy yield under partial shading conditions. The solutions also reduces the losses in bypass diodes during their ON time and forces the shaded modules to produce more power in comparison to TCT interconnections and is capable to provide a complete base for the interconnection of modules in partial shaded photovoltaic arrays.

References

1. BalasubramanianIndu Rani, SaravanaIlangoGanesan and NagamaniChilakapati, "Impact of partial shading on the output power of PV systems under partial shading conditions," IET Power Electronics, vol. 73, pp. 657-666, 2014.
2. Subudhi Bidyadhar and Raseswari Pradhan, "A comparative study on maximum power point tracking techniques for photovoltaic power systems," IEEE transactions on Sustainable Energy, vol. 1.4, pp. 89-98, 2013.
3. Pareek Smita, DahiyaRatna. Simulation and performance analysis of individual module to address partial shading cum parameter variation in large photovoltaic fields. J Energy 2015;2(3):99-104.
4. Pareek, S., Dahiya, R., 2014. Power output maximization of partially shaded 4⁄4 PV field by altering its topology. Energy Procedia 50, 71–78.
5. Pareek, S., Runthala, R., Dahiya, R., 2013. Mismatch losses in SPV systems subjected to partial shading conditions. In: International Conference on Advanced Electronic Systems (ICAES), pp. 343–345.
6. Purohit, J., Naveen, S.P., 2016. Comparison of various techniques to mitigate the effect of partial shading on solar PV systems. IJIREEICE 4 (8), 88–91.
7. Pareek, S., Dahiya, R., 2016. Series-connected Shaded Modules to Address Partial Shading Conditions in SPV Systems. AIP Publishing, 1715, 020020
8. Pareek, S., Dahiya, R., 2015. Power optimization of TCT conFig.d PS-PV fields by forecasting the connection of modules. In: Annual IEEE India Conference (INDICON), pp. 1–6.
9. Pareek, S., Chaturvedi N., Dahiya, R., 2017. Optimal interconnections to address partial shading losses in solar photovoltaic arrays, "Solar Energy, vol. 95,pp. 561–572.
10. Celik Berk, EnginKaratepe, Santiago Silvestre, Nuri Gokmen and AissaChouder, "Analysis of spatial fixed PV arrays configurations to maximize energy harvesting in BIPV applications," Renewable Energy, vol. 75, pp. 534-540, 2015.
11. Pareek, S., Dahiya, R., 2016. Enhanced power generation of partial shaded photovoltaic fields by forecasting the interconnection of modules. Energy 95,561–572.
12. Bastidas-Rodriguez, Juan David, Elisa Franco, Giovanni Petrone, Carlos Andres Ramos-Paja and Giovanni Spagnuolo, "Maximum power point tracking architectures for photovoltaic systems in mismatching conditions: a review," IET Power Electronics, vol. 7, no. 6, pp. 1396-1413, 2014.
13. Bai Jianbo, Yang Cao, YuzheHao, Zhen Zhang, Sheng Liu and Fei Cao, "Characteristic output of PV systems under partial shading or mismatch conditions," Solar Energy, vol., 112, pp. 41-54, 2015.

14. Dolara Alberto, George Cristian Lazaroiu, Sonia Leva and GiampaoloManzolini. "Experimental investigation of partial shading scenarios on PV (photovoltaic) modules," Energy, vol. 55, pp. 466-475, 2013.
15. Kotti R. and W. Shireen, "Efficient MPPT control for PV systems adaptive to fast changing irradiation and partial shading conditions," Solar Energy, vol. 114, pp. 397-407, 2015.
16. Villa Luiz Fernando Lavado, Damien Picault, Bertrand Raison, Seddik Bacha and Antoine Labonne, "Maximizing the power output of partially shaded photovoltaic plants through optimization of the interconnections among its modules," IEEE Journal of Photovoltaics, vol. 2, no. 2, pp. 154-163, 2012.
17. Pareek Smita and RatnaDahiya, "Output Power Comparison of TCT & SP Topologies for Easy-to-Predict Partial Shadow on a 4× 4 PV Field," In Applied Mechanics and Materials, vol. 612, pp. 71-76, 2014.
18. Libing Bai, "An improved MPPT controller for photovoltaic system under partial shading condition," IEEE Transactions on Sustainable Energy, vol. 5, no. 3, pp. 978-985, 2014.
19. TeyKok Soon and SaadMekhilef, "Modified incremental conductance algorithm for photovoltaic system under partial shading conditions and load variation," IEEE Transactions on Industrial Electronics, vol. 61, no. 10, pp. 5384-5392, 2014.
20. Hiren Patel and Vivek Agarwal, "Maximum power point tracking scheme for PV systems operating under partially shaded conditions," IEEE Transactions on Industrial Electronics, vol. 55, no. 4, pp. 1689-1698, 2008.
21. Rani B. Indu, G. SaravanaIlango and ChilakapatiNagamani, "Enhanced power generation from PV array under partial shading conditions by shade dispersion using Su Do Ku configuration, " IEEE Transactions on Sustainable Energy, vol. 4, no. 3, pp. 594-601, 2013.
22. IshaqueKashif and Zainal Salam, "A review of maximum power point tracking techniques of PV system for uniform irradiation and partial shading condition," Renewable and Sustainable Energy Reviews, vol. 19, pp. 475-488, 2013.
23. Qi Jun, Youbing Zhang and Yi Chen, "Modeling and maximum power point tracking (MPPT) method for PV array under partial shade conditions," Renewable Energy, vol. 66, pp. 337-345, 2014.
24. Bidram Ali, Ali Davoudi and Robert S. Balog, "Control and circuit techniques to mitigate partial shading effects in photovoltaic arrays," IEEE Journal of Photovoltaics, vol. 2, no. 4, pp. 532-546, 2012.
25. Rodriguez Cuauhtemoc and G. Amaratunga, "Long-lifetime power inverter for photovoltaic AC modules," IEEE Transactions on Industrial Electronics, vol. 55.7, pp. 2593-2601, 2008.
26. Ji Young-Hyok, Doo-Yong Jung, Jae-Hyung Kim, Chung-Yuen Won and Dong-Sung Oh, "Dual mode switching strategy of flyback inverter for photovoltaic AC modules," IEEE International Power Electronics Conference (IPEC), pp. 2924-2929, 2010.

27. E. Roman, R. Alonso, P. Ibanez, S. Elorduizapatarietxe and D. Goitia, “Intelligent PV module for grid-connected PV systems,” IEEE Transaction on Industrial Electronics, vol. 53, no. 4, pp. 1066–1073, June 2006.
28. Malathy S. and R. Ramaprabha, "A static PV array architecture to enhance power generation under partial shaded conditions," IEEE 11th International Conference on Power Electronics and Drive Systems (PEDS), pp. 341-346, 2015.
29. El-Dein, MZ Shams, Mehrdad Kazerani and M. M. A. Salama, "An optimal total cross tied interconnection for reducing mismatch losses in photovoltaic arrays," IEEE Transactions on Sustainable Energy, vol. 4, no. 1, 99-107, 2013.

IOP Conf. Series: Materials Science and Engineering **331** (2018) 012019 doi:10.1088/1757-899X/331/1/012019

Noise Estimation and Quality Assessment of Gaussian Noise Corrupted Images

V M Kamble and K Bhurchandi

Visvesvaraya National Institute of Technology, Nagpur, India
E-mail: vipinkamble97@gmail.com

Abstract. Evaluating the exact quantity of noise present in an image and quality of an image in the absence of reference image is a challenging task. We propose a near perfect noise estimation method and a no reference image quality assessment method for images corrupted by Gaussian noise. The proposed methods obtain initial estimate of noise standard deviation present in an image using the median of wavelet transform coefficients and then obtains a near to exact estimate using curve fitting. The proposed noise estimation method provides the estimate of noise within average error of +/-4%. For quality assessment, this noise estimate is mapped to fit the Differential Mean Opinion Score (DMOS) using a nonlinear function. The proposed methods require minimum training and yields the noise estimate and image quality score. Images from Laboratory for image and Video Processing (LIVE) database and Computational Perception and Image Quality (CSIQ) database are used for validation of the proposed quality assessment method. Experimental results show that the performance of proposed quality assessment method is at par with the existing no reference image quality assessment metric for Gaussian noise corrupted images.

1. Introduction

Digital images have become an important part of our lives. From CCTV camera to mobile phones, majority devices are equipped with imaging technology. The captured images are expected to be of high quality in accordance with the human perception. An ideal approach to estimate the noise and to evaluate the quality of an image is to score the image by different individuals and then pool the result. But this is a time consuming and subjective procedure and the quality score might not indicate the exact visual quality of an image as it is an average of individual scores. Thus, an algorithmic approach is required for real time quality assessment and noise estimation of images.

The image quality evaluation problem is broadly classified into 3 categories as,

- Full Reference Image Quality Assessment (FR-IQA)
- Reduced Reference Image Quality Assessment (RR-IQA)
- No Reference Image Quality Assessment (NR-IQA)

FR-IQA algorithms require a ground truth or reference image for quality evaluation of the distorted image. RR-IQA assesses quality of the distorted image by comparing only the features of distorted image with that of the reference image. In NR-IQA, the image quality is computed using information available in the distorted image only. This makes NR-IQA a challenging task. The NR-IQA problem is further classified into two sub categories-

- Distortion Specific (DS) NR-IQA
- Non-Distortion Specific (NDS) NR-IQA

Distortion Specific NR-IQA [1-3] refers to the algorithms that are designed only for specific type of distortions and their performance is highly dependent on the type of distortion present in an image. Non-Distortion Specific algorithms are generalized algorithms that can be used to evaluate the quality of an image which is degraded by any type of distortion. Handling all types of distortions in a single algorithm is a challenging task. Few algorithms like BIQI [4] and DIIVINE [5] use a two-step framework to first identify the type of distortion present in an image and then analyze the quality accordingly. BLIINDS-II [6] uses Natural Scene Statistics (NSS) of Discrete Cosine Transform (DCT) coefficients to evaluate the quality of an image. Peng Ye and David Doermann [7] employed Gabor filter features to train a codebook for quality assessment.

The main parameter for distortion of image quality is noise. Estimating this noise can also assist in enhancing the performance of majority denoising algorithms which require a prior estimate of noise present in an image for effective noise removal. This noise estimate information is usually obtained from the user before denoising an image. This makes denoising a user dependent task whose performance is highly influenced by the noise estimate. Hence, an accurate estimate of noise is very important for effective denoising.

In this paper, we present a low complexity based DS NR-IQA algorithm which is optimized for quality evaluation of images corrupted by Gaussian noise. We also present a fairly accurate noise estimator of Gaussian noise present in an image. The proposed methods are based on initial noise estimation in distorted image using wavelet transform [8]. The initial noise estimate is mapped to Differential Mean Opinion Score (DMOS) using Velde's non-linear mapping equation [9] for quality assessment. The same initial estimate is used to obtain the exact noise estimate of an image using curve fitting. A short description of wavelet transform and Velde's non-linear equation is presented in section 2. In section 3, we describe our proposed approach for noise estimation based on initial estimate from wavelet transform. In section 4, we describe the proposed method for image quality assessment. We benchmark the proposed noise estimation method and the proposed image quality metric with existing state of the art algorithms in section 5 and conclude the paper in section 6.

2. Noise Estimate and Mapping Function

We employ the signal noise estimator proposed by D. L. Donoho and I. M. Johnstone [8] to obtain an initial estimate of Gaussian noise using wavelet transform. Performance of this noise estimator is presented in figure 1 for four test images Lena, Barbara, Cameraman and Peppers. Ideally the performance should have been a straight line passing though origin. The important observation of figure 1 is that the performance curve closely matches for all the images. For low values of added noise standard deviation, the estimate is fairly correct.

The estimated noise is then mapped to correlate with human perception using Koen Velde's non-linear equation [9] for image quality assessment. The same noise estimate is used to obtain exact noise present in an image using proposed curve fitting.

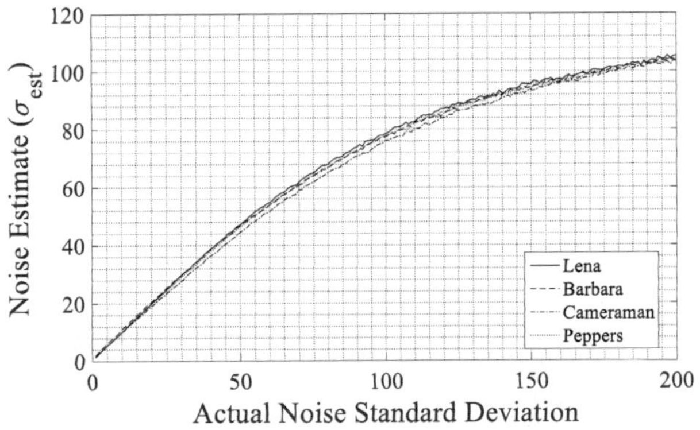

Figure.1. Noise estimation using Wavelet Transform.

2.1 Noise Estimate using Wavelet Transform

Discrete Wavelet Transform (DWT) decomposes a signal into high pass and low pass components. DWT is obtained by passing the signal through series of high pass and low pass filters. DWT for a signal $x(n)$ is defined as in (1)

$$y_{high}[k] = \sum_{n} x[n] \times g[2k-n]$$

$$y_{low}[k] = \sum_{n} x[n] \times h[2k-n] \quad (1)$$

where g and h are high pass and low pass filters respectively. DWT of an image signal results in four sub-band components namely HP-HP, HP-LP, LP-HP and LP-LP (where, LP - low pass, HP – high pass). The noise estimator proposed by D. L. Donoho and I. M. Johnstone [8] uses the median of HP-HP component of wavelet transform to evaluate the noise present in an image. The noise estimate σ_{est} is defined as

$$\sigma_{est} = \frac{median(abs(W))}{0.6745} \quad (2)$$

where, W is the HP-HP component of the finest level of wavelet sub-band. The HP-HP sub-band contains the noise as well as diagonal edges present in an image. The coefficients corresponding to edges are the high vales present in sub-band. Noise is uniformly distributed throughout the sub-band, hence can be estimated using median of the sub-band coefficients.

σ_{est} is able to provides a rough approximation of standard deviation of Gaussian noise present in the image. Though σ_{est} is approximately equal to the noise present in the image, it does not linearly correlate with the human perception of visual quality. Hence, σ_{est} cannot be used independently as a visual quality indicator.

2.2 Mapping Function

Koen Velde [9] proposed a non-linear function to enhance the low value transform domain coefficients. It is observed from heuristic experimentation that the Gaussian noise present in an image and visual perception of human do not correlate linearly. Hence, we use this non-linear mapping function to correlate the noise estimate with the visual quality of an image. The mapping function [9] is defined as

$$y(x) = \left(\frac{m}{x}\right)^{p} \quad (3)$$

where,
y – multiplying parameter for mapping x
m – cross-over parameter
x – input signal intensity
p – degree of non-linearity

'y' produces a value greater than 1 for all x below the cross-over parameter 'm' and vice versa. The values that are lower than m are boosted and the values greater than m are suppressed. y produces a nonlinear curve for boosting and suppressing the input values depending on the cross-over parameter m.

3. Proposed Noise Estimation Method

The noise estimate obtained using (2) provides an approximate value of noise standard deviation of an image. But the error between estimated and actual noise content of the image goes on increasing as the amount of noise increases. To overcome this problem, we have proposed a novel noise estimator which is based on the wavelet domain estimator descried in section 2.1.

The mismatch between actual noise present in an image and the noise estimate using (2) for standard images (*Lena, Barbara, Cameraman* and *Peppers*) is shown in figure 1. It can be observed that the noise estimate is very consistent for different images. But, the difference between actual noise present in the image and estimated noise goes on increasing as the amount of noise increases. This non-linearity of the curve in figure 1 causes the error in estimation of exact noise present in an image.

To obtain the exact estimate of noise from a noisy image, we use curve fitting to the curves in figure 1. After heuristic experimentation, we found that the exact estimate of noise can be obtained by a fifth degree polynomial curve using data from figure 1. The final estimate of noise (σ) using curve fitting is presented in (4).

$$\sigma = p_1.\sigma_{est}^5 + p_2.\sigma_{est}^4 + p_3.\sigma_{est}^3 + p_4.\sigma_{est}^2 + p_5.\sigma_{est}^1 + p_6 \quad (4)$$

The values of polynomial coefficients in (4) are optimized to fit the curves in figure 1. Average of all four curves is taken to obtain a single curve; this curve is then fit using the curve fitting toolbox of MATLAB. Following are the values of coefficients:

$$\begin{aligned} p_1 &= 5.404 \times 10^{-8} \\ p_2 &= -1.092 \times 10^{-5} \\ p_3 &= 8.695 \times 10^{-4} \\ p_4 &= -2.784 \times 10^{-2} \\ p_5 &= 1.383 \\ p_6 &= -1.707 \end{aligned} \quad (5)$$

The polynomial coefficients (p_1 to p_6) are obtained using standard images of denoising experimentation. These images contain all major textures possible in natural images like smooth regions, sharp edges, curves and fine edges. Hence the polynomial coefficients (p_1 to p_6) obtained using these images are robust to the changes in image textures.

4. Proposed Image Quality Assessment Method

The noise estimate and visual perception of humans are not linearly correlated. Hence, we map the noise estimate using a non-linear function to achieve high linear correlation. The noise estimate (σ_{est}) of an image is obtained using (2). The product of σ_{est} and mapping function gives the final quality metric Q as defined in (6)

$$Quality\,Metric(Q) = \left(\frac{m}{\sigma_{est}}\right)^p \times \sigma_{est} = \frac{m^p}{\sigma_{est}^{p-1}} \quad (6)$$

Values of the cross-over parameter (m) and the degree of non-linearity (p) are selected as 65 and 0.65 respectively These values are obtained heuristically by comparing the values of Q with that of DMOS values from LIVE database [10] for varying m and p. At m=65 and p=0.65, the values of Q and DMOS are observed to have a maximum linear correlation. The range of the proposed quality metric is from 0 to 100 where 0 indicates best image quality and 100 indicates poor image quality. A flowchart of proposed quality assessment method is presented in figure 2.

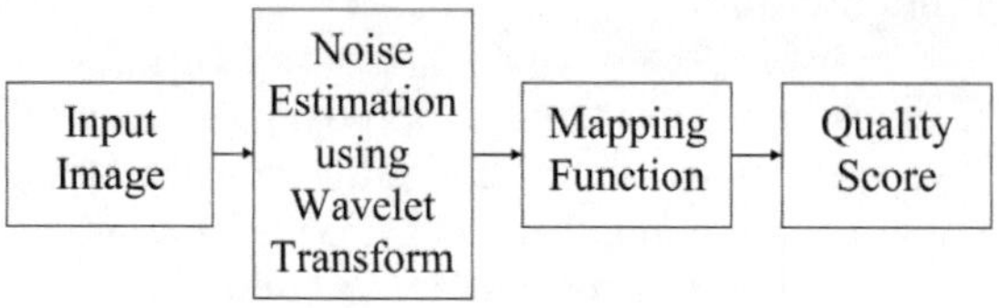

Figure 2. Proposed quality assessment algorithm flowchart.

To establish the consistency of the proposed quality assessment method, we plot the proposed quality metric (Q) values for increasing magnitude of Gaussian noise standard deviation as shown in figure 3. Gaussian noise is added to standard *Parrots.bmp* image from LIVE database. Corresponding quality score obtained using the proposed method is plotted against the Gaussian noise standard deviation. It can be observed from figure 3 that the value of Q is increasing non-linearly with the standard deviation of added noise. This non-linearity is required because the human perception of visual quality and the actual noise present in an image are not linearly related.

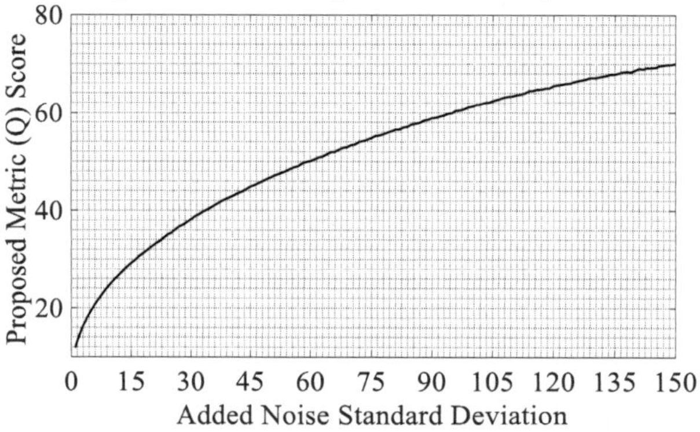

Figure 3. Consistency of proposed quality assessment method.

The proposed quality metric (Q) does not require any training and gives the quality score of an image based on an equation. Hence, it is computationally efficient and easy to calculate by simple coding. The proposed metric (Q) has same computation complexity as that of wavelet transform, i.e. O(N). Average time taken for quality assessment of an image of size 512x512 using proposed metric (Q) is 20ms. The system on which the time computation is evaluated has the following configuration: i7-4770 CPU@3.4GHz, 32GB Ram, Windows 8.1 64 bit operating system.

5. Benchmarking and Discussion

The proposed noise estimation method is benchmarked by comparing the actual (added) noise present in an image with the estimated noise. Noise estimation for standard images with added noise of standard deviation 20 is shown in figure 4.

(a) Lena, σ = 20.25 (b) Barbara, σ = 20.85 (c)Cameraman, σ = 19.27 (d) Peppers, σ = 19.97

Figure 4. Noise estimates (σ) for actual noise standard deviation of 20

The noise estimates and the percentage error in estimation for a wide range of added noise are presented in table 1.

Table 1. Noise estimation using proposed method

Added Noise	**Estimated Noise / % Error**			
	bikes	**caps**	**House**	**lighthouse**
10	11.53/15.30	10.36/3.66	10.48/4.88	10.98/9.84
20	20.81/4.09	20.35/1.79	20.23/1.18	20.55/2.78
30	30.06/0.21	30.30/1.02	29.99/0.01	30.61/2.03
40	39.34/1.63	41.24/3.10	39.69/0.75	41.29/3.23
50	49.09/1.80	52.11/4.23	49.71/0.56	51.72/3.44
60	58.48/2.52	62.29/3.82	59.04/1.59	61.54/2.57
80	77.11/3.60	83.08/3.85	78.60/1.74	83.19/3.99
100	96.59/3.40	103.73/3.73	98.87/1.12	101.80/1.80
120	116.36/3.03	127.23/6.02	118.06/1.61	122.13/1.78
140	134.82/3.7	146.36/4.54	138.69/0.93	147.47/5.34
160	154.30/3.55	168.95/5.59	157.96/1.27	167.92/4.95
180	172.60/4.10	184.19/2.33	176.28/2.06	174.41/2.45
200	194.34/2.82	208.66/4.33	190.14/4.92	200.54/0.27
Avg. % Error	**3.83**	**3.69**	**1.74**	**3.42**

Graphical results for these images are presented in figure 5. The noise estimations are calculated for four popular images (*Lena, Barbara, Cameraman* and *Peppers*) which contain most major textures that might be possible in practical images. The average error in noise estimation over wide range of noise standard deviation is below 4%. Also, in figure 5, it can be observed that the noise estimates obtained using proposed estimator are very close to the ideal value of noise standard deviation which demonstrate the accuracy of the proposed noise estimation method.

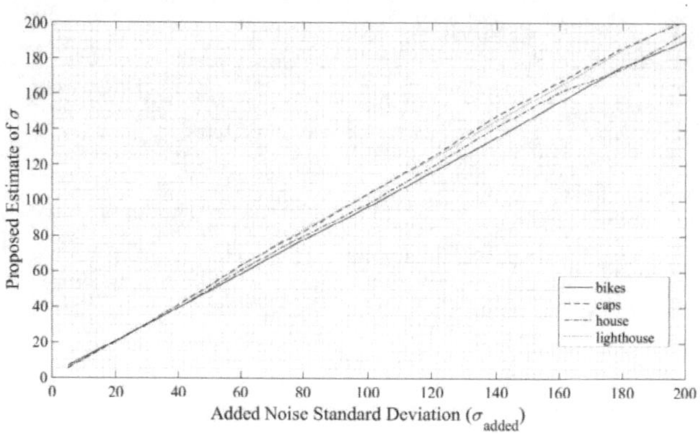

Figure 5. Noise estimation using proposed method for four images

We also present the performance of the proposed noise estimation method on uniform patches of values 0, 128 and 255 in figure 6. The noise estimation for these patches using the proposed method is obtained for linearly increasing noise standard deviation. It can be observed from figure 6 that for the patch of uniform value 128, the estimated noise and the added noise are almost equal. The estimated noise for patches of values 0 and 255 is less than the added noise. This is because of the fact that the pixel values in an image are represented by 8 bits and have values from 0 to 255. So even if noise is added to an image, the deviation in pixel value would not be beyond 0 and 255. Hence effective estimated noise is less than the added noise which can be observed from figure 6.

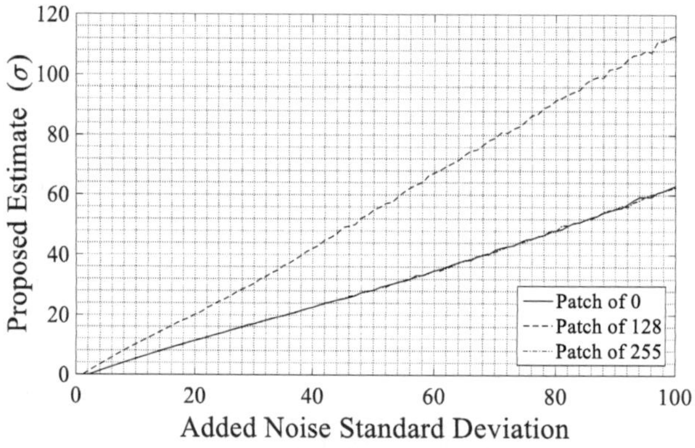

Figure 6. Noise estimation using proposed method for uniform patches

The proposed quality metric (Q) is benchmarked with five state of the art published algorithms. Two databases are used for comparing the performance of proposed quality metric (Q) with these algorithms. Laboratory for Image & Video Engineering (LIVE) database [10] contain 29 reference images. For our experimentation, we have used only the Gaussian noise corrupted images of LIVE database. The 29 reference images are corrupted by five levels of additive Gaussian noise resulting in total 145 images. Categorical Subjective Image Quality (CSIQ) database [11] contain 30 reference images which are corrupted by five levels of Gaussian noise degradation resulting in total 150 images. Total 295 images from two databases are used for benchmarking.

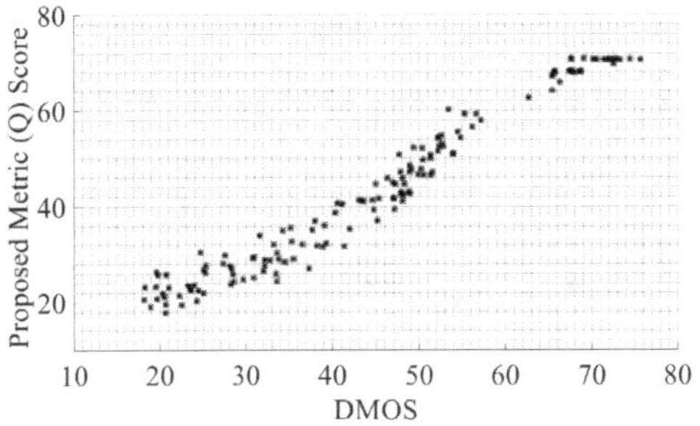

Figure 7(a). Scatter plot of proposed metric (Q) for LIVE database

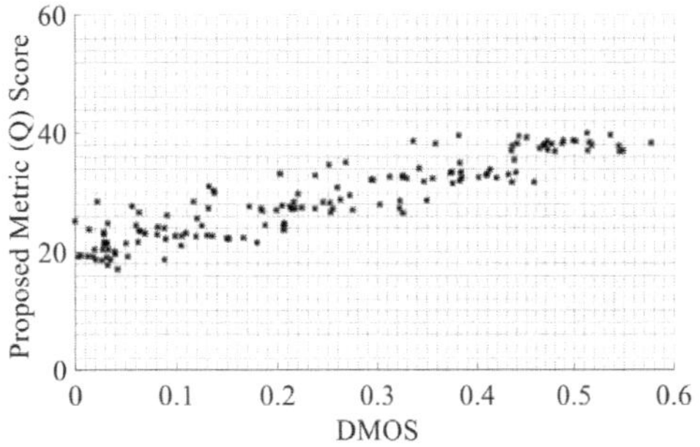

Figure 7(b). Scatter plot of proposed metric (Q) for CSIQ database

Scatter plots of the proposed quality metric (Q) for LIVE and CSIQ databases are shown in figure 7(a) and figure 7(b) respectively. Scatter plot facilitate visual performance evaluation of an algorithm. A dispersed scatter plot indicates poor correlation and a concentrated scatter plot indicates high correlation between two variables. It can be observed from figure 7(a) and figure 7(b) that both the scatter plots can be approximated by a curve which indicates the consistency of proposed quality metric (Q). Also, there are no outliers in the scatter plot which demonstrate the robustness of proposed quality metric (Q). Few sample images from LIVE database are shown in figure 8.

(a) DMOS = 48.86, Q = 48.59 (b) DMOS = 41.16, Q = 40.46 (c) DMOS = 28.32, Q = 27.04

(d) DMOS = 47.80, Q = 47.22 (e) DMOS = 65.53, Q = 67.67 (f) DMOS = 24.37, Q = 22.35

(g) DMOS = 20.30, Q = 21.73 (h) DMOS = 65.42, Q = 64.12 (i) DMOS = 35.15, Q = 35.53

Figure 8. DMOS and proposed quality metric (Q) values for images from LIVE database

Three correlation coefficients are used for quantitative performance comparison of the proposed quality metric (*Q*) with existing algorithms. The three correlation coefficients are Pearson's Linear Correlation Coefficient (LCC), Spearman's Rank Order Correlation Coefficient (SROCC) and Kendall's Rank Order Correlation Coefficient (KROCC). The range of all three correlation coefficients is from -1 to +1 where -1 indicate perfect negative correlation and +1 indicate perfect positive correlation. Values close to -1 or +1 are indicator of better performance of algorithm.

The benchmarking results of the proposed quality metric (*Q*) with state of the art algorithms are presented in table 2. The performance of algorithms is validated on LIVE and CSIQ databases using LCC, SROCC and KROCC parameters. SROCC and KROCC are rank based correlation coefficients whereas LCC calculates correlation between two variables using their standard deviations and covariance resulting in a better indicator of their association.

For LIVE and CSIQ databases, the proposed image quality metric (*Q*) yields high correlation coefficient values which can be observed from table 2. For LIVE database, though the proposed image quality metric (*Q*) does not lead in terms of SROCC and KROCC, LCC value indicates that the performance of proposed quality metric (*Q*) is better than the state of the art algorithms. For CSIQ database, the proposed image quality metric (*Q*) leads in terms of LCC, SROCC and KROCC. The consistently high correlation values of the proposed image quality metric (*Q*) for LIVE and CSIQ databases demonstrate its robustness. Scatter plots in figure 7(a) and figure 7(b) can be approximated by curves which indirectly imply high correlation of the proposed image quality metric (*Q*) with subjective quality scores.

Table 2. LCC, SROCC and KROCC values on LIVE and CSIQ database

Algorithm	**LIVE Database**			**CSIQ Database**		
	LCC	**SROCC**	**KROCC**	**LCC**	**SROCC**	**KROCC**
BIQI [4]	0.9277	0.9600	0.8653	0.8949	0.8816	0.6843
BLIINDS-II [6]	0.9653	0.9596	0.8254	0.7939	0.8004	0.6044
Visual Codebook [7]	0.9435	0.9372	0.7818	0.8692	0.8733	0.6729
DIIVINE [5]	0.9598	0.9769	**0.8976**	0.8724	0.8682	0.6800
SSEQ [3]	0.9693	**0.9801**	0.8762	0.8706	0.8713	0.6795
Proposed(*Q*)	**0.9789**	0.9787	0.8728	**0.9133**	**0.9027**	**0.7179**

DMOS and proposed image quality metric (*Q*) score for few sample images from LIVE database are presented in figure 8. The images are chosen randomly and corresponding quality scores are obtained using the proposed image quality metric (*Q*). It can be observed that the proposed image quality metric value (*Q*) is nearly equal to the DMOS of the image.

The proposed metric (*Q*) is specifically optimized for quality evaluation of images corrupted by Gaussian noise only. This limitation is because of the fact that the estimator is able to gauge the noise present in an image using the median of wavelet domain coefficients. Addition of Gaussian noise raises the median value of HP-HP wavelet sub-band coefficients which is used for noise estimation. Other types of distortions cause different effects on the wavelet coefficients and hence these distortions cannot be handled using the discussed wavelet domain estimator. The methodology used in the proposed quality metric (*Q*) can be extended for quality evaluation of images corrupted by other distortions using a different estimator which has higher sensitivity for those distortions.

6. Conclusion

In this paper, we propose a wavelet transform based noise estimation and image quality assessment method for images corrupted by Gaussian noise. The proposed method for noise estimation obtains the initial estimate of noise standard deviation using wavelet domain high pass sub-band coefficients of the image. This estimate is modified using curve fitting to obtain a near perfect noise estimate of an image. The average noise estimation error using the proposed method is below 4% over a wide range of noise. The proposed method for image quality assessment also uses the same high pass wavelet coefficients for initial noise estimation. The noise estimate is mapped to correlate with human perception using a nonlinear equation. The combined use of noise estimate and a nonlinear mapping yields a robust method to evaluate quality of images corrupted by Gaussian noise. It is a minimum training approach hence it is fast and the performance is consistent on different image databases. The proposed quality metric (Q) is benchmarked on Gaussian noise corrupted images from LIVE and CSIQ databases using three correlation coefficients. The performance of the proposed quality metric (Q) is comparable with existing state of the art algorithms in terms of Spearman's and Kendall's correlation coefficients. It outperforms the existing algorithms in terms of Pearson's correlation coefficient. We plan to extend the concept to tackle distortions other than the Gaussian noise in future work.

7. References

[1] Mittal A, Moorthy A K and Bovik A C 2012 No-reference image quality assessment in the spatial domain *IEEE Transactions on Image Processing* **21** 4695-708

[2] Mittal A, Soundararajan R and Bovik A 2013 Making a completely blind image quality analyzer *IEEE Signal Processing Letters* **22** 209-12

[3] Liu L, Liu B, Huang H and Bovik A C 2014 No-reference image quality assessment based on spatial and spectral entropies *Signal Processing: Image Communication* **29** 856-63

[4] Moorthy A K and Bovik A C 2010 A two-step framework for constructing blind image quality indices *IEEE Signal Processing Letters* **17** 513-6

[5] Moorthy A K and Bovik A C 2011 Blind image quality assessment: From natural scene statistics to perceptual quality *IEEE Transactions on Image Processing* **20** 3350-64

[6] Saad M A, Bovik A C and Charrier C 2012 Blind image quality assessment: A natural scene statistics approach in the DCT domain *IEEE Transactions on Image Processing* **21** 3339-52

[7] Ye P and Doermann D 2012 No-reference image quality assessment using visual codebooks *IEEE Transactions on Image Processing* **21** 3129-38

[8] Donoho D L and Johnstone I M 1994 Ideal spatial adaptation by wavelet shrinkage *Biometrika* **81** 425-55

[9] Velde K V 1999 Multi-scale color image enhancement. In: *International Conference on Image Processing (ICIP)*: IEEE) pp 584-7

[10] Sheikh H R, Wang Z, Cormack L and Bovik A C LIVE image quality assessment database release 2.

[11] Larson E C and Chandler D M Categorical Subjective Image Quality (CSIQ) database.

IOP Conf. Series: Materials Science and Engineering **331** (2018) 012020 doi:10.1088/1757-899X/331/1/012020

Nonlinear Motion Tracking by Deep Learning Architecture

Arnav Verma[1]; Devesh Samaiya[2]; Karunesh K. Gupta[2]

University of Edinburgh, UK.
Birla Institute of Technology & Science, Pilani

kgupta@pilani.bits-pilani.ac.in

Abstract. In the world of Artificial Intelligence, object motion tracking is one of the major problems. The extensive research is being carried out to track people in crowd. This paper presents a unique technique for nonlinear motion tracking in the absence of prior knowledge of nature of nonlinear path that the object being tracked may follow. We achieve this by first obtaining the centroid of the object and then using the centroid as the current example for a recurrent neural network trained using real-time recurrent learning. We have tweaked the standard algorithm slightly and have accumulated the gradient for few previous iterations instead of using just the current iteration as is the norm. We show that for a single object, such a recurrent neural network is highly capable of approximating the nonlinearity of its path.

1. Introduction

The Object tracking is one of the most important problems in the world of artificial intelligence. It is used very commonly in public surveillance systems such as Face Watch in UK to discourage crimes and in the task of counting people indoors for biometric systems [1]. Crowd tracking basically involves segmenting moving people in an image and predicting their future positions based on previous positions-a task commonly called tracking. Major complexities in tracking occur due to lighting and illumination conditions, occlusion, and nonlinearity of motion, resulting in high misclassification rate and misplaced prediction in tracking [2].

Object tracking is performed in two steps; first it is detection then tracking. The detection generally involves, background model development followed by evaluation of deviations in each frame. Pixels that are transformed require further treatment such as background subtraction. Most common challenges faced during detection involve illumination changes which cause false classifications, camouflages resulting in objects similar to background, presence of irrelevant moving objects like leaves, water, flags et cetra, and shadows that may get classified as an object of interest. A more robust technique used for foreground detection is the mixture of Gaussians method [3]. In this algorithm, each pixel is generally modelled as a mixture of Gaussian densities. Based on the local behavior, some of these Gaussians will represent the foreground whereas, the rest will depict the background. Here, using a mixture of Gaussians is necessary because-

- If each pixel represents a single surface, a single Gaussian would have sufficed. However, in practice, multiple surfaces might appear in a single pixel.
- If lighting varied slowly with time, only a single adapting Gaussian would have sufficed. However, lighting conditions may change in many ways, and at varied rates.

Tracking, as the name suggests, refers to tracking the motion of the detected objects, also commonly known as trajectory estimation. In realistic environments, size, shape, and motion of the objects change over time, resulting in one object often getting recognized as a different object, causing the tracker to lose track. Occlusion i.e one object covering another is a very common problem faced in tracking. Occlusion usually occurs in two ways. Inter-object occlusion is when to objects of interest cloud each other, resulting in a single blob of different size than expected. If occlusion occurs in field-of-view, it is very easy to detect. However, if the objects were occluded when they entered the field-of view, as is often common in crowds, then it is very difficult to detect. Occlusion due to scene structure occurs when an object of interest moves behind an irrelevant object such as a car, tree, building et cetra. Additionally, imperfect detection also deteriorates the performance of tracking.

This paves the way for the Kalman filter. A Kalman filter is essentially a state space model for stochastic estimation, composed of predictor-corrector equations and is an optimal linear estimator. Kalman filters are widely used in the fields of aeronautics, orbit determination, speech enhancement, and various other tasks that involve path prediction, or autonomous or assisted navigation. Hence, while Kalman filter is the most widely used tracking technique in object tracking, it does have some shortcomings surrounding the assumptions it makes in arriving at an optimal solution- linearity, and white Gaussian noise. Most tracking systems assume the motion of the object occurs with constant acceleration, resulting in a roughly linear predicted path. When it comes to tracking a nonlinear time-series, which is what the task of tracking boils down to, Kalman filter and its nonlinear variants like Extended Kalman Filter, and Unscented Kalman Filter are inadequate as they require a prior knowledge of the nature of nonlinearity of the data which is to be tracked.

In this paper, the algorithm is proposed to track nonlinear time series using a Recurrent Neural Networks as they are highly efficient at dealing with nonlinearity, even in the absence of external inputs. Now, when we try to track an object, we would essentially be tracking the centroid of the object. This corresponds to online training and for RNNs, this is best carried out by an algorithm known as Real-time Recurrent Learning (RTRL)

2. Mixtures of Gaussians

2.1. Introduction and Initialization

For object detection, we have relied on foreground detection using Mixture of Gaussians due to reasons mentioned earlier [3].

$$\boldsymbol{X_1,\ldots,X_t = I(x_0,y_0,i): 1 \leqslant i \leqslant t}$$

$$\boldsymbol{P(X_t) = \sum_{i=1}^{K} w_{i,t} * n\left(X_t, u_{i,t}, \Sigma_{i,t}\right)}$$

$$\boldsymbol{n\left(X_t, u_{i,t}, \Sigma_{i,t}\right) = \frac{1}{(2\pi)^{n/2}(|\Sigma|)^{1/2}} e^{\frac{-1}{2}(X_t-u_t)^T\Sigma^{-1}(X_t-u_t)}}$$

We choose the number of Gaussians- K, which is usually kept between 3 and 5. The learning rate is generally taken between 0.1 and 0.01. These initializations are done at each pixel with crudely initialized mean vectors and covariance matrices, as the algorithm is expected to evolve and initial observations may be unreliable. At every frame t, we take the feature vector $x_t = [R_t\ G_t\ B_t]$. Now, using Expectation-Maximization algorithm at each pixel in every frame is computationally complex. The algorithm being used is an online K-means approximation.

2.2 The Algorithm

- For each vector x_t in current frame, find which Gaussian matches this observation. This is done by first assuming independence of the three color components and a shared variance and then choosing the Gaussian for which, the observation is within 2.5 of signal mean value. If more than one such Gaussian turns out to be a match, the best such match is considered.

- Now for the matched Gaussian, the weights are updated as shown below and renormalized.
- Additionally, the following updates are also carried out.
- If current pixel value matches none of the K Gaussian pdfs, the distribution with lowest probability is updated to a distribution with the current value as its mean value, an initially high variance, and low prior weight.

$$W_{K,t} = (1-\alpha)W_{k,t-1} + \alpha\left(M_{k,t}\right)$$

$$\mu_t = (1-\rho)\mu_{t-1} + \rho X_t$$

$$\sigma_t^2 = (1-\rho)\sigma_{t-1}^2 + \rho(X_t-\mu_t)^T(X_t-\mu_t)$$

where the second learning rate is-

$$\rho = \alpha\eta(X_t l \mu_k, \sigma_k)$$

The background will theoretically have low variance and high weight. Hence the Gaussians are first ordered based on ratio of weight and standard deviation. Hence, the first B models are chosen as part of background model based on the following constraint where T is a measure of the minimum amount of data that must be accounted for by the background.

$$B = argmin_b\left(\sum_{k=1}^{b} W_k > T\right)$$

Blurring as well as morphological operators such as erosion and dilation to fill holes and get regions corresponding to objects in the scene.

The same process is now repeated for the next frame.

3. Recurrent Neural Network

Tracking nonlinear motion is often carried out with algorithms such as the extended Kalman filter, unscented Kalman filter, particle filter etc. However, these algorithms require in an intrinsic understanding of the inherent nonlinearity of the system, which in turn can often be quite hard to quantify in terms of a simple mathematical function.

Thus, tracking, which is essentially the task of predicting a time-series, seems to demand a certain precognition of the expected path, which defeats the very purpose of tracking. Traditional discrete Kalman filters work very well for linear paths with the assumption of constant acceleration. This, however, does not extend to nonlinear paths as there are infinitely large number of possibilities that might act as an approximation function.

This is where neural networks fill in. We already know that feed-forward neural networks are capable of fitting curves of high amount of nonlinearity. However, predicting a time-series arising from position data of nonlinear motion requires information about history of the time-series to be provided to the network in a manner of online training. Such a system is called a recurrent neural network, a deep learning architecture that is known to do well for time-series data, and has a documented algorithm for online training called the real-time recurrent neural network [5], [6].

3.1 Architecture

- In proposed architecture, Type-I Forward Recurrent Neural Network is used.
- In this architecture, there would be M external inputs, N neurons in the hidden layer, and L output units. Additionally, the output units are a subset of the hidden layer neurons, as seen in Figure 1.

The outputs of the neurons from previous time instant or iteration are sent as inputs to the neurons of the next time instant or iteration. Hence, there are effectively, M+N inputs.

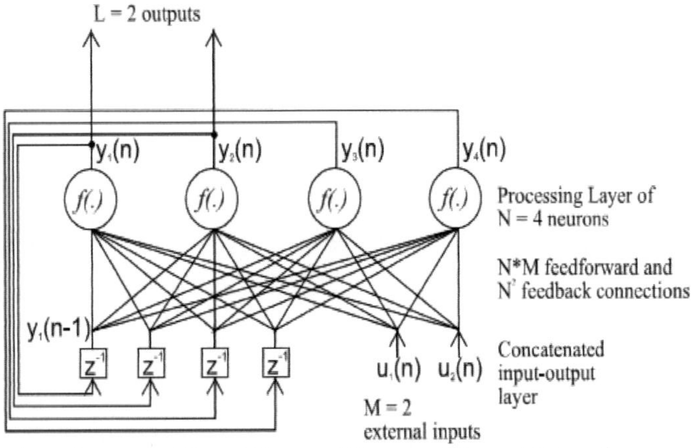

Figure 1. Proposed architecture of Type-I Forward Recurrent Neural Network

3.2 Real-time Recurrent Learning (RTRL)

RTRL is a gradient decent algorithm for training recurrent neural networks in which, an online learning is employed with the error gradient and update weights calculated for each and every time step [4], [6].

Let f be the transfer function of the neurons and z be the concatenated input vector that combines both external inputs as well as feedback of neurons from previous iterations. The algorithm proceeds as follows-

- Let I be the set of indices that correspond to inputs arising from external input. The problem of automatic tracking requires no external inputs and hence, we will not be employing external inputs in our final solution. Similarly, let U be the set of indices that correspond to inputs arising from feedback from neurons of previous iteration.

$$Z_k(t) = \begin{matrix} X_k(t) if k \in I \\ Y_k(t) if k \in U \end{matrix}$$

- If W is the weight matrix with m+n columns and n rows.

$$net_k(t) = \sum_{L \in U \cup I} W_{kL}\, Z_L(t)$$

  Now, if the neural outputs are represented by the vector y,

$$y_k(t+1) = f_k\big(net_k(t)\big)$$

- Let the target values be stored in a matrix d, of vector targets at each time instant. Now, let's define the error for this time instant as:

$$E(\tau) = \frac{1}{2}\sum_{k \in U} e_k\,(\tau)^2$$

where

$$e_k(\tau) = \begin{matrix} d_k(t) - y_k(t) & if\ \ k \in T(t) \\ 0; & otherwise \end{matrix}$$

- If t_o is initial time instant, and t_l is the current time instant, the accumulated error over all time instants, which must be differentiated to obtain gradient for RTRL, and the corresponding gradient are:

$$E_{total}(t_0, t_1) = \sum_{\tau=t_0+1}^{t1} E(\tau)$$

$$\nabla_w E_{total}(t_0, t+1) = \nabla_w E_{total}(t_0, t) + \nabla_w E(t+1)$$

- Let the learning rate be μ. Hence, the weight update required, based on this single step would be (The net update usually employs accumulated weight updates over the previous time steps as well)

$$\Delta W_{i,j} = -\mu \frac{\partial E(t)}{\partial W_{i,j}}$$

$$\sum_{t=t_0+1}^{t1} \Delta W_{i,j}(t)$$

$$\frac{-\partial E(t)}{\partial W_{i,j}} = -\sum_{k \in U} \left(\frac{\partial E(t) \partial Y_k(t)}{\partial Y_{(k)}(t) \partial W_{(ij)}} \right)$$

- Now, we need to evaluate an expression for the partial derivative of the neural outputs with respect to each and every weight matrix element. Combining the expressions for net_k, gradient of error, and output of neurons,

$$\frac{\partial Y_k(t+1)}{\partial W_{i,j}} = f'_k(net_k(t)) \sum_{L \in U} W_{kL} \frac{\partial Y_L(t)}{\partial W_{i,j}} + \delta_{ik} z_j(t)$$

Here, δ is the Kronecker delta function

- The evaluated partial derivative of y involves another intricacy. As external inputs are independent of weights at any time step, their derivatives with respect to the elements of the weight matrix have been taken as 0, i.e-

$$\frac{\partial z_l(t)}{\partial W_{ij}} = 0 for l \in L$$

- Our starting state at time step to is again independent of the weights. Hence, for k spanning all neural outputs at this time instant, i spanning over all neurons, and j spanning over all inputs (internal, as well as external),

$$\frac{\partial y_k(t_0)}{\partial W_{ij}} = 0$$

- Let p be the matrix containing values of derivatives of neural outputs:

$$p_{ij}^k(t) = \frac{\partial y_k(t)}{\partial W_{ij}}$$

Hence,

$$\Delta W_{ij}(t) = \mu \sum_{k \in U} e_k(t) p_{ij}^k(t)$$

In our approach, we will slightly tweak the RTRL algorithm to accumulate error over few previous iterations, instead of just the current iteration, as is the norm.

4. Experimental Results

The method was tested using fabricated nonlinear centroid data, which was then normalized as the output of the tan sigmoid function that we used as the activation function ranges from 0 to 1 on MATLAB ver.12. We used one hidden layer with 10 hidden units, 300 frames, and normalized, symmetric, square, data to train the RNN. The optimum number of hidden units, and number of previous iterations over which we accumulate the error for a particular dataset maybe fixed by using Genetic Algorithm to optimize the total error over all iterations. Throughout all the tests, the learning rate was maintained at 0.1.

Table 1 shows a decreasing trend for the total error with increasing number of iterations for which, error gradient is accumulated until this measure reaches 7 for the case of symmetric, normalized, square centroids. However, even with accumulation over 6 iterations, the RNN tracks the initial data poorly.

Table 2 shows the variation of accumulated error with change in the number of units in the hidden layer. For the case of symmetric, normalized, square centroid, 8 hidden units provide optimal performance for prediction, for the RNN. While using 3, 6, and 7 hidden units provide lower total error, all these cases show better tracking for starting data and poorer tracking for later data, which is undesirable as opposed to using 8 hidden units, in which case, the tracking improves with arrival of new data.

Table 1. Iterations and accumulated error.

No: of iterations for which error is accumulated	Accumulated error
1	0.6188
2	0.2272
3	0.1443
4	0.988
5	0.746
6	0.718
7	0.1267

Table 2. Hidden Units and Accumulated Error

Number of hidden layers	Accumulated error
3	0.0868
4	0.1165
5	0.0925
6	0.0844
7	0.0640

8	0.0893
9	0.0943

Figure 2 (a) shows the recurrent neural network's attempt to track linear motion of Figure 2(b) without any prior knowledge of the path.

Figure 3 (a) & 3 (b) represent the coordinates of the centroid of the object along the X and Y axes respectively for 10 hidden units, and error gradient accumulation over 4 iterations. This network, when trained, predicts the X and Y coordinates as shown in Figure 4 (a) & (b) respectively, again without any prior indication of the nature of motion.

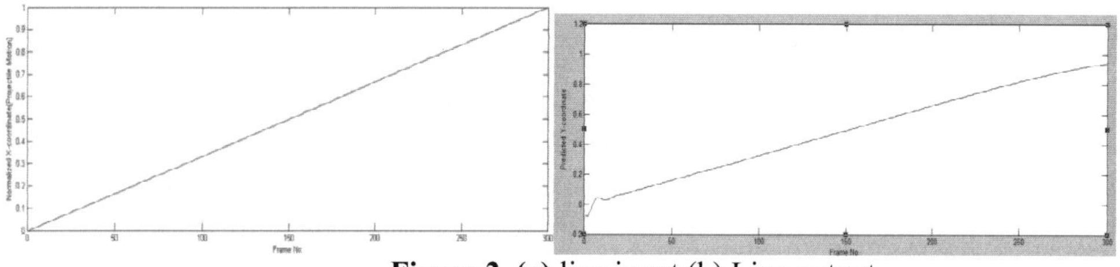

Figure 2. (a) line input (b) Line output

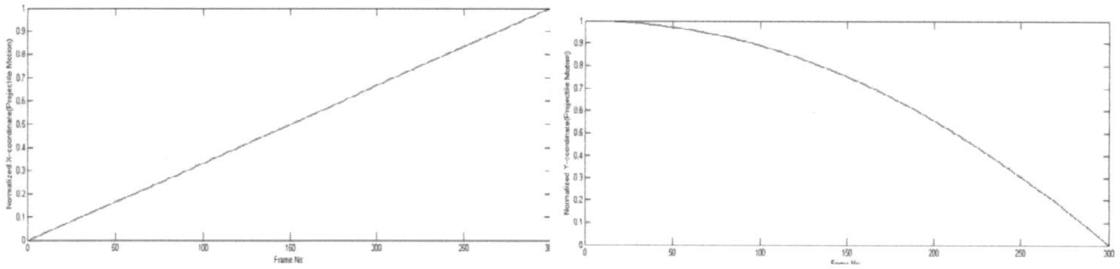

Figure 3. (a) x coordinate of projectile (b) y coordinate of projectile

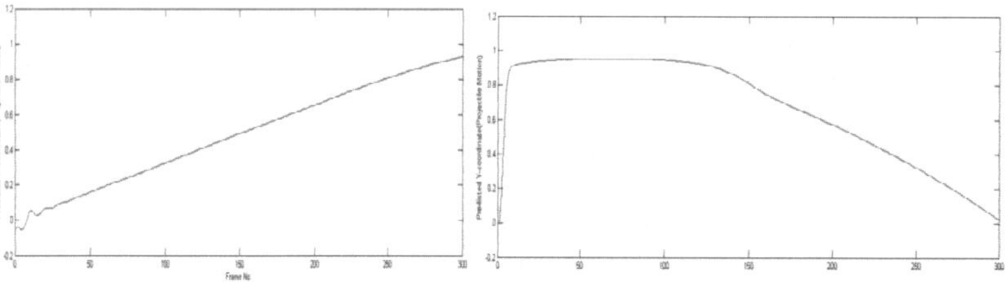

Figure 4. (a) projectile predicted X coordinate (b) Projectile Predicted Y coordinate

Finally, Figure 5(a) & (b) show the tracking performance for the case of sinusoidal data with low frequency for the first few frames with the tracked response and the target function plotted respectively. Highly chaotic data proved to be difficult to track in real-time using real-time recurrent learning.

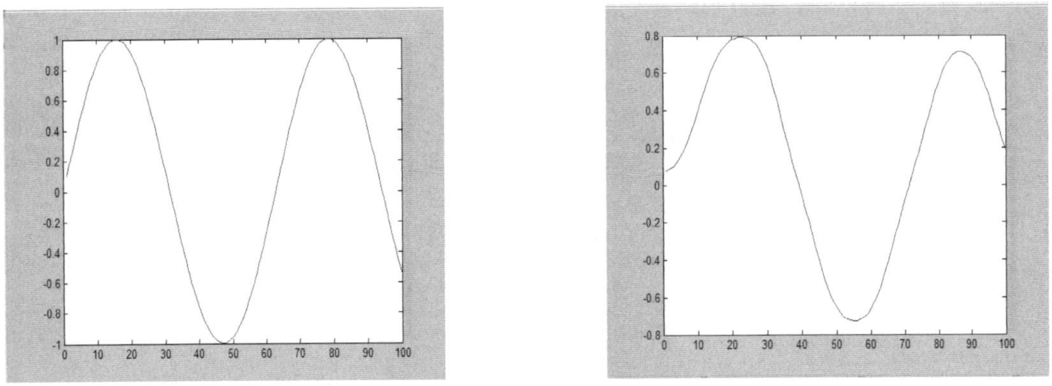

Fig 5. (a) target sinusoidal data (b) predicted sinusoidal data

5. Conclusions and Future work

The proposed algorithm based on deep learning architecture is highly capable of predicting nonlinear tracking. It has proven to be highly capable of predicting nonlinear paths of a single object without any previous knowledge or assumption about the path the object is going to take, albeit with some amount of deviation. This is an improvement over nonlinear tracking algorithms like Extended Kalman Filter and Unscented Kalman Filter. While the complexity of real-time recurrent learning is O(N4) for every time-step in the absence of external input, for fewer number of hidden units, the algorithms computes very quickly.

However, there are many aspects of this method to improve upon in the future, the first of which is to handle multi object tracking efficiently, even in the presence of occlusion. Occlusion has adverse effects on the tracking algorithm's performance. There is also an observably larger error in prediction when the coordinates of centroids have different natures. This may be dealt with by using different RNNs for each coordinate. There is also work to be done in identifying the apt number of hidden layers and units within these layers that satisfy the trade-off between complexity and tracking performance.

References

[1] Milan Sonka, Vaclav Hlavac and Roger Boyle, "Image Processing, Analysis and Machine Vision" International Student Edition, Third Edition, 2008.
[2] Chris Stauffer, W.E.L.Grimson, Adaptive Background Mixture Model for real-time tracking, IEEE Computer Society Conference on Computer Vision and Pattern Recognition. pp. 246-252, Santa Barbara, CA, June 1998.
[3] Stauffer, C., and Grimson, E. Adaptive background mixture models for real-time tracking. In Proceedings of the IEEE International Conference on Computer Vision and Pattern Recognition, 1999.
[4] Esko O. Djik, Analysis of Recurrent Neural Networks with Application to Speaker Independent Phoneme Recognition, Laboratory of Signals & Systems ñ Network theory, University of Twente Enschede, The Netherlands, 1999.
[5] Prof. C.Anderson, rtrl.m,rtrlinit.m
[6] R.J.Williams, D.Zisper, Gradient Based Learning Algorithms for Recurrent Networks and their Computational Complexity,1995.

Numerical analysis of THz radiation wave using upper hybrid wave wiggler

Pratibha Malik*[1], Suresh C. Sharma[1], Jyotsna Panwar[1] and Rinku Sharma[1]

[1]Department of Applied Physics, Delhi Technological University, Shahbad Daulatpur, Bawana Road, Delhi-110042, India

*pratibhamalik27@gmail.com

Abstract. A theory for upper hybrid wave induced by relativistic electron beam in magnetized plasma emits tuneable and coherent terahertz radiation. The nonlinear interaction with REB is used to generate terahertz radiation. The enhancement in the amplitude of THz wave is also observed when pre-bunched REB is used. The ponderomotive force applied on beam electrons due to radiation wave and upper wave wiggler modifies the dispersion relation. By solving the dispersion relation, we have derived the growth rate of the radiation wave. Numerical studies indicate that by increasing the beam energy the growth rate of the radiation wave decreases, while it increases with wiggler frequency. Besides this, the growth rate of the radiation wave increases with beam density and decreases with radiation frequency and static magnetic field.

1. Introduction

For past few decades the production of terahertz (0.3- 10 THz) radiation has been witnessed by the researchers in the broad range of spectroscopy, military security, sensing and communications [1-3]. Since several approaches have been demonstrated to generate the terahertz radiation such as laser based sources, electron beam based sources. In electron beam based sources: surface plasmons [4] and free electron lasers (consist of an electron beam and electromagnetic wiggler i.e., provides magnetic field) [5-6], which are the perfect sources for the generation of tuneable high frequency radiation. On account of high cost and larger size the implication of such kind of wigglers are replaced by the plasma wigglers [7]. In this scheme, the upper hybrid wave is employed as an electrostatic wiggler to generate the THz radiation.

2. Enhancement of THz radiation

We consider an upper hybrid wave of potential $\phi = \phi \exp(-i(\omega t - \vec{k}.\vec{r}))$ propagates in x-z plane is being dipped in magnetized plasma (B_1). The interaction of relativistic electron beam $(v_{bo}\hat{z})$ with upper hybrid wave $(\vec{E}_{\omega})$ produces the nonlinear current density and responsible to obtain the nonlinear dispersion relation. The ponderomotive force and nonlinear REB induced upper hybrid wave wiggler is the main mechanism for the enhanced THz radiation emission. We have derived the results using fluid equation model and the equations as mentioned as below.

The interaction of REB with upper hybrid wave produces the electron oscillatory velocity $\vec{v}_0$ and taking into account the equation of motion as

1. $$\frac{\partial}{\partial t}(\gamma_1\vec{v})+(\vec{v}.\nabla)(\gamma_1\vec{v})=-\frac{e}{m}\vec{E}_\omega-\vec{v}_0\times(\omega_c\vec{z}) \tag{1}$$

2. Equation of continuity

$$\frac{\partial n}{\partial t}+\nabla.(n\vec{v})=0 \tag{2}$$

3. The magnetic field of the radiation wave

$$\vec{B}=\frac{c}{\omega_1}(\vec{k}_1 x\vec{E}) \tag{3}$$

Here the terms n, $\vec{v}$ and m are defined as density of electrons, velocity and mass of the electron respectively, relativistic gamma factor $\gamma_1=\gamma_{b0}(\approx(1-v_{bo}^2/c^2)^{-1/2})+\gamma_{10}$ and beam velocity $\vec{v}=v_{bo}\hat{z}+\vec{v}_0$. However, the cyclotron frequency is represented as $\omega_c(=eB_1/mc)$.
The electromagnetic wave with electric field $\vec{E}=\hat{x}E\exp(-i(\omega_1 t-\vec{k}_1.\vec{z}))$ is assumed the perturbation field for the equilibrium. In addition, the potential offered by negatively charged beam wave is $\phi_b=\phi_b\exp(-i(\omega_b t-\vec{k}_b.\vec{z}))$.

The frequency of THz radiation wave is calculated as $\omega_1=2\gamma_{b0}^2(k_z v_{b1}+\omega)-2\gamma_{b0}^{1/2}\omega_{pb}$, where $\omega_{pb}=\left(4\pi n_{b1}e^2/m\right)^{1/2}$ is the beam plasma frequency. We obtain the relations as $\omega_1=k_1 c$ and $\omega_b=k_{bz}v_{bo}-\omega_{pb}/\gamma_{b0}^{3/2}$.

However, the beam wave is characterized as $n=n_{b1}+n_{b2}\exp[-i\omega_m(t-z/v_{b0})]+n_{b3}\exp[-i(\omega_1 t-\vec{k}_1.\vec{x})]$ and $n_{b2}=\Delta n_{b1}$, where n_{b1} is the beam velocity at equilibrium. However, n_{b3} is the perturbed density due radiation wave and Δ is the modulation index.

Therefore, the ponderomotive force [8] produced by radiation wave $(\omega_1,\vec{k}_1)$ and wave wiggler $(\omega,\vec{k})$ on beam electrons at $(\omega_b,\vec{k}_b)$ is obtained as

$$\vec{F}_p=-\frac{m}{2}[\vec{v}.\nabla(\gamma_1\vec{v})]-\frac{e}{2c}(\vec{v}_o\times\vec{B}) \tag{4}$$

and $\vec{F}_p=-e\nabla\phi_p$

After using Eqs. (1), (2) and Eq.(4) and solving them , the nonlinear current density due to ponderomotive force and non-linearity, becomes

$$\vec{J}^{nl}_{(\omega_1,\vec{k}_1)}=-\frac{1}{2}n_{b3}e\vec{v}_0 \tag{5}$$

Now, solving Maxwell's wave equation and simplifying the terms after putting the value of current density and approximate value of static magnetic field, we get instability of the radiation wave and derived the dispersion relation. Thus, the amplification of the radiation wave is perceived as long as the beam velocity is larger than the phase velocity of the radiation wave. In this case, we get the small frequency mismatch i.e., the imaginary part of the radiation frequency $\omega_1=k_1c+\delta$. This leads to the enhancement in the radiation wave.
The growth rate of the radiation wave i.e., the imaginary part of δ

$$\alpha=\text{Img }\delta=\frac{\sqrt{3}}{2}\left[\frac{\omega^2 v_{os}^2\omega_{pb}[(1+\Delta k_{bz})+\Delta k_{bzo}]|k_z|}{32\omega_1\omega_c\gamma_{b0}^{1/2}}\right]^{\frac{1}{3}}. \tag{6}$$

where $v_{os}(=e\phi k_x/m\omega\gamma_{bo})$ is the measurement of beam oscillatory velocity due to wiggler wave along x-direction. The Equation (6), represents the relation for the growth rate α(in sec^{-1}) of the unstable mode of THz radiation wave.

3. Results and Discussions

The typical parameters are used to analyse the behaviour of terahertz radiation wave in the magnetized plasma: electronic mass m= 9.1x10^{-28} g, electronic charge e = 4.8 x10^{-10} Statcoulomb, electron plasma density $n_p = 4.1 \times 10^{18}$cm^{-3}, wiggler wave frequency $\omega = 1.6 \times 10^{14}$ rad/sec, electron cyclotron frequency $\omega_c = 0.2\omega$, 0.15ω, and 0.09ω etc, static magnetic field $B_1 = 100$kG, 105 kG and 111 kG etc. The modulation index Δ ranges from 0 to 1 in the steps of 0.1.

Using Equation (6), we have plotted Figure1, which illustrates the variation of growth rate α(in sec^{-1}) of the radiation wave with beam energy γ_{b0} and radiation frequency ω_1(in rad/sec). This figure reveals that the growth rate decreases with beam energy and radiation frequency. The physical mechanism may be: As long as the beam velocity is larger than the phase velocity of the radiation wave, the amplification in the growth rate of unstable radiation mode keeps on rising.

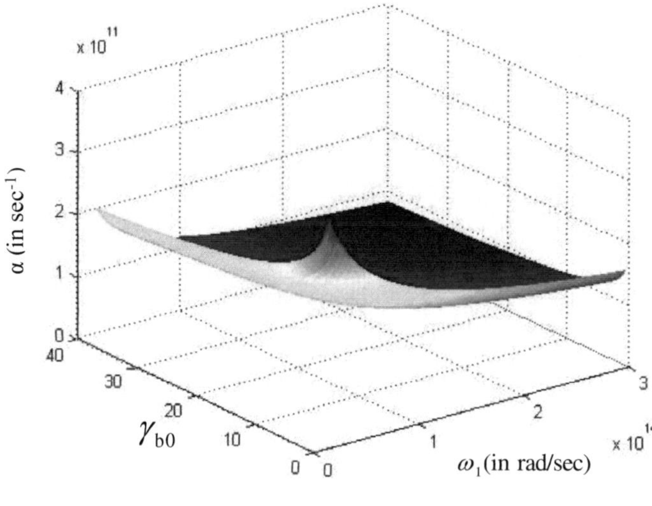

Figure 1: Variation of growth rate α(in sec^{-1}) of radiation wave with radiation frequency ω_1(in rad/sec) and beam energy γ_{b0} at modulation index $\Delta = 0.8$.

However, Figure 2 depicts the variation of growth rate α(in sec^{-1}) of the unstable radiation mode with static magnetic field ω_c(in rad / sec) and beam density n_{b1}(in cm^{-3}) of electrons. In this sketch, we see that the growth rate of radiation unstable mode increases with beam density but decreases as static magnetic field increases [9]. The reason may be understood as: for the increased values of magnetic field, the amplitude of the radiation wave is suppressed.

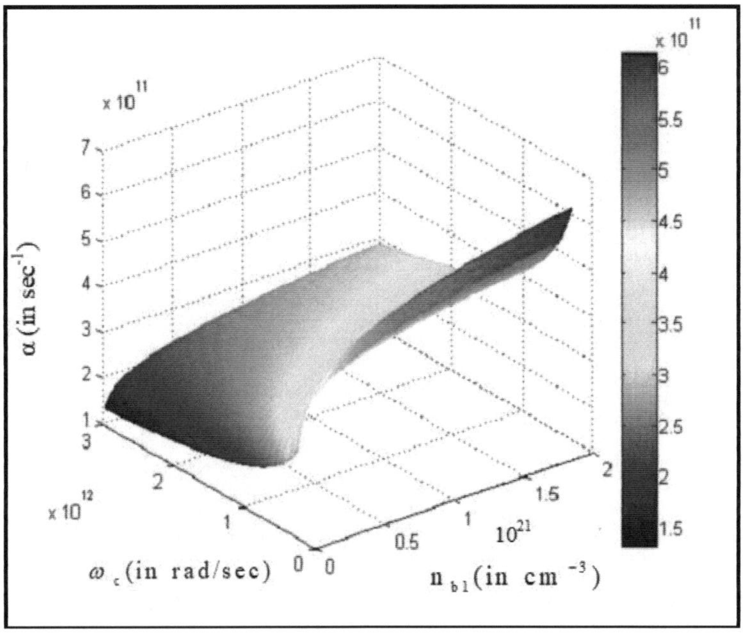

Figure 2: Variation of growth rate $\alpha(\text{in sec}^{-1})$ of radiation wave with cyclotron frequency $\omega_c(\text{in rad / sec})$ or magnetic field B_1 and beam density $n_{b1}(\text{in cm}^{-3})$ at modulation index $\Delta = 0.6$.

4. Conclusion

The present work shows the possibility of upper hybrid wave as a wiggler for producing the coherent terahertz radiation. The interaction of relativistic electron beam with such wiggler is analogous to free electron laser, which shows the coherence and tuneability of the radiation in THz frequency range, if phase matching conditions are satisfied. The nonlinear current is involved to obtain the nonlinear dispersion relation and that leads to the instability of growth rate of the radiation wave.

References

[1] Mittleman D M, Jacobsen R H, and Nuss M C 1996 *IEEE J. Sel. Top. Quantum Electron* **2** 679
[2] Ferguson B and Zhang X C 2002 *Nat. Mater.* **1** 26
[3] Pickwell E and Wallace V P 2006 *J. Phys. D* **39,** 301
[4] Sharma S C and Malik P 2015 *Phys. Plasmas* **22** 043301
[5] Dattoli G , Ottaviani P L and Renieri A 2005 *Laser Part. Beams* **23** 303
[6] Geddes C G R, Toth C, Van Tilborg J, Esarey E, Schroeder C B, Bruhwiler D, Nieter C, Cary J and Leemans W P 2004 *Nature* **431** 538
[7] Joshi C, Katsouleas T, Dawson J M, Yan Y T and Slater Jack M 1987 *IEEE J. Quant. Electron* **23,** 1571
[8] Liu C S and Tripathi V K, 1994 *Interaction of Electromagnetic Waves with Electron Beams and Plasmas* (World Scientific publishing Co. Pvt. Ltd., Singapore)
[9] Namiot V A and Shchurova L Y 2011 *Phys. Lett.* A **375** 2759

Offset fed slot antenna for broadband operation

K Ritish[1], S Piyush[2] and A V Praveen Kumar[3]

[1, 2, 3] Department of Electrical & Electronics Engineering,
Birla Institute of Technology and Science, Pilani, Rajasthan, India

E-mail: kr.ritish@gmail.com[1], f2013440@pilani.bits-pilani.ac.in[2],
praveen.kumar@pilani.bits-pilani.ac.in[3]

Abstract. In this paper, a microstrip fed rectangular slot antenna with wideband characteristics is proposed. Both the impedance and radiation characteristics of the proposed antenna are presented. It is shown that a properly offset feed can give a dual resonance nature, which can be optimized to enable wideband behavior. From HFSS simulation, an impedance bandwidth (-10 dB) of 49.92 % (2.51 GHz to 4.18 GHz) about the center frequency of the band is obtained. Prototype measurement demonstrates a bandwidth of 45.30 % (2.51 GHz to 3.98 GHz). Simulated radiation patterns show bidirectional behavior, which is stable in the band with a peak gain of 5.7 dBi and a gain variation of 2 dBi.

1. Introduction

In the past several years, multiband and wideband wireless technologies received considerable attention. Hence the development of antennas with multiple / wide bandwidth, small size, ease of fabrication and low cost are required. One of the good candidates bearing wideband properties is the slot antenna and the characteristics of microstrip fed slot antennas have been extensively studied [1][2][3]. Slot antennas dominate patch antennas in several aspects such as (i) larger bandwidth (ii) bidirectional radiation pattern (iii) low cross-polarization and (iii) less sensitive to manufacturing tolerances etc [3]. Also various investigations on enhancing the bandwidth of wide slot antenna are reported [4][5][6][7]. The bandwidth enhancement is based on the principle of merging of dual resonances produced by the offset fed rectangular slot antenna.

In the present paper, a compact wideband slot antenna, excited by a simple microstrip transmission line is proposed. The wideband performance is achieved by generating dual-resonances of a slot antenna, offset-fed with a microstrip line. ANSYS HFSS has been used for modeling the antenna and optimizing the design. A prototype is fabricated and the bandwidth characteristics are measured and compared with the simulation result.

2. Antenna Configuration

The geometry of the proposed antenna is shown in Figure 1. As shown in the figure, the antenna has overall size L = W = 100 mm. and the rectangular slot has size Ls= 34 mm., Ws= 17 mm. The antenna is printed on FR4 microwave substrate of thickness 1.6 mm, relative permittivity of 4.4 and loss tangent of 0.002. This slot is excited by a 50 Ω microstrip line with a simple tuning stub having a length G = 13.5 mm and offset parameter d =12.58 mm. The microstrip is printed on the back side of the substrate. A conventional center-fed (i.e. d =0) printed slot antenna with a simple tuning stub has been taken for comparison. All the design parameters of center-fed antenna are the same as those of the offset-fed antenna except the length of stub (G) that need to be varied for impedance matching.

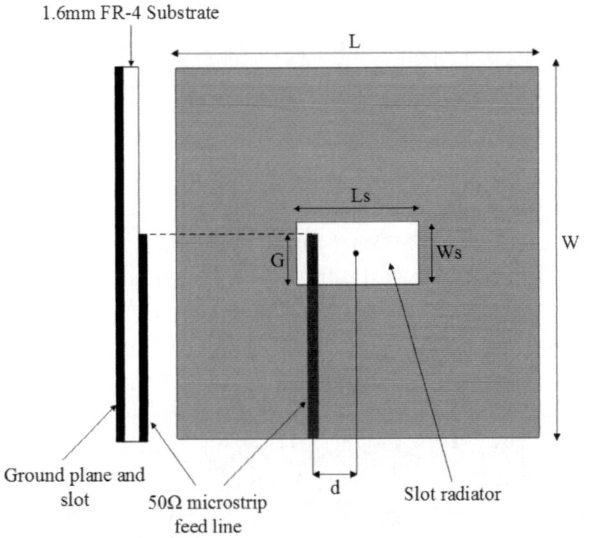

Figure 1. Geometry of microstrip line fed printed rectangular slot antenna with offset parameter of 'd'

3. Results and discussion

3.1. Impedance Characteristics:

By optimizing the off-centered parameter 'd' in Figure 1, the measured reflection coefficient of proposed offset-fed antenna and center-fed antenna are shown in Figure 2. A comparison study of the design parameters and corresponding simulated data are listed in Table 1. From the obtained results, it is seen that the proposed antenna which is fed by (d = 0.37Ls mm) off-centered microstrip feed has an impedance bandwidth as large as 1.67 GHz or 49.92% about the center frequency of the band. This is nearly about seven times that (230 MHz) that of the center-fed antenna. This bandwidth is increased because a second resonant mode in addition to the fundamental mode can be excited by the off-centered microstrip line fed rectangular slot antenna. It has been found that dual-resonant characteristics are enhanced as the microstrip line is moved off center i.e. towards the edge of the slot. And these dual resonances can be merged below -10 dB reflection coefficient together to give wide bandwidth behavior by properly optimizing the off-centered microstrip line feed location.

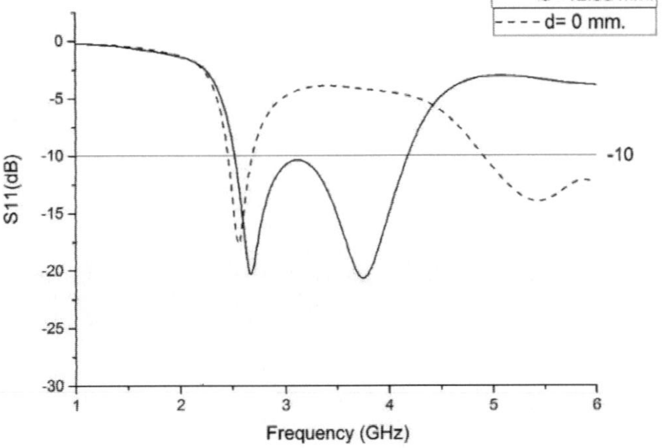

Figure 2. Reflection coefficient versus frequency, for the designs shown in Table 1.

IOP Conf. Series: Materials Science and Engineering **331** (2018) 012022 doi:10.1088/1757-899X/331/1/012022

Table 1. Comparison between the center-fed and the offset-fed proposed antennas with parameters.

Antenna	d (mm.)	G(mm.)	f_L(GHz)	f_H(GHz)	Impedance B.W.(GHz)	Percentage B.W. (%)
(a) Centre-fed	0	10.5	2.46	2.69	0.23	8.93
(b) Offset-fed (Proposed)	12.58	13.5	2.51	4.18	1.67	49.92

Once optimized the geometry of the slot antenna, a prototype has been fabricated and measured. Figure 3 shows the top and bottom views for the fabricated prototype of the offset-fed slot antenna with SMA connector. The Keysight N-9928A Vector Network Analyzer is used for measurement of impedance characteristics. Figure 4 shows the measured and simulated reflection coefficient against frequency for the proposed antenna. Measured bandwidth is 1.47 GHz or 45.30 % about the center frequency. The difference is attributed to the fabrication tolerances.

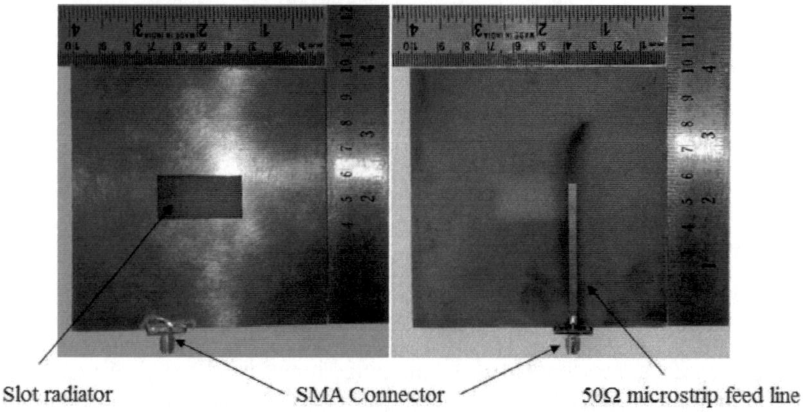

Figure 3. Photograph of the antenna prototype for $\varepsilon r = 4.4$, $h = 1.6$ mm, $L=W=100$ mm, $Ls = 37$ mm, $Ws = 17$ mm, $G=13.5$ mm, and $d=12.58$ mm.

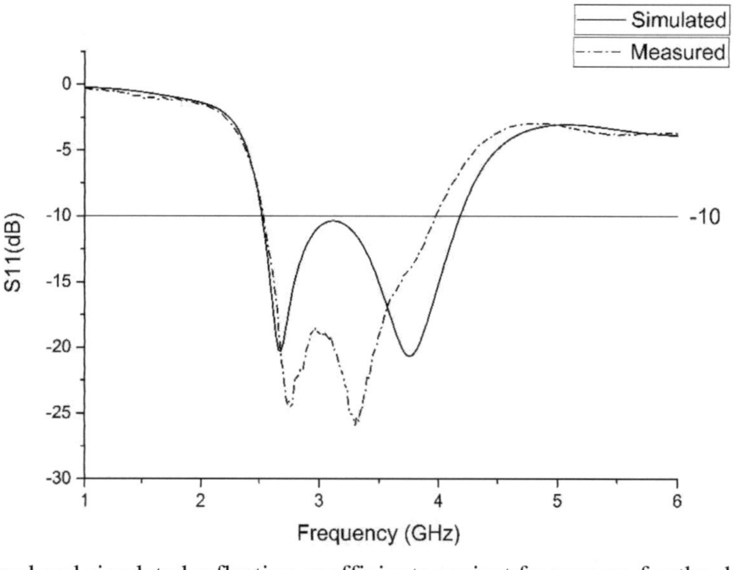

Figure 4. Measured and simulated reflection coefficients against frequency, for the design shown in Figure 3.

3.2. Radiation characteristics:

Figure 5 shows the radiation patterns of this antenna in the two principal planes (XZ and YZ) at three different frequencies in the band 2.68 GHz, 3.12 GHz and 3.74 GHz. The patterns are broadsided, having very little variation with frequency indicating the usefulness of the design. Also note that the slot antenna being open on both sides radiates bi-directionally. Figure 6 represents the antenna gain within the operating bandwidth of this antenna. The gain varies from 3.9 dBi to 5.7 dBi in the impedance band showing a variation less than 2 dB.

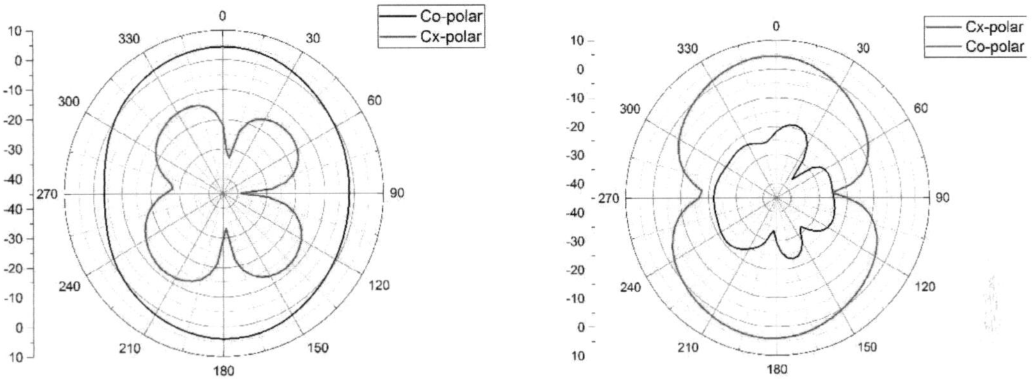

(i) (ii)

(a)

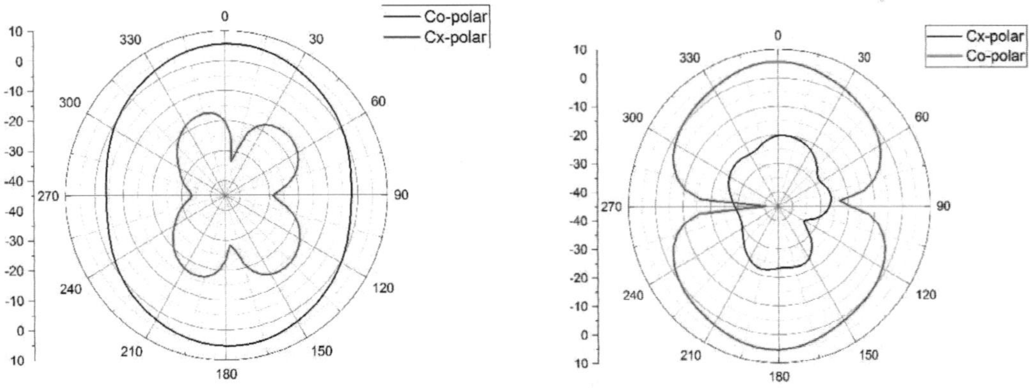

(i) (ii)

(b)

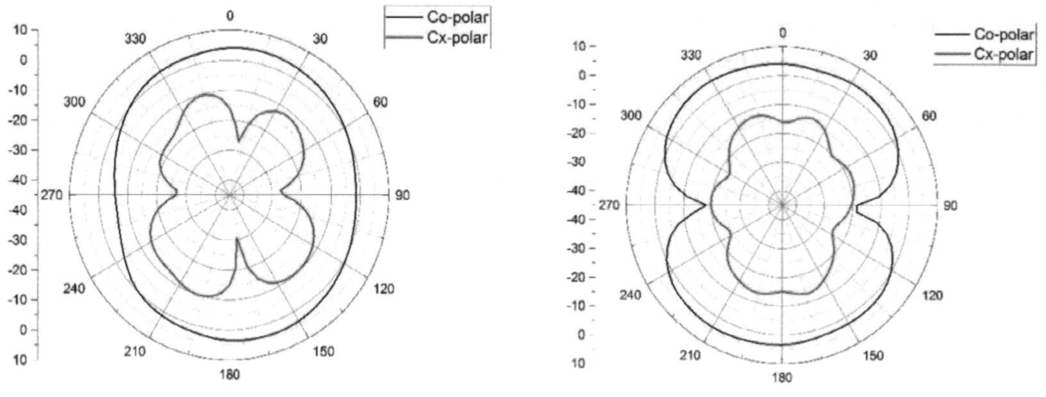

(i) (ii)

(c)

Figure 5. Simulated (i) XZ-plane and (ii) YZ-plane radiation patterns of proposed slot antenna (a) $f = 2.68$ GHz (b) $f = 3.12$ GHz and (c) $f = 3.74$ GHz, for the design shown in Figure 3.

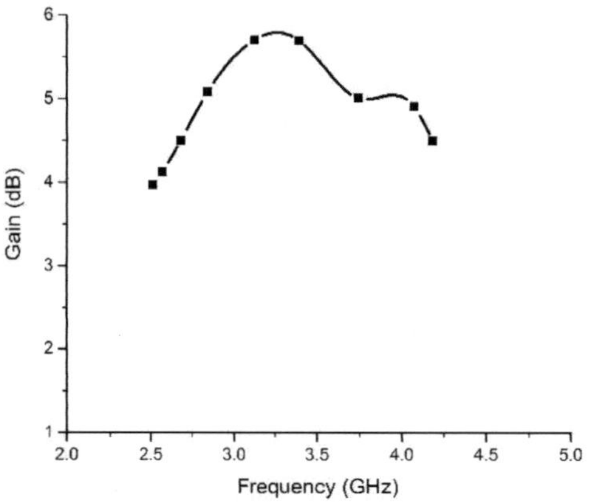

Figure 6. Simulated antenna gain against frequency for proposed slot antenna design shown in Figure3.

4. Conclusion

A printed rectangular slot antenna fed by a 50 Ω offset microstrip line for bandwidth enhancement was investigated. Results show that the -10 dB impedance bandwidth can be significantly improved by selecting suitable offset location of the microstrip feed. For the optimized offset parameter (d/Ls = 0.37, d=12.58 mm), simulated bandwidth becomes nearly 1.67 GHz or 49.92 % while the measured bandwidth is 1.47 GHz or 45.30 %. This is about seven times that of a centre-fed slot antenna. Within this wide impedance band, the peak antenna gain is 5.7 dBi with a gain variation of less than 2 dB.

Acknowledgement

Authors acknowledge DST-FIST grant [Grant Ref: SR/FST/ETI-346/2013] for financial support. The first author also acknowledges University Grants Commission, India for the UGC-NET-JRF/SRF fellowship.

References

[1] Jia-Yi Sze and Kin-Lu Wong 2001 "Bandwidth enhancement of a microstrip-line-fed printed

wide-slot antenna," in *IEEE Transactions on Antennas and Propagation*. **49**, no. 7,1020-24.

[2] Qing X., Chia M. Y. W., and Wu X. 2003 "*Wide-slot antenna for UWB applications*," in *Proc. IEEE Antennas and Propagation Soc. Int. Symp.*, vol.1, pp. 834–837.

[3] R. Garg P. Bhartia I. Bahl and A. Ittipiboon 2001 *Microstrip Antenna Design Handbook MA* (Norwood:Artech House).

[4] Jan J.Y. and Su J.W. 2005 "Bandwidth enhancement of a printed wide-slot antenna with a rotated slot," *IEEE Trans. Antennas Propag.*, **53**, no. 6, 2111–14.

[5] Sung Y. 2012 "Bandwidth Enhancement of a Microstrip Line-Fed Printed Wide-Slot Antenna with a Parasitic Center Patch," in *IEEE Transactions on Antennas and Propagation*, **60**, no. 4, 1712-16.

[6] Fan S. T., Yin Y. Z., Lee B., Hu W. and Yang X. 2012 "Bandwidth Enhancement of a Printed Slot Antenna with a Pair of Parasitic Patches," in *IEEE Antennas and Wireless Propagation Letters*, **11**,1230-33.

[7] Tan Y., Yan L., Zhao X., Liu C., and Huang K.M. 2014 'Bandwidth enhancement of printed slot antenna with a diamond-shaped tuning stub', *Prog. Electromagn. Res. C*, **50**, 87.

IOP Conf. Series: Materials Science and Engineering **331** (2018) 012023 doi:10.1088/1757-899X/331/1/012023

Optimal placement of FACTS devices using optimization techniques: A review

Dipesh Gaur and Lini Mathew
Department of Electrical Engineering, NITTTR, Chandigarh, India

Email: gaur.dipesh@gmail.com

Abstract Modern power system is dealt with overloading problem especially transmission network which works on their maximum limit. Today's power system network tends to become unstable and prone to collapse due to disturbances. Flexible AC Transmission system (FACTS) provides solution to problems like line overloading, voltage stability, losses, power flow etc. FACTS can play important role in improving static and dynamic performance of power system. FACTS devices need high initial investment. Therefore, FACTS location, type and their rating are vital and should be optimized to place in the network for maximum benefit. In this paper, different optimization methods like Particle Swarm Optimization (PSO), Genetic Algorithm (GA) etc. are discussed and compared for optimal location, type and rating of devices. FACTS devices such as Thyristor Controlled Series Compensator (TCSC), Static Var Compensator (SVC) and Static Synchronous Compensator (STATCOM) are considered here. Mentioned FACTS controllers effects on different IEEE bus network parameters like generation cost, active power loss, voltage stability etc. have been analyzed and compared among the devices.

1. Introduction

Today's power system becomes more complex interconnected system due to alarming increase in load demand and dynamic load pattern which affect severely on transmission lines. They are operating either overloaded or under loaded. The uneven load distribution affects voltage profile and makes system voltage security vulnerable to the fault. It becomes difficult to maintain power system security and reliability. Conventional approach of add new transmission lines in the system and build new power generation facilities is bound with certain factors such as technical and economical bounds. So the best and necessary solution left is to make optimal use of existing generation and transmission network. FACTS controllers are the best and effective alternative for power system performance improvement like voltage security, transfer capability and reduction in losses etc. instead of making complex new transmission corridor. These devices can be connected in series, shunt, series-series and series-shunt. It is important to decide FACTS devices type according to the purpose of need. For voltage control at the point, shunt controllers are desirable and power flow in the line can be controlled through series controllers [1].

In 1999, Hingorani and Gyugyi introduced the concept of FACTS. Modulation and alteration of line power flow becomes accurate, fast and precise manner is attainable with FACTS concept [2]. The core of FACTS controllers is basic power electronics devices. FACTS devices applications include enhancement of transmission lines power transfer capacity and regulate different parameters in transmission network such as current, impedance, phase angle and voltage. Power flow can be made flexible or controllable using these devices. FACTS devices helps to increase loadability of the network through reduction of power flow in overloaded lines and line losses also reduced [3].FACTS devices effectively tackle the problem of voltage collapse and system security. These devices help in the problem of Congestion management. System accommodated the changes easily with FACTS devices [4].

Optimal location and settings of FACTS controllers play an important role for enhancement of system performance and economic benefits. In the past, several approaches are proposed by researchers to work out the problem of optimal location of FACTS devices. Common techniques of placement of devices are categorized into analytical, linear programming and heuristic search methods. The problem of optimal location is considered as combinatorial analysis and heuristic search methods are best tools for such problems as they are robust, fast and best suited for real problems of the power system. Common heuristic search methods proposed for optimal placement in research are Genetic Algorithm (GA), Differential Evolution (DE), Particle Swarm Optimization (PSO), Harmony Search Algorithm (HSA), and Ant Colony Optimization (ACO) [5].

In this paper, different optimization approaches especially heuristic approach for different parameters and their results are reviewed and compared. Literature regarding optimal location and ratings of SVC, TCSC and STATCOM using optimization technique especially GA are reviewed. Different objective functions and bus networks are considered and separately each FACTS devices effects on certain parameters of bus network such as generation cost, losses, voltage profile etc. are represented in tabular form. Comparison of different FACTS devices on the ground of parameters affected like reduction in active losses, improvement in voltage profile, minimization of cost etc. This paper compares different search techniques and review of recent techniques in literature and show effectiveness of individual FACTS devices for various parameters for a given bus network.This paper has following sections: II. FACTS devices and their mathematical models, III. Recent optimization techniques, IV. FACTS Applications using GA, V. Conclusion, VI. Research Scope.

2. Recent Optimization Techniques

2.1 Particle Swarm Optimization(PSO)

PSO method was introduced in 1995 by Dr. Kennedy and Eberhart. PSO is motivated by swarm intelligence like fish school and bird flocks etc. In birds, social interaction with each other and with environment to search the food is the basis of PSO. In PSO, compared to GA few parameters need to be changed and it is easy to implement. PSO has been used to loading of branch and voltage stability maximization and minimization of losses on IEEE 30 bus network using TCSC and SVC together at optimal location. In [6], Result of PSO method has been compared with Evolutionary Programming (EP) and it was concluded that loading factor (182%) increased more with minimum number of FACTS devices in PSO as compared to EP. PSO is faster than EP.

Kavitha et al. in [7] presented the comparison of WIPSO (improved weight PSO), BBO (Biogeography optimization) and PSO methods for number of FACTS devices, optimal location, type, rating of SVC and TCSC. Objective functions which consist of load voltage deviation, cost and line loadings are considered. Implementation has been done on IEEE 30, 57 and 14 bus systems. Analysis of result shows that WIPSO performs better than PSO and BBO perform best compared to WIPSO and PSO. SVC and TCSC optimally placed together gives better voltage profile security compared to when placed individually.

PSO is more popular compared to other due to its advantages such as easy implementation, low computational time, robustness and rapid convergence. Comparison of PSO with different methods through pie chart shown in Figure 1 [8].

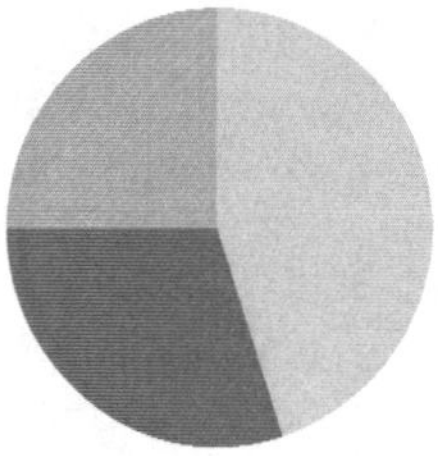

Figure 1. Pie chart showing use of methods

Inkollu et al. in [9], proposed new technique which is a hybrid of PSO and Gravitational Search Algorithm (GSA). PSO has been used for optimizing the gravitational constant of GSA to improve its search performance and it has been used to find the optimal location and rating of FACTS devices. Interline Power Flow Controller (IPFC) and Unified Power Flow Controller (UPFC) is considered. The objective function has been made on the basis of voltage collapse and power loss. Technique has been implemented on IEEE 30 bus network. Comparison made between PSO-GSA and GA-GSA technique and it has been observed that computation time is less in former technique.Venkatesh et al. in [10], presented PSO method used for optimal location of FACTS controller for enhancing the Available Transfer Capability (ATC) of the system. FACTS devices considered are SSSC, STATCOM, UPFC on six bus network.

2.2 Genetic Algorithm

It is a natural genetic based evolutionary technique inspired by Darwin theory of survival of fittest. It comprises three operators: selection, crossover and mutation applied at each iteration.Adebayo et al. presented new method called Network Structural Characteristics Techniques (NSCT) to search optimal location of FACTS controllers such as TCSC and UPFC and compared it with GA. Both methods are implemented on IEEE 14 bus network. Generation and devices cost function minimization has been taken as the objective. It has been observed from simulation that NSCT is more superior than GA due to less time consumption and no iteration for searching optimal location of FACTS controllers. Total computational time for GA is 6.73 sec and for NSCT is 1.23 sec [11].

Khandani et al. proposed hybrid of GA and Sequential Quadratic Programming (SQP) to find optimal location of SVC and solution of Optimal Power Flow problem for enhancement of TTC. GA method is relatively slow but does not depend on initial point while SQP depends on initial point and cannot handle discrete variables, hybrid model takes advantage of both. This hybrid method has been implemented on IEEE 5 bus system. It has been shown that SVC at position 2 is the optimal place for enhancement of TTC.[12]

Rashed et al. discussed the enhancement of loadability of the system by optimally placed multiple TCSC in the system with the help of PSO and GA. Minimizing investment cost of devices taken as objective function with voltage limits and thermal limits of lines as constraint. IEEE 6 and 14 bus networks are used for simulation results. In 6 bus network, active power flow in line by installing 3 TCSC's optimally improves 39% with GA and 43% with PSO. Similarly, in 14 bus network, with 5 TCSC, 20% improvement with GA and 29% with PSO. Maximum loadability increase in 6 and 14 bus system are 15% and 22% respectively. By the comparison of two methods, it has been shown that PSO is faster than GA. It is found that TCSC is the one of most effective device to increase the system loadability [13].

2.3 Brainstorm Optimization Algorithm (BSOA)

BSOA technique was proposed by Shi in 2011 and its idea comes from brainstorming procedure in human species. BSOA development based on the notion that human beings are considered most intellect creatures so algorithm motivated by their approach of problem solving is considered best over algorithms inspired by ants, birds, bees etc. it contains four operations: initialization, clustering, cluster center perturbation, individual perturbation.

A. R. Jordehi in [14] presented BSOA approach with eight other approaches i.e. PSO, GA, Simulated Annealing (SA), Backtracking Search Algorithm (BSA), DE, hybrid of Genetic Algorithm and Pattern Search (GA–PS), Asexual Reproduction Optimization (ARO) and GSA are used for optimal allocation of FACTS controllers such as SVC and TCSC on IEEE 57 bus network and comparisons are made. BSOA emerged best among other techniques to minimize power loss and voltage deviation of the system. In context of TCSC, it has been observed that the main work of TCSC helps to control power flow but also shows the effectiveness to enhance voltage security of the network.

2.4 Gravitational Search Algorithm (GSA)

GSA is a meta-heuristic technique developed in 2011 based upon Newton's law of gravity and motion. In this technique, each agent is associated with mass. Agent with heaviest mass among other attract low mass agents according to the Newton law of gravitation and the position of heaviest mass agent considered as optimal solution in the search area.

Bhattacharyya et al. in [15] applied GSA method for optimal placement of SVC and TCSC to increase the system loadability and power transfer capability of generators. The proposed approach is compared with DE, PSO and GA. The techniques are applied on IEEE 57 and 30 bus network. It has been shown that GSA is most effective in minimizing operating cost and active power loss compared to others.

2.5 Adaptive Evolutionary Algorithm (AEA)

Evolutionary methods are potent algorithms for multi-objective optimization problem where optimal solutions can be found in one simulation run. EA's does not required problem information that is being solved unlike classical methods [16].

The main challenge evolutionary method is dealt with maintaining its population diversity to avoid premature convergence. The proposed AEA method in [17] provides control of population diversity to prevent premature convergence and maintain the global search. In the paper, AEA method performance compared with PSO, SA and EA on the basis of L-index, voltage deviation, and reactive power loss. AEA emerged superior compared to other methods on IEEE 57 bus network and it outperforms for voltage deviation minimization in IEEE 14 bus system. It was shown that AEA is better than standard EA.

2.6 Moth Flame Algorithm (MFA)

MFA technique is proposed by Seyedali Mirjalili, inspired by moth navigation in environment called transverse orientation. In night, moth travel in straight direction by maintaining specific angle with moon for long distances but these insects trapped in circular or spiral paths around artificial lights. Mathematical modeling of this travelling behavior is done for optimization.

M. Ebeed et al. in [18] presented MFA method for optimal setting of STATCOM for voltage profile and stability improvement and loss minimization. Comparison of MFO with PSO has been done and validated on IEEE 30 bus network. MFA emerged more superior than PSO in stabilizing voltage deviation and minimize power losses with STATCOM like power loss minimized by 6 and 4 percent respectively by MFO and PSO respectively. It has been shown that STATCOM is effective for voltage profile improvement and loss minimization.

2.7 Imperialistic Competitive Algorithm (ICA)

Atashpaz-Gargari and C. Lucas developed novel meta-heuristic technique ICA which is motivated by the imperialism and socio-political development of human beings. Each candidate is treated as country and become either a colony or an imperialist.

A. R. Jordehi in [19] discussed TCSC and Thyristor Control Phase Shifting Transformer (TCPST) optimal placement by ICA on IEEE 14 bus system.

ICA outperforms some state of art methods like Evolutionary Programming (EP), Artificial Bee Colony (ABC), Bat Swarm Optimization (BSO), GSA. It has been shown that FACTS devices installation made the system secure at the time of contingencies and load growth with ICA.

2.8 Adaptive Cuckoo Search Algorithm (ACSA)

Suash Deb and Xin-She Yang introduced meta-heuristic CSA technique in 2009 inspired from search method of cuckoo bird species for laying eggs. Cuckoo bird lays eggs on other birds nest and they search the nest through levy flights or random walks. Mathematical modeling including step size parameter and probability inspired from above behavior leads to CSA. In ACSA, only step size parameter is removed for search of next nest instead value of best nest updated based on current best nest.

Taleb et al. proposed ACSA for the optimal location of TCSC in IEEE 9 bus network for minimum bus bar voltage and minimizing active and reactive power loss. Evaluation of result of ACSA with PSO and GA has been done. It has been found out that minimum bus voltage reduced by 0.62% in comparison of PSO and GA. Overall active and reactive power loss is more reduced than PSO and GA [20].

3. FACTS applications using GA

In this section, overview of FACTS devices optimal placement, type and size using Genetic Algorithm with consideration of various objective functions (single and multi-objective) with different equality/inequality constraints are analyzed and study the effect of FACTS on different parameters of power system. The device considered mainly is SVC and TCSC. Table.1 and Table.2 shows Multi and single objective function optimization using GA with different bus system.

In [21], [23, [26], [30], [35] from table.1 and [38], [43], [46], [49] from table.2, it is inferred that TCSC has been used to enhance the performance of system by affecting various parameters like reducing generation cost, reducing active and reactive power loss, improving voltage stability by decreasing L-index, improving total transfer Capability (TTC), controlling the active power flow, enhancement of stability margin (SM) etc. It is found in [26], that cost of 10 MVAR TCSC installed in IEEE 30 bus network for voltage stability and reduction in losses is 1,53,000 US$. Similarly for 13 MVAR, cost of TCSC is 1,881,500 US$.

In [22], [24], [32], [36], [39], it is observed that SVC has been used to reduce generation cost reduction, active and reactive power loss reduction. STATCOM has been used in [40], [48], [50] for improvement in voltage stability, loadability and in loss reduction. It is seen in [48] that 60 MVAR STATCOM cost 10.8 million US$ in IEEE 14 bus network for improving voltage security.

It is observed from [44], that active power loss increases when SVC and TCSC are used together in IEEE 30 bus network but reactive power loss decreases significantly. In [37], it has been concluded that TCSC is more effective than SVC for enhancing the voltage stability of the 14 bus power system and generation cost reduction is nearly same in both the cases. It is inferred from [29] and [31] that active power loss reduces more in case of SVC than TCSC for IEEE 5 bus system and TCSC is more effective to reduce reactive power loss in IEEE 30 bus system. TCSC is best suited FACTS device among other for loadability enhancement.

In single type optimization, it has been shown that TCSC is the most effective device over Thyristor Controlled Phase Shifting Transformer (TCPST) and Thyristor Controlled Variable Reactor (TCVR) for enhancement of system loadability. TCSC drives the power in other direction to reduce loading on the line. [25].Nireekshana et al. shown that TCSC is better than SVC for ATC enhancement when the same bus–to–bus transactions are chosen [47].Karami et al shown the cost benefit analysis of STATCOM in which active power losses reduction saves 80 million$ and this study shows that tradeoff between STATCOM initial cost and saving from losses gives benefit of 69.3 million $.[50].Rashed et al presented Differential Evolution (DE) technique and compared it with GA. DE technique has superiority over GA in some features like high quality solution, stable convergence etc. [46].

Table 1. Multi-objective optimization

Objective functions	Multi-objective function	Equality/Inequality Constraints	Test Bus and Optimal location	Rating of device	Parameters affected
Investment cost function $C_1(f)$= depends on type of FACTS device **Bid function** $C_2(P_G) = P_{min}^T P_G$ [21]	Total cost function $C_T=C_1(f)+C_2(P_G)$ Overall obj. function m-C_{Total} m= 4000 US$/hour to convert obj. function into max. one	Equality const. E(f,g)= 0 Inequality const. $B_1(f)<b_1$, $B_2(g)<b_2$	*10 bus system* TCSC between bus 4-5	capacitive 70% of X_{line}	Generation cost reduced from 1117.75 to 915.90 US$/hour
For real power loss $F_1(u,v) = \sum_{k=1}^{NTL}[g_k[v_i^2 + v_j^2 - 2v_iv_j\cos(\delta_i - \delta_j)]$ **For voltage deviation** $F_2(u,v) = \sum_{k=1}^{NL}\lvert(v_k - v_k^{ref})\rvert$ **For SVC size** $F_3(SVC\ size) =$ Rating of SVC in p.u. [22]	F(u,v) = $F_1(u,v)+F_2(u,v)$ $+F_3(SVC\ size)$	Equality Const. $P_{Gi} - P_{Di} - P(V,\delta) = 0$ $Q_{Gi} - Q_{Di} - Q(V,\delta) = 0$ Inequality Const. $Q_{Gi}^{min} \leq Q_{Gi} \leq Q_{Gi}^{max}$	*30 bus system* SVC at bus 30 (for LO 36) 28(for LO 5) 28(for LO 15) LO = line outage	capacitive 11.04 MVAR Inductive 1.55 MVAR Inductive 15.42 MVAR	
Total cost function $C_T=C_1(f)+C_2(P_G)$ $C_1(f)$= FACTS cost function $C_2(P_G)$= generation cost function [23]	Fitness= $\frac{1}{C_T+W(\sum_i^N P_{Gi}-P_{Di}-P_{Li})}$	Equality const. E(f,g)= 0 Inequality const. $B_1(f)<b_1$, $B_2(g)<b_2$	*9 bus system* TCSC 5th bus OL TL 3 TL 4 9th bus OL TL 2 OL=overloading	(In MVAR) 10.166 17.977 62.989	
For real power losses $O_1 = \sum_{l=1}^{b}[(v_i^2 + v_j^2 - 2v_iv_j\cos(\delta_i - \delta_j)]y_{ij}cos\theta_{ij}$ **For voltage deviation** $O_2 = \sum_{i=1}^{n}(u_{iref} - u_i/u_{iref})^2$ **For installation cost** $O_3 = 0.0003Q^2 - 0.3051Q + 127.38$	$f(x) = 0.4\frac{O_i}{\sum \Delta loss_{base}} + 0.4\frac{O_2}{\sum \Delta v_{base}} + 0.2\frac{O_3}{C_{max}} - \xi\sum_{i=1}^{nr} bal_i - \varsigma\sum_{k=1}^{n} thermal_k - \varrho\sum_{k=1}^{n} voltage_k$ S_l=power of line l U_i=voltage of bus i	Equality Const. $P_{Gi} - P_{Di} - P(V,\delta) = 0$ $Q_{Gi} - Q_{Di} - Q(V,\delta) = 0$ Inequality Const. $S_l \leq S_l^{max}$ $U_{imin} \leq U_{nom} \leq U_{imax}$	*13 bus system* SVC at 10th bus	143 MVAr	Loss drop from 25.886 to 22.432

[24]					
For BL $Ovl_{line} = e^{\lambda_{out}(100-BL)}$, >100% loading 1, < 100% loading **For BVL** $Vtg_{bus} =$ 1, $0.95 < Vt < 1$ $e^{\lambda_{vtg}(\|(1-VL)-0.05\|)}$, otherwise BL= branch loading BVL= bus voltage loading [25]	Obj= $\prod_{line} Ovl_{line} +$ $\prod_{bus} Vtg_{bus}$ λ_{out} and λ_{vtg} are coefficients to adjust slope of the exponentials	Equality Const. $P_{Gi} - P_{Di} - P(V,\delta) = 0$ $Q_{Gi} - Q_{Di} - Q(V,\delta) = 0$ Inequality Const. $S_l \le S_l^{max}$ $U_{imin} \le U_{nom} \le U_{imax}$	*118 bus system* TCSC at branch 5-8 23-24 92-100 SVC at branch 43-44 95-96		15 devices used ,TCSC emerged as most efficient device for loadability enhancement
Reactive power loss $f_1(x)=\sum_{i=1}^{N_{line}} Q_{iloss}$ **Voltage stability margin (L index)** $f_2(x)= max(L_j)$ **Cost of FACTS controllers** $f_3(x)= C_{upfc}+C_{tcpst}+C_{tcsc}$ [26]	Min f(x)= $[f_1(x),f_2(x),f_3(x)]$	Equality const. $\sum_{i=1}^{N_{bus}} P_G - P_D - P_L = 0$ $\sum_{i=1}^{N_{bus}} Q_G - Q_D - Q_L = 0$ Inequality const. $P_{Gimin}<P_{Gi}<P_{Gimax}$ $Q_{Gimin}<Q_{Gi}<Q_{Gimax}$ $S_{ij}<S_{ijmax}$ $V_{imin}<V_i<V_{imax}$ $\delta_{imin}<\delta<\delta_{imax}$	*30 bus system* For TCSC line 10-22 line 10-20	(In MVAR) 1 13	L index drop to 0.072 from 0.136 Q_{loss} reduced to 17 from 22MVAR Cost of 10 MVAR TCSC 153000 US$ 13 MVAR 1881500 US$
Generation cost function $C_{Gi}(P_G)=a_{Gi}+b_{Gi}P_G+C_{Gi}P_G^2$ **Cost of FACTS controllers** $C_2(f)= C_{svc}+C_{upfc}+C_{tcsc}$ [27]	objective function= min $[C_{total}=C_{Gi}(P_G)+ C_2(f)]$	Equality const. E(f,g)= 0 Inequality const. (power flow eqn.) $B_1(f)<b_1,\ B_2(g)<b_2$	*30 bus system* Bus 2 load increment SVC at line 36 Bus 15 load increment TCSC at line 10	73.38 MVAR 72.90 MVAR	System cost in SVC case 881.87 $/hr System cost in TCSC case 849.41 $/hr
Fuel cost $C_{Gi}(P_G)=a_{Gi}+b_{Gi}P_G+C_{Gi}P_G^2$ **Power loss** $P_{loss} = \sum_{j=1}^{m} loss_j$ [28]	Obj. function F(x,u) = min.($C_{Gi}(P_G)+P_{loss}$)	Equality const. (active and reactive power balance) g(x,u)= 0 Inequality const $P_{Gimin}<P_{Gi}<P_{Gimax}$ $Q_{Gimin}<Q_{Gi}<Q_{Gimax}$ $S_{ij}<S_{ijmax}$ $V_{imin}<V_i<V_{imax}$ $-0.5X_L \le X_{TCSC} \le 0.5X_L$	*30 bus system* For TCSC Line 27 for minimum cost between line 12 & 14 for min. loss For SVC Bus 24 for min. cost Bus8 for min. loss	capacitive 30.6% X_{line} inductive 29% X_{line} 4.2 MVAR 50 MVAR	In TCSC and SVC case, Loss reduced from 2.01 to 1.953 MW and 1.934 MW respectively Fuel cost reduced from 576.60 to 575.99 and 575.94($/hr) respectively

VSI (L index) $L_j = \left\|1 - \sum_{i=1}^{g} F_{ji}\frac{V_i}{V_j}\right\|$ **FACTS cost fn.** $C_{tcsc}=0.015s^2-0.7130s+153.75$(US$/kvar) $C_{svc}=0.0003s^2-0.3051s+127.38$(US$/kvar) [29]	Fitness fn = a_1*(max L_j) + a_2*(cost fn) + a_3* (losses) $a_1 = 2.78$ $a_2 = 0.1$ $a_3 = 2.05$ VSI = voltage stability index	Equality const. $S^k(\|V\|^k,\theta,u^k)= 0$ Inequality const. $h^k((\|V\|^k,\theta,u^k) \le 0$	*30 bus system* TCSC Bus 2-4 SVC Bus 18	13 MVA 97.18 MVA	In TCSC and SVC case, loss reduced from 28.36 to 24.91 MVA and 26.3 MVA respectively Overall cost 9197500 US$ and 4521600 US$ respectively
Cost function $g_1= \sum_{i=1}^{t} f_i(S_i)$ **Line flow limits** $g_2= \prod_{i=1}^{l} f_2^k(p_i)$ **Active power losses** $g_3= p_{sl}= Re\{V_{sl}[\sum Y_{sli}.V_i]^*\}$ g_4= ATC= TTC-ETC [30]	Objective function= min(g_1,g_2,g_3,g_4) TTC calculated by RPF method in which maximize λ ATC= available transfer capability TTC = total transfer capability ETC= existing transfer commitment	Equality const. $S^k(\|V\|^k,\theta,u^k)= 0$ Inequality const. $h^k((\|V\|^k,\theta,u^k) \le 0$ u is control variables	*30 bus system* TCSC Line 2-6 Line 27-28	19.097 MVAR 10.327 MVAR	Loss reduced from 3.006 to 2.231 MW TTC increase from 70.25 to 86.85 MW
Generation cost function $C_{Gi}(P_G)=a_{Gi}+b_{Gi}P_G+C_{Gi}P_G^2$ **Benefit function** $B_{Dj}(P_D)=d_{Dj}P_D-e_{Dj}P_D^2$ [31]	Objective fn $\max(\sum_{i=1}^{N_D} B_{Dj}(P_D) - \sum_{i=1}^{N_G} C_{Gi}(P_G) + cost(FACTS))$ N_D= no. of loads N_G=no. of generators	Equality const. $\sum_{i=1}^{N_{bus}} P_G - P_D - P_L = 0$ Inequality const. $P_{Gimin}<P_{Gi}<P_{Gimax}$	*5 bus system* TCSC bus 3-5 For SVC 3rd bus		In TCSC and SVC case, losses reduced from 21.95 to 20.303 MW and 19.81 MW respectively Generation cost reduced from 6126.50 to 6095.20 and 6078.6 US$/hr respectively
Generation cost function $C_{Gi}(P_G)=a_{Gi}+b_{Gi}P_G+C_{Gi}P_G^2$ **FACTS devices cost functions** $C_{svc}=0.0003s^2-0.3051s+127.38$(US$/kvar) [32]	Min F = min($C_{Gi}(P_G)+C_{svc}(g)$)	Power flow equations E(f,g)= 0 Inequality const. $B_1(f)<b_1$, $B_2(g)<b_2$	*14 bus system* SVC at bus 5 *30 bus system* SVC at bus 5	51.31 MVAR 136.973 MVAR	Fuel cost reduced to 829.93 US$/hr
Generation cost function $C_{Gi}(P_G)=a_{Gi}+b_{Gi}P_G+C_{Gi}P_G^2$ **FACTS devices cost functions** $C_{tcsc}=0.015s^2-0.7130s+153.75$(US$/kvar) $C_{svc}=0.0003s^2-0.3051s+127.38$(US$/kvar) [33]	Total cost function $C_T=C(f)+C_{Gi}(P_G)$ Overall obj. function = m-C_{Total} m= 4000 US$/hour to convert obj. function into max. one	Equality const. E(f,g)= 0 Inequality const $B_1(f)<b_1$, $B_2(g)<b_2$	*14 bus system* For SVC 5 -6 bus For TCSC 2 -5 bus	10.5 MVAR Capacitive 22%X_{line}	

IOP Conf. Series: Materials Science and Engineering **331** (2018) 012023 doi:10.1088/1757-899X/331/1/012023

Voltage stability index (VSI) $F_1 = L_{max}$ $\lvert L_j \rvert = \left\lvert 1 - \frac{\sum i\epsilon\alpha_G C_{ij} V_i}{V_j} \right\rvert j\varepsilon\alpha_L$ **Generation cost function** $F_2 = C_{Gi}(P_G) = a_{Gi} + b_{Gi}P_G + c_{Gi}P_G^2$ **For real power loss** $F_3 = \sum_{k=1}^{NTL}[g_k[v_i^2 + v_j^2 - 2v_i v_j \cos(\delta_i - \delta_j)]$ [34]	$F = h_1F_1 + h_2F_2 + h_3F_3$ $h_1 + h_2 + h_3 = 1$ $h_1 = 0.35$ $h_2 = 0.3$ $h_3 = 0.35$	Equality const. $\sum_{i=1}^{Nbus} P_G - P_D - P_L = 0$ Inequality const. $P_{Gimin} < P_{Gi} < P_{Gimax}$	*14 bus system* For SVC Bus 4 For TCSC 9-14 bus	 18.94 MVAR 23% X_{line}	In SVC and TCSC case, VSI reduced from 0.0783 to 0.0772 and 0.0732 respectively Loss reduced by 0.0946 and 2.0378 MW respectively Generation cost reduced by 0.5611 and 0.2278 $/hour respectively
Generation cost function $C_{Gi}(P_G) = a_{Gi} + b_{Gi}P_G + C_{Gi}P_G^2$ **Benefit function** $B_{Dj}(P_D) = d_{Dj}P_D - e_{Dj}P_D^2$ [35]	Objective fn $min\ [\ \sum_{i=1}^{N_G} C_{Gi}(P_G) - \sum_{i=1}^{N_D} B_{Dj}(P_D) + cost(FACTS)]$ N_D= no. of loads N_G=no. of generators	Equality const. $\sum_{i=1}^{Nbus} P_G - P_D - P_L = 0$ Inequality const. $P_{Gimin} < P_{Gi} < P_{Gimax}$ $Q_{Gimin} < Q_{Gi} < Q_{Gimax}$ $S_{ij} < S_{ijmax}$ $V_{imin} < V_i < V_{imax}$ $X_{tcscmin} < X < X_{tcscmax}$	*14 bus system* For TCSC Between 6-13 bus	23.98 % X_{line}	Generation cost reduced from 1407.154 to 1344.47 $/hr
Generation cost function $C_{Gi}(P_G) = a_{Gi} + b_{Gi}P_G + C_{Gi}P_G^2$ **FACTS device cost function** $C_{svc} = 0.0003s^2 - 0.3051s + 127.38$(US$/kvar) [36]	Overall obj. function $min(C_T)$ $= C(f) + C_{Gi}(P_G)$	Equality const.(active and reactive power balance) E(f,g)= 0 Inequality const. (power flow eqn.) $B_1(f) < b_1,\ B_2(g) < b_2$	*14 bus system* SVC at 5th bus	 51.66 MVAR	Fuel cost 1064 $/hour
System security $F_t = \sum_{l=1}^{a} w_1 \left(\frac{s_l}{s_{lmax}}\right)^{2q} + \sum_{m=1}^{b} w_m \left(\frac{V_{mref} - V_m}{V_{mref}}\right)^{2r}$ $F_e = C_{svc} + C_{tcsc}$ [37]	Fitness fn = Min F(x) = [$F_t(x)$, $F_e(x)$]	Equality Const. E(f, g) = 0 Inequality const. $B(f) > b_1,\ B(f) < b_2$ f = variables of FACTS devices g = power systems operating state	*14 bus system* TCSC Bus 9-14 SVC Bus 7-9	 Capacitive 0.996 MVAR 22.91 MVAR	

Table 2. Single objective optimization

Objective functions	Equality/Inequality Constraint	Test bus and Optimal location	Rating	Parameters affected
Obj. fcn = max $\boldsymbol{\lambda}$ $\boldsymbol{\lambda}$ is system loading factor [38]	Equality const. $P_{gi}–P_{di} - \sum \mid V_i \mid * \mid V_j \mid * (G_{ij} \cos\delta_{ij} + B_{ij} \sin\delta_{ij}) = 0$ $Q_{gi}–Q_{di} - \sum \mid V_i \mid * \mid V_j \mid * (G_{ij} \sin\delta_{ij} – B_{ij} \cos\delta_{ij}) = 0$ Inequality const. $V_{imin}<V_i<V_{imax}$ $- 0.5 * X_{line} < X_{TCSC} < 0.5 * X_{line}$	*4 bus sytem* TCSC at line 3	Capacitive 23% X_{line}	TTC improved from 489.6 to 813.6 MW
Susceptance equation $B^{i+1}_{SVC} = B^{i}_{SVC}+(\Delta B_{SVC}/B_{SVC})B^{i}_{SVC}$ [39]	Equality const. $\sum_{i=1}^{Nbus} P_G - P_D - P_L = 0$ Inequality const. $P_{Gimin}<P_{Gi}<P_{Gimax}$ $Q_{Gimin}<Q_{Gi}<Q_{Gimax}$ $S_{ij}<S_{ijmax}$ $V_{imin}<V_i<V_{imax}$ $X_{tcscmin}<X<X_{tcscmax}$	*9 bus system* SVC at 9th bus(NL) *30 bus system* SVC at 1st bus(NL) NL= normal loading	B = 46.77 siemens B = 9.6 siemens	Q loss reduced from 79.55 to 4.73 Q loss reduced from 68.68 to 19.79
$P_{li} = \boldsymbol{\lambda} P_{0li}$ & $Q_{li} = \boldsymbol{\lambda} Q_{0li}$ $P_{gi} = \boldsymbol{\lambda} P_{0gi}$ Obj.fcn= max($\boldsymbol{\lambda}$) $\boldsymbol{\lambda}$ is system loading factor [40]	Inequality const. $P_{Gimin}<P_{Gi}<P_{Gimax}$ $S_{ij}<S_{ijmax}$ $\lvert V_{bi} \rvert \le 0.05$	*57 bus system* STATCOM with SMES At Bus 38	Capacitive 67.528 MVAR	$\boldsymbol{\lambda}$ improved 14% Loss reduced by 12 MW
For voltage stability $\sum abs\lvert v_i - v_{refi} \rvert^3$ [41]	Equality const. $\sum_{i=1}^{Nbus} P_G - P_D - P_L = 0$	*68 bus system* For SVC 40 49 50	(in MVAR) 145 150 150	
Obj. = Maximize λ λ= system loading margin [42]	Equality const. $\sum_{i=1}^{Nbus} P_G - P_D - P_L = 0$	*14 bus system* For SVC at bus 9 *30 bus sytem* For SVC at bus 30		λ improved from 3.9752 to 4.11 λ improved from 3.01 to 3.30
min F = $\sum_{k=1}^{nl} Q_{lk} + \sum_{k=1}^{nl} P_{lk} + \sum_{n=1}^{ng} Q_{gn}$ [43]	Equality const. $P_{gi}–P_{di} - \sum \mid V_i \mid * \mid V_j \mid * (G_{ij} \cos\delta_{ij} + B_{ij} \sin\delta_{ij}) = 0$ $Q_{gi}–Q_{di} - \sum \mid V_i \mid * \mid V_j \mid * (G_{ij} \sin\delta_{ij} – B_{ij} \cos\delta_{ij}) = 0$ Inequality const. $P_{Gimin}<P_{Gi}<P_{Gimax}$ $Q_{Gimin}<Q_{Gi}<Q_{Gimax}$	*30 bus sytem* TCSC at Line 2 Line 3 Line 6	capacitive 40% X_{line} inductive 16.7 X_{line} Capacitive 2.8% X_{line}	Active flow Increase by 10 MW Decrease by 3 MW Increase by 0.64 MW

	$X_{mintcsc}<X<X_{maxtcsc}$ $V_{imin}<V_i<V_{imax}$ $\delta_{min}<X<\delta_{max}$			
Min(f,u) = $P_L(V,\delta,S)=\sum_{i=1}^{N}[P_{Lt}\times E_{loss}\times \Delta T-C_{in}]$ [44]	E(b,v)= 0 $B_1(s)<b_1$, $B_2(v)<b_2$	*30 bus system* For SVC Bus 24th For SVC and TCSC together at bus 24th and line 1-3 respectively	 25 MVAR 25 MVAR & 53.99X_{line}	**SVC case** Real loss reduced from 17.869 to 17.695 MW & Q loss reduced from 24.862 to 22.3 MVAR **TCSC and SVC** Real loss reduced from 17.869 to 17.979 MW & Q loss reduced from 24.862 to 14.145 MVAR
OVL(k)=1, if $S_{pq}<S_{pqmax}$ $e^{\lambda\left\|1-\frac{\|S_{pq(k)}\|}{S_{pqmax(k)}}\right\|}$, if $S_{pq}>S_{pqmax}$ [45]	Inequality const. $Q_{Gimin}<Q_{Gi}<Q_{Gimax}$ $S_{ij}<S_{ijmax}$ $V_{imin}<V_i<V_{imax}$	*30 bus system* TCSC at line 11 SVC at bus 22	 Capacitive 32.12X_{line} 19.28 MVAR	29% reduction in losses
Real power loss minF = $\sum_{k=1}^{ntl}P_{lk}$ [46]	Equality const. $P_{gi}-P_{di}-\sum_{j=1}^{n}V_iV_jY_{ij}(x_{tcsc})\cos(\delta_{ij}+\gamma_j-\gamma_i)=0$ $Q_{gi}-Q_{di}-\sum_{j=1}^{n}V_iV_jY_{ij}(x_{tcsc})\sin(\delta_{ij}+\gamma_j-\gamma_i)=0$ Inequality const. $P_{Gimin}<P_{Gi}<P_{Gimax}$ $Q_{Gimin}<Q_{Gi}<Q_{Gimax}$ $V_{imin}<V_i<V_{imax}$ $\delta_{ijmin}\le\delta_{ij}\le\delta_{ijmax}$	*3 bus system* TCSC bus 1-3 *5 bus system* TCSC bus 1-2 *14 bus system* TCSC bus 2-3	 Capacitive 2.20% X_{line} Capacitive 0.1% X_{line} Capacitive 4.2% X_{line}	
Obj. = max ATC ATC = TTC-TRM-(ETC+CBM) ATC= available transfer capability TTC = total transfer capability ETC= existing transfer commitment [47]	Equality const. $\sum_{i=1}^{N}P_{Gi}-P_D-P_{Loss}=0$ Inequality const. $P_{Gimin}<P_{Gi}<P_{Gimax}$ $Q_{Gimin}<Q_{Gi}<Q_{Gimax}$ $V_{imin}<V_i<V_{imax}$ CBM= capacity benefit margin TRM= transmission reliability margin (TRM),	*14 bus system* TCSC at line 9 SVC at bus 10 *30 bus system* TCSC at line 28 SVC at bus 20	 Capacitive 8.8%X_{line} 8.1 MVAR Capacitive 1.03%X_{line} 9.9 MVAR	ATC increase by15.5 by 7 by 40 by 20

Z =MANP×VANP MANP=$\frac{1}{n}\sum_{j=1}^{n} NP_j$ VANP= $\sqrt{\frac{1}{n}\sum_{j=1}^{n}(NP_j - MANP)^2}$ Z= objective function NP= nodal price of active power VANP= variance of nodal price [48]	Equality const. G(X,P,Q) =0 Inequality const. H(X,P,Q)<0	*14 bus system* STATCOM at line 5 *30 bus system* STATCOM at line 15 *118 bus system* STATCOM at line 35	30.02 MVAR capacitive 28.75 MVAR capacitive 27.92 MVAR capacitive	**14 bus case** Fuel cost reduced from 1265643 to 12644.53 $/hr **30 bus case** Fuel cost reduced from 14444.88 to 14278.6 $/hr **118 bus case** Fuel cost reduced from 83311.63 to 83189.6 $/hr
J= $\sum\left\|V_j - V_{jref.}\right\|^2$+ (1-SM)+$n_f$ [49]	Equality const. $\sum_{i=1}^{N} P_{Gi} - P_D - P_{Loss}=0$ Inequality const. $X_{min}<X<X_{max}$ $0.95<V_{bus}<1.05$	*AEP 14 bus system* TCSC between bus 1-2 1-5 2-3	 Capacitive 1.7 X_{line} 11.91X_{line} 8.53 X_{line}	SM increase from 0.1166 to 0.2188 SM= security margin
Max F(u) = (**λ**-**λ**$_0$) **λ** = distance between operating and collapse point.[50]	Equality const. f(x, **λ**, p) = 0 $D_x^T F(x, \lambda, p_0)w = 0$ D_x^T = jacobian matrix	*14 bus system* STATCOM line 9-14	 60 MVAR	Increase voltage security by 44.2 MW Loss reduced by 50.83 MW 10.8 million $ investment cost

4. Conclusion

In this paper, recent optimization techniques are discussed such as Particle Swarm Optimization (PSO), Genetic Algorithm, Brainstorm Optimization Algorithm (BSOA), Gravitational Search Algorithm (GSA), Adaptive Evolutionary Algorithm (AEA), Moth Flame Algorithm (MFA), Imperialistic Competitive Algorithm (ICA), Adaptive Cuckoo Search Algorithm (ACSA) and compared among each other. It has been observed that recent methods are generally inspired by human behavior like BSOA, ICA etc and show good performance to solve practical problems. They outperformed conventional heuristic methods like GA and PSO in speed.

The other part of the paper presents analytical review on optimal location of FACTS devices mainly SVC, TCSC and STATCOM. Different bus system and objective function for various parameters optimization has been studied e.g. for minimization of generation cost, active power losses and investment cost of FACTS and regarding voltage stability.

Power system parameters changes with and without FACTS devices has also been considered. It can be concluded that FACTS devices enhance system performance and maintain system security at the time of contingencies. This paper will be helpful for choosing specific type of FACTS devices on the basis of different objective.

5. Future Scope

It has been observed that hybrid optimization techniques are more effective, robust and fast to search the particular solution. More work can be done to develop the hybrid optimization technique from recent methods.

References

[1] Kavitha K and Neela R 2016 Comparison of BBO , WIPSO & PSO techniques for the optimal placement of FACTS devices to enhance system security *Glob. Colloq. Recent Adv. Eff. Res. Eng. Sci. Technol. (RAEREST 2016)* **25** 824–37

[2] Hingorani N G and Gyugyi L 2000 *Understanding FACTS: concepts and technology of flexible AC transmission systems*

[3] Medhi B K and Bhuyan S 2014 Performance Analysis of Some FACTS Devices Using Newton Raphson Load Flow Algorithm *2014 First Int. Conf. Autom. Control. Energy Syst. (ACES), Hooghy* 1–6

[4] Kulkarni P P and Ghawghawe N D 2013 A Review Paper On Optimal Location And Parameter Setting Of Facts To Improve The Performance Of Power System *Int. J. Electr. Electron. Data Commun.* **1** 1–5

[5] Kavitha K and Neela R 2017 Optimal allocation of multi-type FACTS devices and its effect in enhancing system security using BBO , WIPSO & PSO *J. Electr. Syst. Inf. Technol.* 1–17

[6] Venugopal D and Jayalaxmi A 2015 Optimal Placement of Facts Devices Using Particle Swarm Optimization Technique for the Increased Loadability of a Power System *IOSR J. Electr. Electron. Eng.* **10** 20–6

[7] Kavitha K and Neela R 2016 Comparison of BBO , WIPSO & PSO techniques for the optimal placement of FACTS devices to enhance system security . *Glob. Colloq. Recent Adv. Eff. Res. Eng. Sci. Technol. (RAEREST 2016)* **25** 824–37

[8] Dixit S, Agnihotri G, Srivastava L and Singh A 2014 An Overview of Placement of TCSC For Enhancement of Power System Stability *2014 Int. Conf. Comput. Intell. Commun. Networks, Bhopal* 1186–9

[9] Inkollu S R and Kota V R 2016 Optimal setting of FACTS devices for voltage stability improvement using PSO adaptive GSA hybrid algorithm *Eng. Sci. Technol. an Int. J.* **19** 1166–76

[10] Bavithra K, Charles Raja S and Venkatesh P 2016 Optimal Setting of FACTS Devices using Particle Swarm Optimization for ATC Enhancement in Deregulated Power System *Int. Fed. Autom. Control.* **49** 450–5

[11] Adebayo I G, Arun Bhaskhar M, A. Yusuff A and A.Jimoh A 2016 Optimal Location Identification of FACTS Devices through Genetic Algorithms and the Network Structural Characteristics Techniques *5th Int. Conf. Renew. Energy Res. Appl. 20-23 Birmingham, UK* **5** 5–9

[12] Khandani F, Soleymani S and Mozafari B 2011 Optimal Allocation of SVC to Enhance Total Transfer Capability Using Hybrid Genetics Algorithm *8th Electr. Eng. Electron. Comput. Telecommun. Inf. Technol. Assoc. Thail. - Conf. 2011, Khon Kaen* 861–4

[13] Rashed G . I, Shaheen H . I and Cheng S J 2007 Optimal Location and Parameter Settings of Multiple TCSCs for Increasing Power System Loadability Based on GA and PSO Techniques *Third Int. Conf. Nat. Comput. (ICNC 2007), Haikou* 335–44

[14] Jordehi A R 2015 Brainstorm optimisation algorithm (BSOA): An efficient algorithm for finding optimal location and setting of FACTS devices in electric power systems *Int. J. Electr. Power Energy Syst.* **69** 48–57

[15] Bhattacharyya B and Kumar S 2016 Loadability enhancement with FACTS devices using gravitational search algorithm *Int. J. Electr. Power Energy Syst.* **78** 470–9

[16] Abido M A 2006 Multiobjective Evolutionary Algorithms for Electric Power Dispatch Problem *IEEE Trans. Evol. Comput.* **10** 315–29

[17] Nascimento S do and Gouvêa Jr. M 2017 Voltage stability enhancement in power systems with automatic facts device allocation *3rd Int. Conf. Energy Environ. Res. ICEER 2016, 7-11 Sept. 2016, Barcelona, Spain* **107** 60–7

[18] Ebeed M, Kamel S and Youssef H 2016 Optimal Setting of STATCOM Based on Voltage Stability Improvement and Power Loss Minimization Using Moth-flame Algorithm *2016 Eighteenth Int. Middle East Power Syst. Conf. (MEPCON), Cairo* 815–20

[19] Jordehi A R 2016 Optimal allocation of FACTS devices for static security enhancement in power systems via imperialistic competitive algorithm (ICA) *J. Appl. Soft Comput.* **48** 317–28

[20] Taleb M, Salem A, Ayman A and Azma M A 2016 Optimal Allocation of TCSC Using Adaptive Cuckoo Search Algorithm *2016 Eighteenth Int. Middle East Power Syst. Conf. (MEPCON), Cairo* 387–91

[21] Cai L J, Erlich I and Stamtsis G 2004 Optimal Choice and Allocation of FACTS Devices in Deregulated Electricity Market using Genetic Algorithms *IEEE PES Power Syst. Conf. Expo.* **1** 201–7

[22] Dixit S, Srivastava L and Agnihotri G 2014 Optimal placement of SVC for minimizing power loss and improving voltage profile using GA *2014 Int. Conf. Issues Challenges Intell. Comput. Tech.* 123–9

[23] Vijayakumar K, Kumudinidevi D R P and Suchithra D 2007 A Hybrid Genetic Algorithm for Optimal Power Flow incorporating FACTS Devices *Int. Conf. Comput. Intell. Multimed. Appl. (ICCIMA 2007), Sivakasi, Tamil Nadu* 463–7

[24] Pisica I, Bulac C, Toma L and Eremia M 2009 Optimal SVC Placement in Electric Power Systems Using a Genetic Algorithms Based Method *2009 IEEE Bucharest PowerTech, Bucharest* 1–6

[25] Gerbex S, Cherkaoui R and Germond A J 2001 Optimal Location of Multi-Type FACTS Devices in a Power System by Means of Genetic Algorithms *IEEE Trans. Power Syst. Aug 2001* **16** 537–44

[26] Tiwari R, Niazi K R and Gupta V 2012 Optimal Location of FACTS Devices for Improving Performance of the Power Systems *2012 IEEE Power Energy Soc. Gen. Meet. San Diego, CA* 1–8

[27] Tiwari P K and Sood Y R 2009 Optimal location of FACTS devices in power system using Genetic Algorithm *2009 World Congr. Nat. Biol. Inspired Comput. (NaBIC), Coimbatore* 1034–40

[28] Malakar T, Sinha N, Goswami S K and Saikia L C 2010 Optimal Location and Size Determination of FACTS Devices by using Multiobjective Optimal Power Flow *TENCON 2010 - 2010 IEEE Reg. 10 Conf. Fukuoka* 474–8

[29] Baghaee H R, Jannati M, Vahidi B, Hosseinian S H and Rastegar H 2008 Optimal Multi-type FACTS Allocation using Genetic Algorithm to Improve Power System Security *2008 11th Int. Conf. Optim. Electr. Electron. Equipment, Brasov* 209–14

[30] Nguyen T T and Yousefi A 2010 Multi-Objective approach for optimal location of TCSC using NSGA II *2010 Int. Conf. Power Syst. Technol. Hangzhou* 1–7

[31] Balamurugan K, Muralisachithanandam R and Dharmalingam V 2015 Performance comparison of evolutionary programming and differential evolution approaches for social welfare maximization by placement of multi type FACTS devices in pool electricity market *Int. J. Electr. Power Energy Syst.* **67** 517–28

[32] Swarnalatha K and Amaresh K 2013 Optimal Location of Static Var Compensator in Power System Using Genetic Algorithm *Int. J. Electr. Electron. Eng. Telecommun.* **2**

[33] Cai L J and Erlich I 2004 Optimal Choice and Allocation of FACTS Devices using Genetic Algorithms *Power Syst. Conf. Expo. IEEE PES.10-13 Oct* 1–6

[34] Dheebika S K and Kalaivani R 2014 Optimal location of SVC, TCSC and UPFC devices for voltage stability improvement and reduction of power loss using genetic algorithm *2014 Int. Conf. Green Comput. Commun. Electr. Eng. (ICGCCEE), Coimbatore* 1–6

[35] Hosseinipoor N and Nabavi S M H 2010 Optimal locating and sizing of TCSC using genetic Algorithm for congestion management in deregualted power markets *2010 9th Int. Conf. Environ. Electr. Eng. Prague, Czech Republic,* 136–9

[36] Metwally M M E, Emary A A E, Bendary F M E and Mosaad M I 2008 Optimal Allocation of Facts Devices in Power System Using Genetic Algorithm *2008 12th Int. Middle-East Power Syst. Conf. Aswan* 3–6

[37] Radu D and Bésanger Y 2006 A Multi-Objective Genetic Algorithm Approach to Optimal Allocation of Multi-Type FACTS Devices for Power Systems Security *2006 IEEE Power Eng. Soc. Gen. Meet. Montr. Que.* 1–8

[38] Fozdar M 2007 GA based optimisation of thyristor controlled series capacitor *2007 42nd Int. Univ. Power Eng. Conf. Bright.* 392–6

[39] Bhandari M and Gurav S S 2015 Genetic Algorithm Based Optimal Allocation of SVC for Reactive Power Minimization in Power Systems *2015 Int. Conf. Ind. Instrum. Control (ICIC), Pune, 2015* 1651–6

[40] Ghahremani E and Kamwa I 2014 Optimal Allocation of STATCOM with Energy Storage to Improve Power System Performance *2014 IEEE PES T&D Conf. Expo. Chicago, IL, USA* 1–5

[41] Farsangi M M, Nezamabadi-pour H, Song Y, Member S and Lee K Y 2007 Placement of SVCs and Selection of Stabilizing Signals in Power Systems *IEEE Trans. Power Syst.* **22** 1061–71

[42] Skaria N A, Baby S and Anumodu D M 2014 Genetic algorithm based optimal location of SVC in power system for voltage stability enhancement *2014 Annu. Int. Conf. Emerg. Res. Areas Magn. Mach. Drives (AICERA/iCMMD), Kottayam* 1–6

[43] Padhy N P, Praveen Kumar B J and Abdel-Moamen M A 2004 Optimal Location and Initial Parameter Settings of Multiple TCSCs for Reactive Power Planning Using Genetic Algorithms *IEEE Power Eng. Soc. Gen. Meet. 2004., Denver, CO* **1** 1110–4

[44] Nikoukar J and Jazaeri M 2007 Genetic Algorithm Applied to Optimal Location of FACTS Devices in a Power System *Proc. 3rd IASME/WSEAS Int. Conf. Energy, Environ. Ecosyst. Sustain. Dev. Ag. Nikolaos, Greece,July 24-26* 526–31

[45] Shah I, Srivastava N and Sarda J 2016 Optimal Placement of Multi-Type Facts Controllers Using Real Coded Genetic Algorithm *2016 Int. Conf. Electr. Electron. Optim. Tech. (ICEEOT), Chennai,* 482–7

[46] Rashed G I, Sun Y and Shaheen H . I 2012 Optimal Location and Parameter Setting of TCSC for Loss Minimization Based on Differential Evolution and Genetic Algorithm *2012 Int. Conf. Med. Phys. Biomed. Eng. Optim.* **33** 1864–78

[47] Nireekshana T, Rao G K and Naga S S 2012 Electrical Power and Energy Systems Enhancement of ATC with FACTS devices using Real-code Genetic Algorithm *Int. J. Electr. Power Energy Syst.* **43** 1276–84

[48] Rahimzadeh S, Tavakoli M and Viki B A H 2010 Simultaneous application of multi-type FACTS devices to the restructured environment: achieving both optimal number and location *IET Gener. Transm. Distrib.* **4** 349–62

[49] Gitizadeh M and Kalantar M 2007 A new approach for optimal location of TCSC in a power system to enhance voltage stability: Steady state studies *2007 42nd Int. Univ. Power Eng. Conf. Bright.* 570–5

[50] Karami A, Rashidinejad M and Gharaveisi A A 2006 Optimal Location of STATCOM for Voltage Security Enhancement via Artificial Intelligent *2006 IEEE Int. Conf. Ind. Technol. Mumbai,* 2704–8

Optimized Latching Control of Floating Point Absorber Wave Energy Converter

Chaitanya Gadodia[1],Shubham Shandilya[1], Hari Om Bansal[2]

[1] *Undergraduate Student , Department of Electrical and Electronics, BITS Pilani, Pilani-333031, India.*
[2] *Associate Professor, Department of Elctrical and Electronics, BITS Pilani, Pilani -333031, India.*
E-mail: f2014177@pilani.bits-pilani.ac.in

Abstract. There is an increasing demand for energy in today's world. Currently main energy resources are fossil fuels, which will eventually drain out, also the emissions produced from them contribute to global warming. For a sustainable future, these fossil fuels should be replaced with renewable and green energy sources. Sea waves are a gigantic and undiscovered vitality asset. The potential for extricating energy from waves is extensive. To trap this energy, wave energy converters (WEC) are needed. There is a need for increasing the energy output and decreasing the cost requirement of these existing WECs. This paper presents a method which uses prediction as a part of the control scheme to increase the energy efficiency of the floating-point absorber WECs. Kalman Filter is used for estimation, coupled with latching control in regular as well as irregular sea waves. Modelling and Simulation results for the same are also included.

1.Introduction

Sea waves are gigantic, to a great extent undiscovered energy asset, and the potential for separating energy from waves is significant. The energy contained in sea surface waves is wind energy gathered over an endless range of the sea's surface and transported to the shorelines in the concentrated type of gravitational waves.

Progressing endeavors are made keeping in mind the end goal to decide reasonable strategies to discover areas where the harvest of this energy is beneficial and actually feasible. These requirements limit's the choices of appropriate areas down to the most part of the Atlantic and Pacific seas, where there is high density of waves along with consistency, particularly in the winter months is common[1].

This paper presents the general status of wave energy and assesses the devices that represent current wave energy converter (WEC), and a portion of the control techniques to improve efficiency of point absorbers. The dimensions of point absorbers are smaller than the sea wavelengths, making them less demanding to handle. For the most part, point absorbers work best when in resonance with the approaching waves, i.e., when the angular frequency of device is same as that of incoming wave. As a result of the randomness of sea waves and their large frequency and amplitude spectrum, control strategies are required to extract maximum power from incoming wave.

2.Wave Energy Converter

There are various techniques for wave energy conversion, the devices used in energy conversion are called as wave energy converters. There are various types of wave energy converters [1], [2]. Some of the predominant types of wave energy converters are as follows:

2.1. Point Absorber

A point absorber is a wave energy converter which has smaller wavelength respect to incident wavelength of sea waves. They can be floating structure that hurl all over on the surface of the water or submerged beneath the surface depending on pressure differential. In view of their small size, the direction of incoming wave is not essential for these devices.

A typical kind of point absorbers, which are studied here are heaving float WECs. They get excited by the heave motion of waves to produce electrical energy. In general, these gadgets include a kind of floating body, the float, whose motion is the function of incoming wave. Most devices of this kind have a second body, the spar, which is more inert to the incoming waves than the buoy, so that the difference in motion of the two bodies can be utilized to generate power. The second body is generally much heavier than the float and creates a lot of drag force. Moreover, it is moored to the seabed to limit its movement and to keep it on position.

2.2. Attenuators
It has various floaters on its arms. The energy of the movement of the arms is again used in hydraulic line which can be converted into electric current. This device has a high endurance for rough sea conditions. A significant advantage of these sorts of devices is the negligible contact with water, preventing corrosion's and preventing electrics far from any of the waves.

2.3. Terminators
Terminator devices lie parallel to the wave front and physically intercept waves. These terminator devices incorporate a stationary part and a moving part that moves due to incoming waves. The "stationary" part could be tied to the ocean bottom or shore. It must stay still, as opposed to the mobile part, to produce sufficient drag.

2.4. Submerged pressure differential
The submerged pressure differential converters are a submerged point safeguard that uses the pressure difference over the device between incoming wave's peaks and troughs. It involves two fundamental parts: an ocean bed fixed air-filled cylinder-shaped chamber with a moveable upper cylinder. As a crest passes over the converter, the water pressure over then pressurizes the air inside the chamber, causing the movement of the upper cylinder downward. As a trough passes over the converter, the water pressure on the device decreases and the upper chamber rises. Favorably this converter is completely submerged, it is not exposed to the hazardous pummeling forces experienced by floating wave energy converters, and decreases the visual effect of the device.

2.5. Overtopping devices
An overtopping device can capture ocean water of incoming waves in a reservoir over the ocean level, then discharges the water back to ocean through turbines. A case of such a device is the Wave Dragon. These converters utilize a couple of expansive reflectors to assemble waves into the central part, where they stream up a slope and over the top into a raised reservoir, from which the water is permitted to come back to the ocean by means of various low-head turbines.

2.6. Oscillating water columns
The working of the oscillating water columns wave energy converters (OWCs) is fundamentally the same as that of a wind turbine, being founded on the functioning of wave induced air pressurization. The device is set upon an air chamber, which is closed and is set over the water. The section of waves changes the water level inside the closed chamber and the rising and falling water level increments and declines the pneumatic force inside the chamber, causing a bidirectional wind stream. By setting a turbine on top of this chamber air will go all through it with the changing air pressure levels. There are two choices to isolate the bi-directional stream: a) Wells turbine to make suction or b) the pressure producing valves. OWC converters can be moored seaward or be set on the shoreline where waves break.

3. Sea Waves
Sea waves with consistent wave amplitudes and periods are rarely found. Normal sea conditions are more mathematically represented by Stochastic-wave time series that considers the superposition of various wave forms with different amplitudes and periods. This superposition of waves is portrayed by an ocean spectrum. Through factual investigation, spectra are described by particular parameters, for example, significant wave amplitude, peak period, wind speed, fetch length, and others. The general kind of spectrum

that are utilized for sea waves are as follows. The Eq. (1). gives the general equation of the sea wave spectrum for irregular sea.

$$S(f) = Af^{-5}exp[-Bf^{-4}] \quad (1)$$

3.1. Pierson-Moskowitz spectrum

It is one of the simplest spectra, proposed by Pierson and Moskowitz [4]. It assumed that after the wind blew relentlessly for quite a while over a substantial territory, the waves would come into harmony with the wind. The spectrum is calculated from Eq. (2).

$$S(f) = \frac{\alpha_{PM} g^2}{(2\pi)^4} f^{-5} \, exp\left[-\frac{5}{4}\left(\frac{f_p}{f}\right)^4\right] \quad , A = \frac{\alpha_{PM} g^2}{(2\pi)^4} , B = \frac{5}{4}(f_p)^4 \quad (2)$$

Where $\alpha_{PM} = 0.0081$ was calculated such that the required wave height is met, g is gravity and f_p is peak frequency of the spectrum.

3.2. Bretschneider spectrum

This is a two-parameter spectrum, based on peak wave frequency and significant wave height. The spectrum is calculated from Eq. (3).

$$S(f) = \frac{H_{m0}^2}{4}(1.057f_p)^4 f^{-5} \, exp\left[-\frac{5}{4}\left(\frac{f_p}{f}\right)^4\right] \quad (3)$$

Here $A = \frac{H_{m0}^2}{4}(1.057f_p)^4 \approx \frac{5}{16} H_{m0}^2 \, f_p^4$, $B = (1.057f_p)^4 \approx \frac{5}{4} f_p^4$.

Where 'H_{m0}' is significant wave height, which is generally defined as the mean height of the one third highest waves.

3.3. JONSWAP (Joint North Sea Wave Project) Spectrum

The spectrum was proposed by Hasselmann et al. [5], and the original formulation of the spectrum is given in Eq. (4). Which was later on changed at International Towing Tank Conference (ITTC).

$$S(f) = \frac{\alpha_j g^2}{(2\pi)^4} f^{-5} \, exp\left[-\frac{5}{4}\left(\frac{f_p}{f}\right)^4\right]\gamma^{\tau} \tau = exp\left[-\left(\frac{\frac{f_p}{f}-1}{\sqrt{2\sigma}}\right)^2\right], \sigma = \begin{cases} 0.07 & f \leq f_p \\ 0.09 & f > f_p \end{cases} \quad (4)$$

Here $A = \frac{\alpha_j g^2}{(2\pi)^4}$, $B = \frac{5}{4} f_p^4$.

Where α_j is the non-dimensional variable, that is a function of wind speed and fetch length.

3.4. White Gaussian Noise

White Gaussian noise can be combined with the system to make system more random. Gaussian noise is statistical noise having a probability density function (PDF) equal to that of the normal distribution, which is also known as the Gaussian distribution. In other words, the values that the noise can take on are Gaussian-distributed. Gaussian noise has a fixed power spectral density and it is also sometimes referred as white noise because it affects all frequency component equally, like a white light, which basically has all the frequency components. Since it has Gaussian distribution its probability density function depends on mean and variance of the random variable. The sea wave spectrums generated by simulation for regular as well as irregular sea is as shown in figure 1.

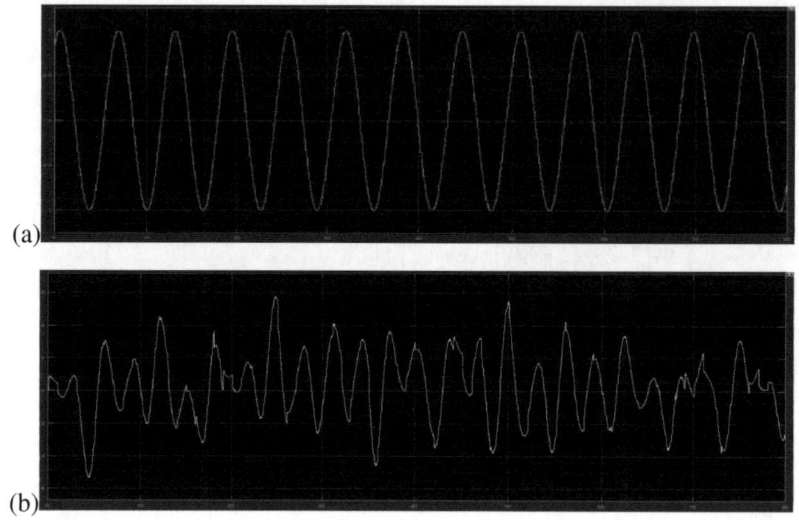

Figure 1. Sea Wave Spectrums, as generated by simulation. (a)Regular Sea Wave (b) Irregular Sea Wave.

4. Floating Point Absorbers

4.1. Coordinate System and Dynamics

The coordinate system of a floating-point absorber subject to incoming waves in water, is so chosen that the X-axis is in the direction of wave propagation, with the wave heading angle equal to zero. The Z-axis is in the vertical upwards direction, and the Y-axis direction is defined by the right-hand rule. Surge (x), sway (y), and heave (z) correspond to the first, second and third position respectively. Roll (Rx), Pitch (Ry), and Yaw (Rz) correspond to the fourth, fifth and sixth position respectively. This is illustrated in figure 2(a).

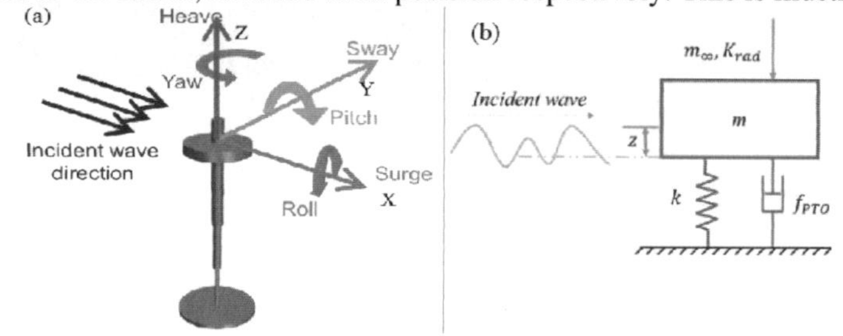

Figure 2. Floating Point Absorbers: (a) Coordinate System and the six directions of motion (b) Dynamic Model.

While considering hydrodynamics and wave-float interaction in preparatory hypothetical analysis, we normally consider its vertical movement i.e. only one degree of motion of wave-energy converter and that is in Z direction [3]. As illustrated in figure 2(b). The Force equation of the system in time domain can be obtained as below

$$(m + m_\infty)\ddot{z} + (b + c_{PTO})\dot{z} + (\rho g S + k_{PTO})z = f_{ext} \tag{5}$$

Where 'm' is mass of the system, 'm_∞' is the added mass, 'k_{PTO} ' is ' f_{ext}' is the external force because of sea waves, 'c_{PTO}' is the power take off constant, 'ρ' is sea water density,'g' is gravity, 'S' is surface area of the buoy, 'b' is viscous damping constant, and the Power Take Off force is considered to be linear

$F_{PTO} = -K_{PTO}X_{rel} - C_{PTO}\dot{X}_{rel}$. Here k_{PTO} is stiffness of the Power Take Off and X_{rel} relative motion of the two bodies. The power consumed by the Power Take Off is given by Eq. (6).

$$P_{PTO} = -F_{PTO}\dot{X}_{rel} = (K_{PTO}X_{rel}\dot{X}_{rel} + C_{PTO}\dot{X^2}_{rel}) \quad (6)$$

The displacement of the float in frequency domain can be written as in Eq. (7). The average power consumed by the Power Take Off can be formulated as in Eq. (8). This gives us a condition for maximizing the power out, Eq. (9).

$$Z = \frac{F_{ext}}{(\rho g S + k_{PTO}) - (m + m_\infty)\omega^2 + (b + c_{PTO})j\omega} \quad (7)$$

$$P = \frac{1}{2}c_{PTO}\omega^2|Z|^2 = \frac{1}{8B}F_{ext}^2 - \frac{B}{2}\left|i\omega Z - \frac{F_{ext}}{2B}\right|^2 \quad (8)$$

$$\omega = \sqrt{\frac{\rho g S + k_{PTO}}{m + m_\infty}} \quad and \ c_{PTO} = b. \quad (9)$$

Hence relative motion and velocity between two bodies is out of phase by 90 degrees. The conditions thus obtained are condition for maximum power transfer between incoming wave and wave energy converter.

4.2. Control Strategy for Point Absorbers

There are various control strategies which can also be used, like latching control, declutch control, latching operating declutch control, Reactive control, Model predictive control. We have used the latching control method [4]. Latching is sub-optimal kind of phase control [6]. The principle behind is to latch and unlatch the relative velocity at appropriate time instances for achieving optimal phase and also optimal output power. The optimum power output is achieved if the motion is unlatched in a time instance such that the oscillatory motion is at its maximum at the same time as the excitation force is at its maximum [7]. By using this we can force our wave energy converter in the resonance with the incoming wave. The resonance maximizes the power output of the device. Latching is done when the relative velocity is zero. If a wave is periodic like sinusoidal, we can easily find the appropriate latching time, but if wave is irregular then, the knowing the appropriate latching time is not trivial. We have used Kalman filter to find the optimal latching time of WEC.

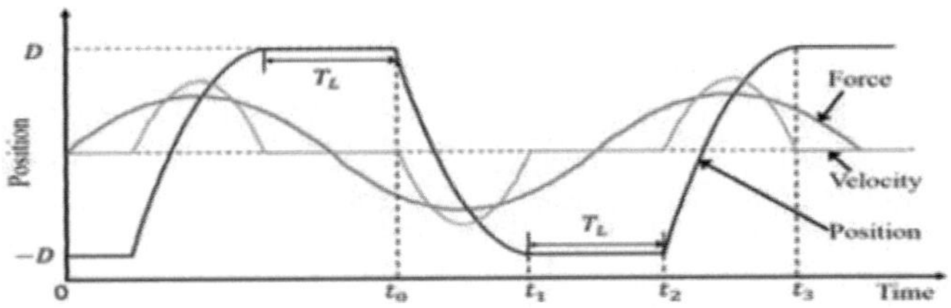

Figure 3. The incoming wave (red), latched WEC (blue) and the latching time.

4.3. Kalman Filter

Kalman filter is a predictor for linear quadratic problem, which is basically problem of predicting the next or instantaneous state of a linear dynamic system, Table 1. Which can have white Gaussian noise, it is statistically optimal than any quadratic function for error estimation. Kalman filter provides a means to get any missing piece of information in noisy environments [8]. The solution of the Kalman filter gives the dependence of the output on the input and it also the functionally depended on the coefficients of the system.

Table 1. Mathematical model of the dynamic systems

	Continuous	Discrete
Time invariant		
Linear	$\dot{x}(t) = Fx(t) + Cu(t)$	$x_k = \Phi x_{k-1} + \Gamma u_{k-1}$
General	$\dot{x}(t) = f(x(t), u(t))$	$x_k = f(x_{k-1}, u_{k-1})$
Time varying		
Linear	$\dot{x}(t) = F(t)x(t) + C(t)u(t)$	$x_k = \Phi_{k-1} x_{k-1} + \Gamma_{k-1} u_{k-1}$
General	$\dot{x}(t) = f(t, x(t), u(t))$	$x_k = f(k, x_{k-1}, u_{k-1})$

4.4. Linear Kalman Filter

Input output model of linear dynamic system, inputs of the system are assumed to be in control, or at least could be measured by us. The algorithm used to predict the output is as given in figure 4.

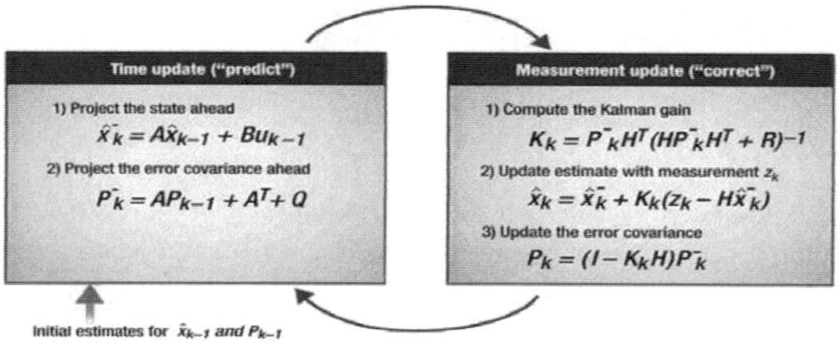

Figure 4. Kalman Filter recursive algorithm

State variables can't be measured directly and could be inferred from the variables which could be measured and output of the system could be measured.

4.5. Extended Kalman Filter

The principle downside when utilizing a linear Kalman filter estimator on the nonlinear model of the WEC is the quick deviation of the linearized mooring power from the real one over the happening scope of plant states. As a well-known instrument for noisy systems of loud frameworks, different expansions and changes of the Kalman filter exist. It basically linearizes the nonlinear system before applying state estimation by a linear Kalman filter.

5. Modelling and Simulation Results

The WEC Parameter values used for simulation are available in [8] and given in Table 2 below. The plant model was made using Solidworks [9], the testing of the model was done using MATLAB SimMechanics [10] and the simulation was done on MATLAB Simulink [9].

Table 2. The Characteristics of the Wave Energy Converter

Property	Value	Property	*Value*
Buoy mass	1000 kg	Pole width	50 mm
Translator weight	*1000 kg*	Winding resistance	0.44 Ω
Buoy Type	Cylinder	Vertical stator length	1264 mm
Buoy Radius	1.5m	Vertical translator length	1867 mm
Buoy Height	0.8m	Winding inductance	11.7 mH
Buoy draft	0.4m		

The regular and irregular wave spectrum, were fed into the system and the response of the system was measured, with and without the use of Kalman Filter. The efficiency is calculated on the basis of the average power of the input wave spectrum and the response wave generated by the system. The simulation results for Latching control without using Kalman Filter is as illustrated in figure 5. The input wave to the system is denoted by the blue curve and the response of the system by the yellow curve. Next the Kalman Filter was used for estimation of sea waves, the results obtained are as illustrated in figure 6. The input wave to the system is denoted by the blue curve, the Kalman Filter predicted wave by the Yellow curve and the orange curve depicts the response of the system.

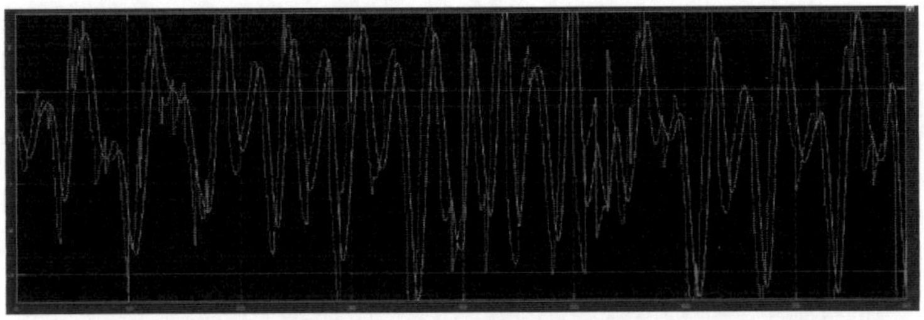

Figure 5. The input and output wave of the system without latching control.

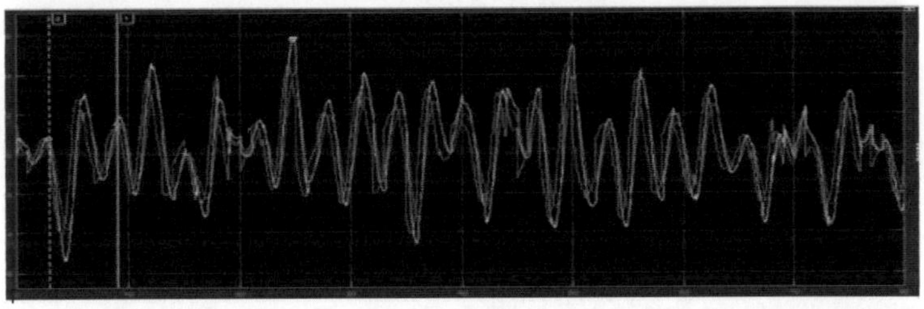

Figure 6. The input and output wave of the system with latching control and Kalman filter.

The simulation results show that a 100% power efficiency for regular wave spectrum and an efficiency of 74% for irregular wave spectrum is obtained. Thus, the adaptive control technique of prediction using Kalman Filter gives better results than the traditional Latching Control techniques.

6. Conclusion

The Sea wave energy is one of alternative energy sources which can be exploited for meeting the energy demand. This paper describes a novel method to increase the power efficiency of a floating-point absorber wave energy converter. Latching control is used to increase the gain of bring to wave energy converters. In this paper, Latching Control coupled with Kalman filter are employed to estimate the sea wave spectrum. This method was applied to both regular and irregular sea waves and an increase in the power efficiency was observed in both the cases.

Simulation results show that the Latching Control and Kalman Filter combination has increased the power conversion significantly in irregular waves, whereas in regular waves the technique can achieve perfectly optimal power conversion. Thus, the adaptive control technique of prediction using Kalman Filter gives better results than the traditional Latching control techniques. Further work on real time implementation of this method in a full-scale device can be done.

References

[1] B Drew, A. R Plummer, and M N Sahinkaya. A review of wave energy converter technology. Proc. IMechE Vol. 223 Part A: J. Power and Energy.Pg 887-902.

[2] Iraide López, Jon Andreu, Salvador Ceballos, Iñigo Martínez de Alegría, Iñigo Kortabarria. Review of wave energy technologies and the necessary power-equipment. Renewable and Sustainable Energy Reviews 27 (2013) 413–434.
[3] Jingjin Xie · Lei Zuo. Dynamics and control of ocean wave energy converters. Int. J. Dynamic Control (2013) 1:262–276.
[4] W. J. Pierson and L.A. Moskowitz. Proposed Spectral Form for Fully Developed Wind Seas Based on the Similarity Theory of S. A. Kitaigorodskii, Geophysical Research, 69:5181-5190, 1964.
[5] K. Hasselman, T.P. Barnett, E. Bouws, H. Carlson, D.E. Cartwright, K. Enke, J.A. Ewing, H. Gienapp, D.E. Hasselmann, P. Kruseman, A. Meerburg, P. Miller, D.J. Olbers, K. Ritcher, W. Sell, and H. Walden. Measurement of wind-wave growth and swell decay during the Joint North Sea Wave Project (JONSWAP). Technical Report12, German Hydrographic Institute, 1973.
[6] J. Falnes. Optimum Control of Oscillation of wave-energy converters. June 1993.
[7] Annette Kristin Brask. Control and Estimation of Wave Energy Converters. June 2015.
[8] http://www.cs.unc.edu/~welch/kalman/
[9] http://www.solidworks.in/
[10] https://www.mathworks.com/products/simmechanics.html
[11] https://www.mathworks.com/products/simulink.html

Performance Analysis of Three-Phase Induction Motor with AC Direct and VFD

Dinesh Kumar

Department of Electrical Engineering, Sekhawati Institute of Engineering and Technology, Sikar

E-mail: saini.dinesh92@gmail.com

Abstract. The electrical machine analysis and performance calculation is a very important aspect of efficient drive system design. The development of power electronics devices and power converters provide smooth speed control of Induction Motors by changing the frequency of input supply. These converters, on one hand are providing a more flexible speed control that also leads to problems of harmonics and their associated ailments like pulsating torque, distorted current and voltage waveforms, increasing losses etc. This paper includes the performance analysis of three phase induction motor with three-phase AC direct and variable frequency drives (VFD). The comparison has been concluded with respect to various parameters. MATLAB-SIMULINK™ is used for the analysis.

1. Introduction

VFD is variable frequency drive. VFD used to control the speed of electric induction motor. Induction motors are fixed speed motors which are used in most industries because of its reduced cost, reliability and rugged nature. The speed of the induction motor can be changed by various methods such as poles changing, voltage changing, connecting the resistance in rotor circuit etc., but the most efficient method is changing the supply frequency and voltage to the motor. Speed is directly proportional to the frequency of supply. The variable frequency drive (VFD) varies the frequency and hence varies the speed or torque of the induction motor as per the requirements of the load. As VFD provides the power efficient operation, smooth speed and torque control but also leads the harmonics in the system. Harmonics are generated due to use of power electronic switching devices. These harmonics affect the performance of the induction motor. This paper describes the effect of harmonics on motor performance and also how to improve the performance of induction motor by using appropriate LC filter with the help of MATLAB/Simulink. This paper has detail study of motor performance with different cases those showing performance of induction motor. First, we study the parameters when three phase induction motor is directly supplied by three phase AC supply. Then these parameters are compared with the parameters those are getting when same three phase induction motor is supplied by VFD without a filter. Then the performance is improved with the injection of LC passive filters in the system.

IOP Conf. Series: Materials Science and Engineering **331** (2018) 012025 doi:10.1088/1757-899X/331/1/012025

2. Three Phase Induction Motor

Three phase induction motor is widely used in any kind of industry like automobile, cement, fertilizer, chemical, Yarn production etc. because of its simple and rugged construction. A three phase induction motor is singly excited machine. Induction motor works on faraday law of electromagnetic induction. In this paper behaviour of an induction motor is studied when supplied with or without VFD by using MATLAB/Simulink. In this paper studied a three phase induction motor with rated parameters listed in table1. [1]

Table 1 Design parameters of three phase induction motor for proposed MATLAB/Simulink

S. No.	Parameters
Power	5.4 HP
Voltage	400 V
Frequency	50 Hz
Stator resistance and	1.405Ω ,
Inductance	0.005839H
Rotor resistance and inductance	1.395Ω, 0.005839H
Mutual inductance	0.1722 Ω
Inertia	0.0131kg-m^2

3. Induction Motor Supplied from 3-phase AC Direct

In this paper, the performance analysis of three phases IM is discussed in details. The behaviour of stator current, speed and torque characteristics are explained with respect to time.

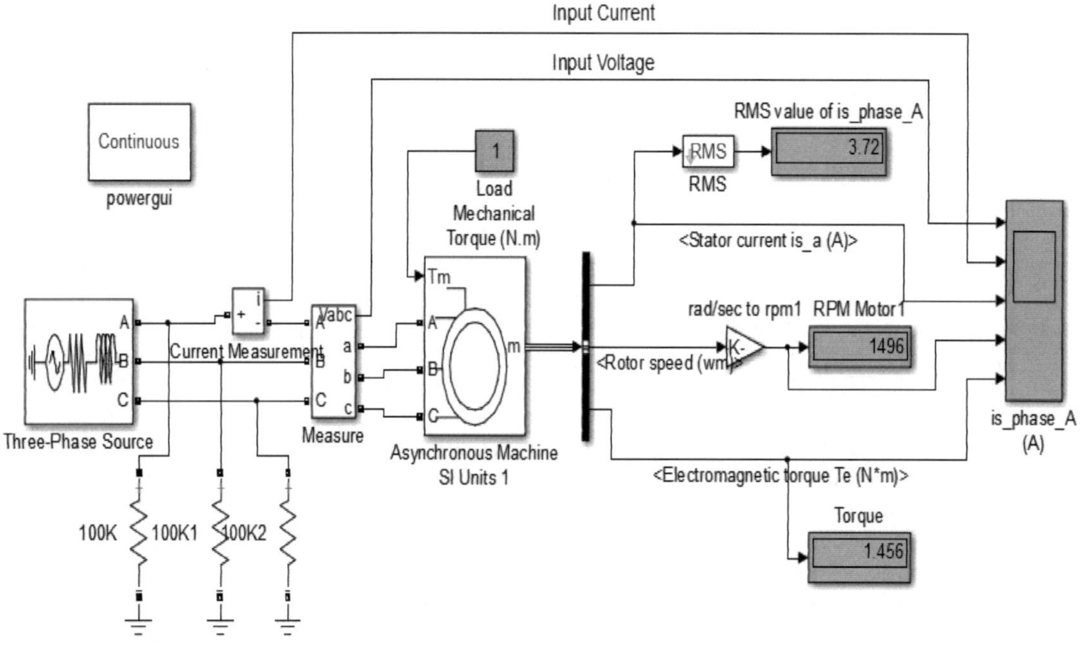

Figure 1 Simulation Circuit of Three Phase Induction Motor when Supplied from Three Phase AC Direct

Figure 1 shows simulation model of three phase induction motor which is directly connected to three phase supply. Performance analysis simulation is done using MATLAB/Simulink for a total time of 0.5 sec.

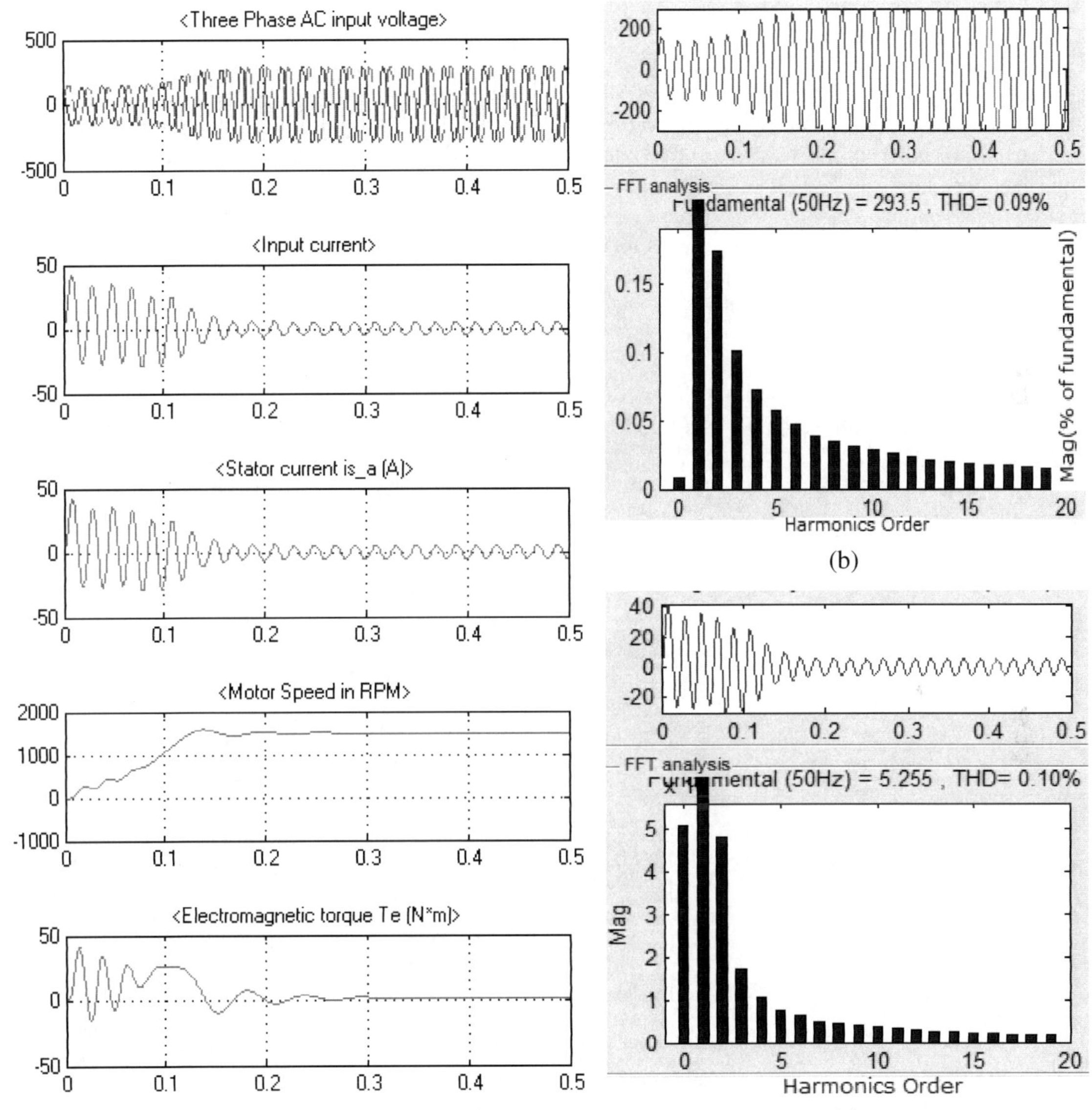

(a) (c)

Figure 2 (a) Input voltage, input current, stator current, rotor speed, electromagnetic torque waveforms (b) THD of input voltage waveform (c) THD of input current waveform when three phase induction motor is supplied by AC direct

From all curves, the nature of the implemented induction motor is explained in transient as well as steady state condition. The parameter of induction motor is containing in table 1. When this motor is operated at

1 Nm mechanical torque then it draw of 3.72 A per phase at 1496 rpm and develop 1.456 Nm electromagnetic torque on armature conductor. As shown in figure 2 (a) the input voltage and current are sinusoidal in nature. The input current shows some transient effects. The THD in input voltage waveform is 0.09% and in input current waveform is 0.10% which is shown in figure 2(b) and 2(c) respectively. These values of harmonics are too less due to sinusoidal input to the motor. Now motor works more efficiently and give good performance.

4. Operating principle and components of VFD

The variable speed drive has a rectifier, inverter, DC link and control components. In VFD a solid state three phase full wave diode rectifier convert three phase 400V 50 Hz supply to DC voltage. This DC output has very high ripples. To minimize these ripples a DC link having low pass filter is connected to get smooth DC output. The harmonic generated are filter out by the low pass filter. After removal of ripple from DC voltage the DC output goes to inverter section of VFD. The inverter converts DC supply to three phase AC supply at any desired frequency with the help of gate pulse to control the speed of three phase induction motor. Inverter consists of power electronics devices like IGBT, Thyristors, MOSFET, power transistors etc. In this paper I use inverter with PWM (Pulse Width Modulation) technique.

5. Induction motor supplied from VFD without filter

The use of power electronics devices in VFD led harmonics in the system. As shown in figure 3 that a three phase induction motor is connected to supply through VFD. In the inverter, the triggering pulse is given by the discrete six pulse generator now the output values are changed as shown in results and waveforms. The resulting torque is decreased from 1.456 Nm to 1.375 Nm due to harmonics effect on induction motor generated by VFD.

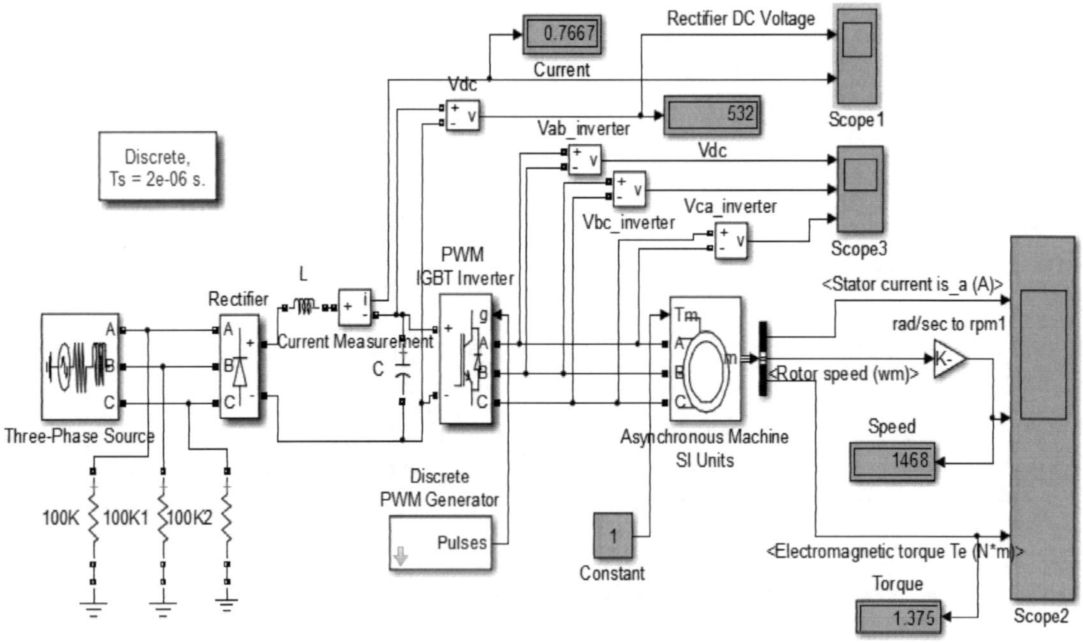

Figure 3 Simulation circuit of three phase induction motor when supplied by VFD without filter

Table 2 VFD parameters used in MATLAB/Simulink

Inverter	IGBT
Snubber resistance and capacitance	1e5,inf
Rectifier	Diode bridge
Inductance (DC Link)	150e-3
Capacitance (DC Link)	100e-6
Initial condition	1 0 0 0 0 0 0 0

Due to the power electronics switching harmonics are increased in the system. The output of an inverter is ac but not sinusoidal due to switching time taken by the Switches [3]. Harmonics are increased in the system so that the motor torque becomes pulsating. These harmonics reduce the performance of three phase induction motor. As a percentage of harmonics increases in the system then distortion of input voltage and input current get increase. This distorted nature of current lead pulsation in torque on rotor of induction motor.

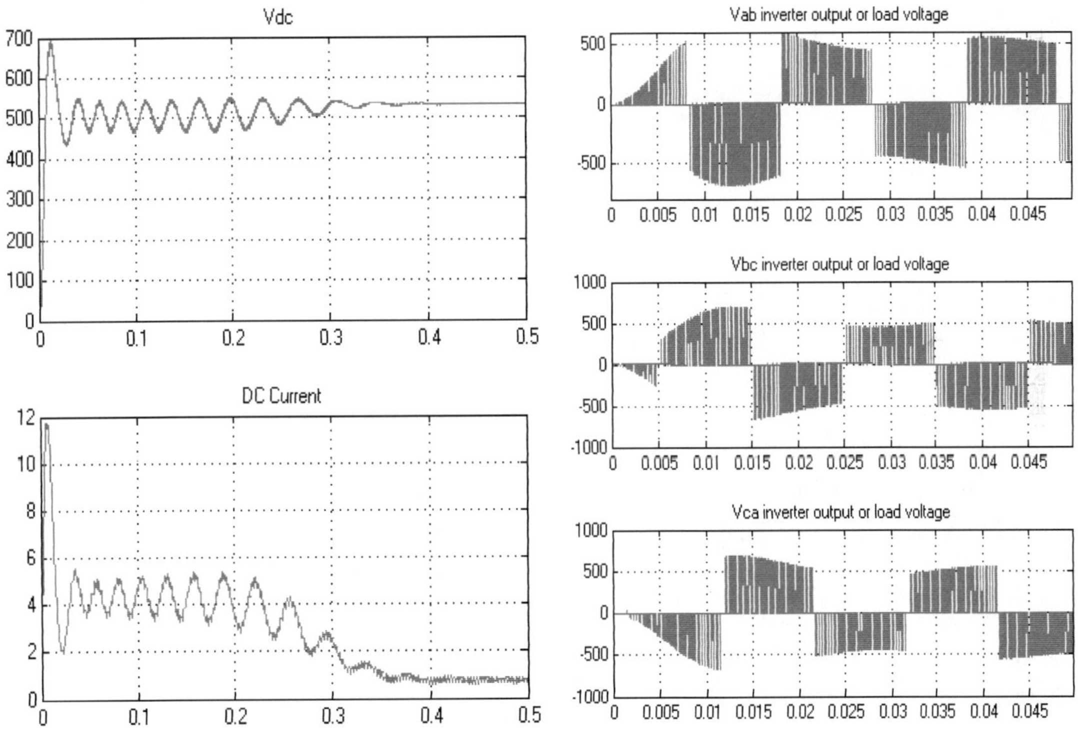

(a) (b)

Figure 4 (a) Rectifier output waveforms (b) inverter outputs waveforms without filter

Figure 4 (a) and 4 (b) shows the DC voltage, DC current characteristics of rectifier section and three phase AC voltage output waveforms of inverter section respectively. The ripple from DC supply can be minimised by using an appropriate LC filter after the rectification. This LC filter is known DC link in VFD.

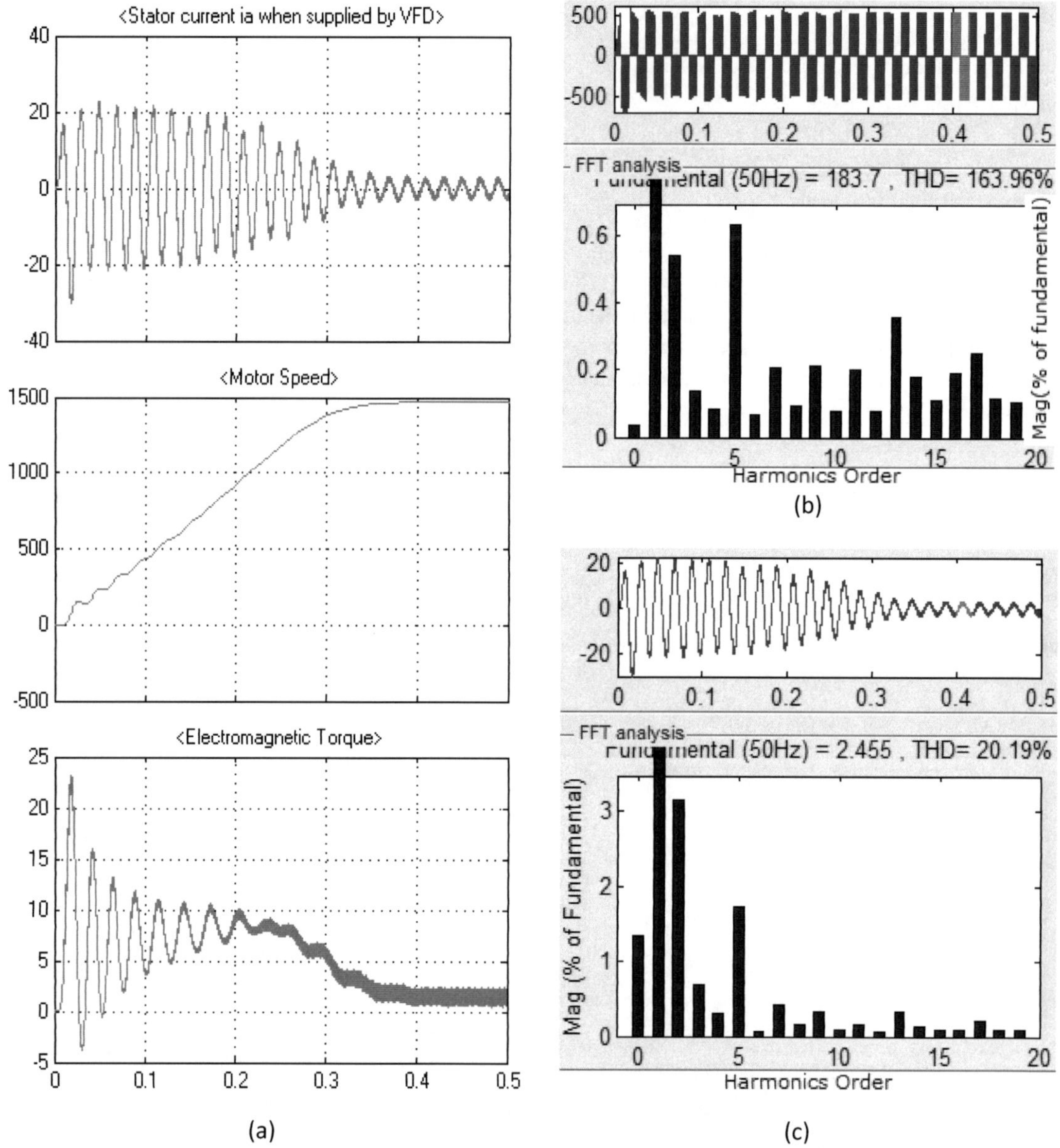

Figure 5(a) Stator current, motor speed and electromagnetic torque waveforms (b) THD in voltage waveform (c) THD in current waveform when motor is supplied by VFD without filter

As shown in figure 5(b) THD in the voltage waveform is 163.96 % and in figure 5(c) THD in the current waveform is 20.19 %. These harmonic inputs decrease the performance of induction motor. Figure 5(a) electromagnetic torque waveform shows the pulsation torque nature even after steady state condition reached.

6. Induction motor supplied from VFD with LC filter

The performance of induction motor can be improved by removing high frequency power signals from the input supply of induction motor.

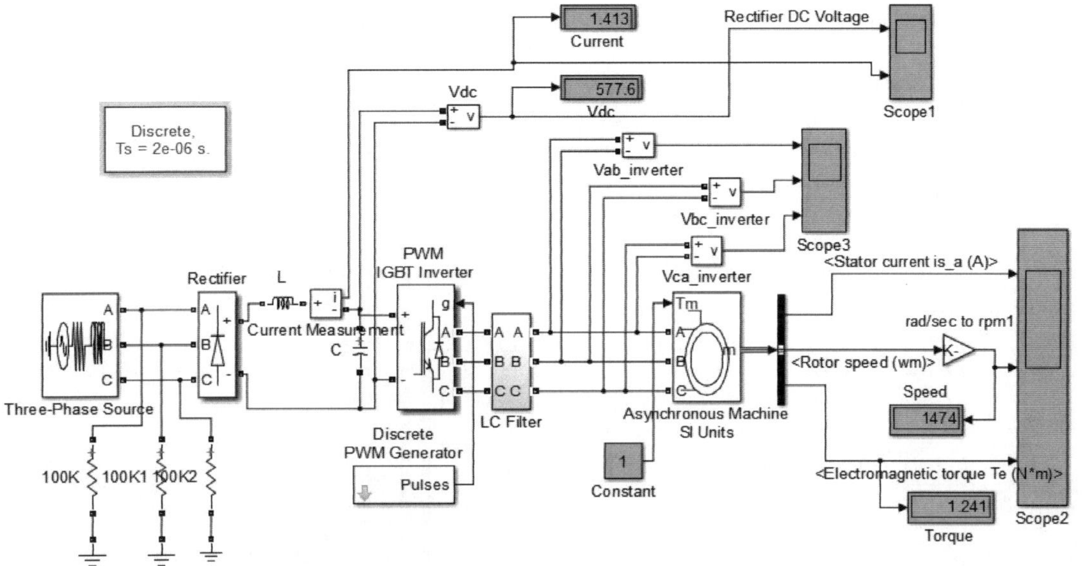

Figure 6 Simulation circuit of three phase induction motor when supplied by VFD with filter

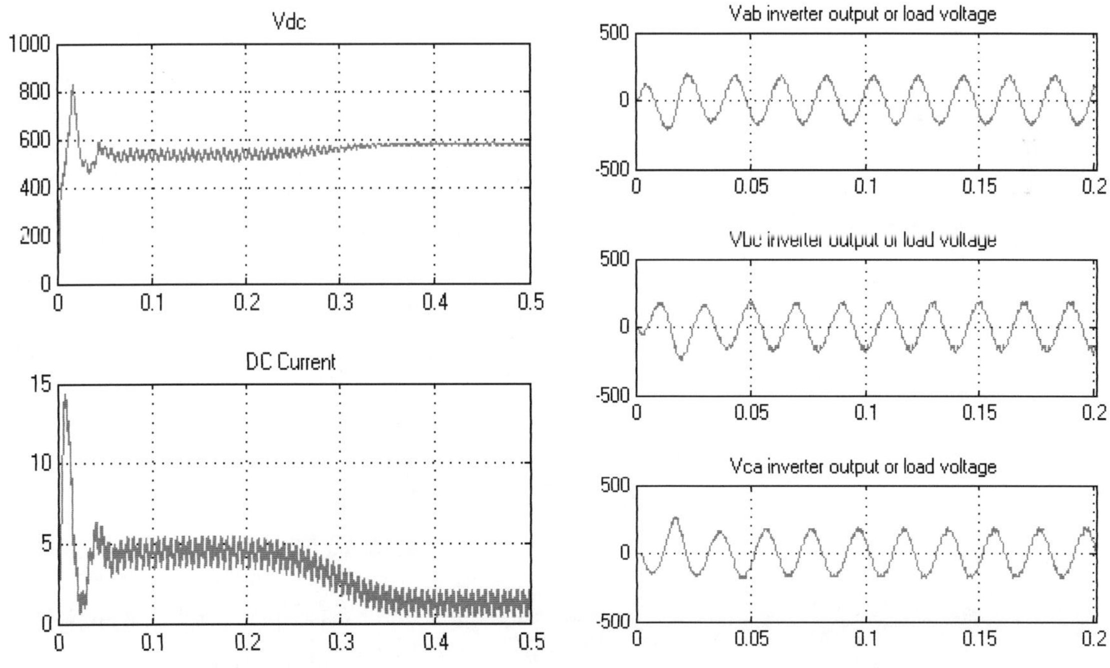

(a) (b)

Figure 7 (a) Rectifier output waveform (b) Inverter output waveform with filter

These harmonics can be removed by using an appropriate passive filter after the output of an inverter. The aim of the filter is to ensure sinusoidal stator current and voltage waveforms for the induction motor. In figure 6 a passive filter is connected after the inverter section of VFD. By connecting appropriate inductance and capacitance the inverter output waveforms can be improved.

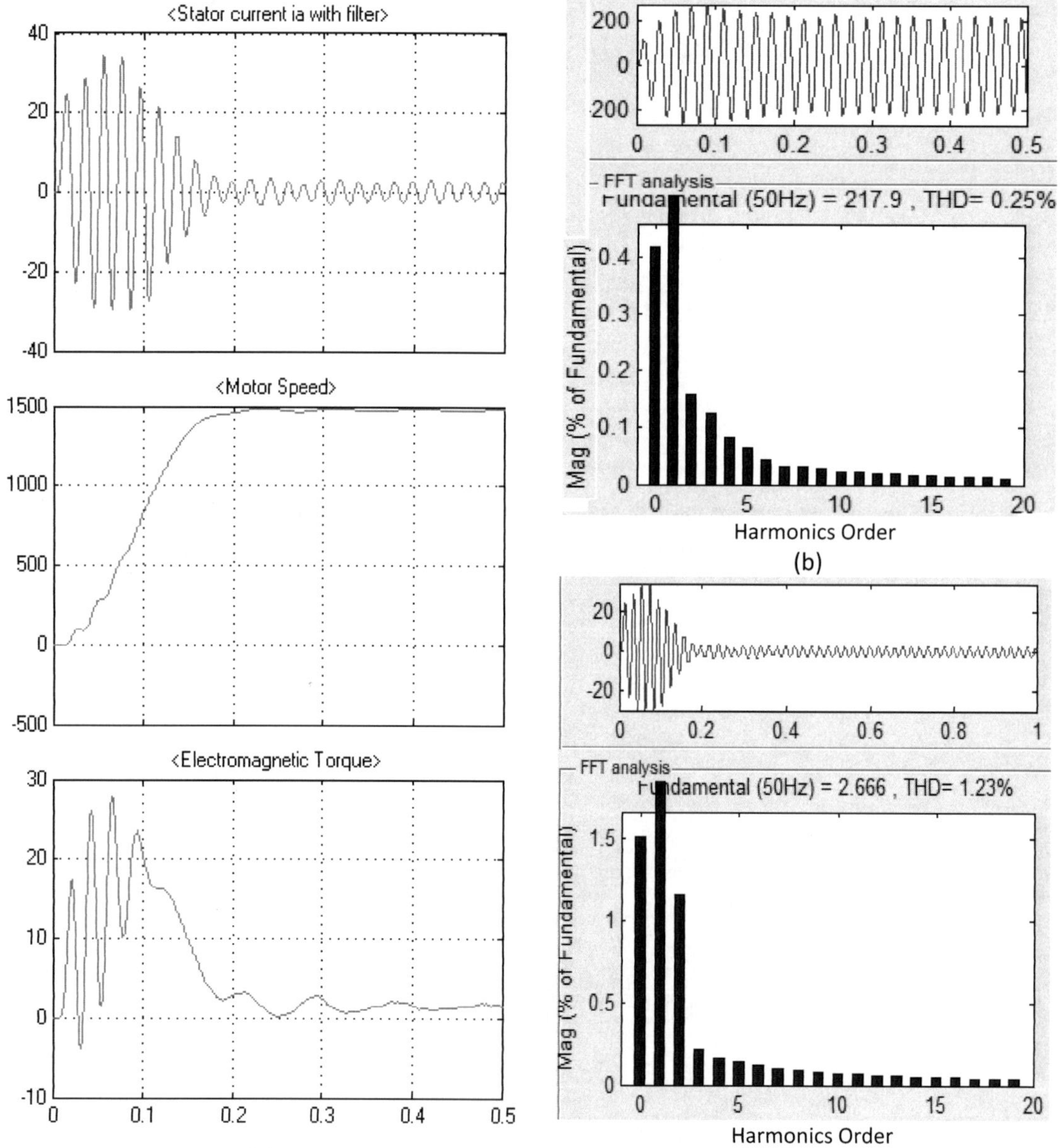

(a) (c)

Figure 8 (a) Stator current, motor speed, electromagnetic torque waveforms (b) THD in voltage waveform (c) THD in current when induction motor is supplied by VFD with filter

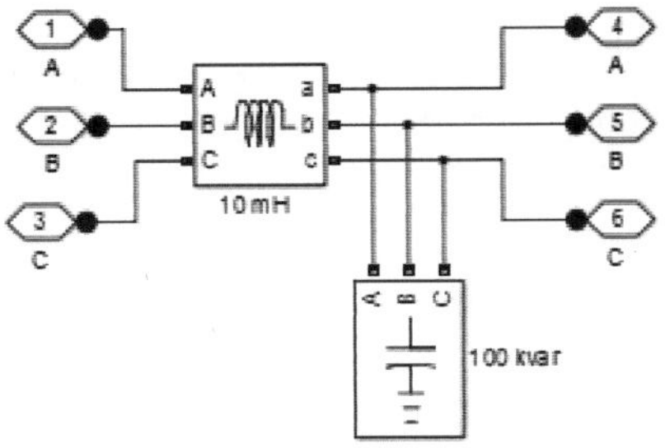

Figure 9 LC passive filters

As figure 7 (b) shows the inverter output voltage waveform with minimum harmonics. When this output is supplied to induction motor then the performance of motor improved which is shown in figure 8 (a), 8 (b) and 8 (c). Figure 8 (a) shows improved stator current, motor speed and electromagnetic torque waveform with less harmonic distortion in steady state condition which results less pulsation electromagnetic torque has on the rotor. Now harmonics are reduced as shown in figure 8 (b) and 8 (c) shows that voltage THD is 7.12 % and current THD is 1.23 % respectively which is less than the previous results when the induction motor is supplied by VFD without a filter. This passive filter consists of LC elements with value of L= 25 mH and Qc= 100 Kvar as shown in figure 9. These values are calculated by random variation in values for a specific motor of 5.4 HP, whose parameters are shown in table 1.

7. Conclusion

Operation of three phase induction motor, when connected to three phase ac source is ideally very efficient. The only problem is that speed control is not easily possible. VFD is used for speed control of induction motor however injects harmonics within the circuit and harmonics lead to peaky rotor current, oscillations in torque, which do not die and generate additional heating. When an appropriate LC filter is added after the VFD converter the waveforms of rotor current improve, torque oscillations are damped after initial transients and converter output waveform becomes more sinusoidal.

References

[1] Pratyusha Biswas Deb, Sudhangshu Sarkar, "Dynamic Model Analysis of Three Phase Induction Motor Using Matlab/Simulink" International Journal of Scientific & Engineering Research, Volume 7, Issue 3, March-2016 572 ISSN 2229-5518

[2] B. Ozpineci, L. M. Tolbert, "Simulink implementation of induction machine model – A Modular approach", IEEE, pp 728-734, 2003.

[3] Naveed Rahaman, H.V. Govindraju "Modelling &Simulation of a Three-Phase Electric Traction Induction Motor Using Matlab Simulink", IJEECS, ISSN : 2347-2820, June, 2014.

[4] Enemuoh F. O.1, Okafor E. E.2, Onuegbu J. C.3, Agu V. N.4 "Modelling, Simulation and Performance Analysis of A Variable Frequency Drive in Speed Control Of Induction Motor" International Journal of Engineering Inventions, e-ISSN: 2278-7461, p-ISSN: 2319-6491, Volume 3, Issue 5 (December 2013) PP: 36-41, www.ijeijournal.com.

[5] Jigar N. Mistry1 , Hetal D. Solanki2, Tejas M. Vala3, "VARIABLE FREQUENCY DRIVE" Research Expo International Multidisciplinary Research Journal Available online at www.researchjournals.in Volume – II , Issue – III September – 2012, ISSN : 2250 – 1630.

[6] Mr. Amit Kale1, Mr. Nikhil R. Kamdi2, Ms. Priya Kale3, Prof. Ankita A. Yeotikar4 "A REVIEW PAPER ON VARIABLE FREQUENCY DRIVE" International Research Journal of Engineering and Technology (IRJET) e-ISSN: 2395 -0056, Volume: 04 Issue: 01 I Jan -2017 www.irjet.net p-ISSN: 2395-0072, © 2017, IRJET I Impact Factor value: 5.181 I ISO 9001:2008 Certified Journal I Page 1281.

[7] Thida Win, Nang Sabai, and Hnin Nandar Maung, "Analysis of Variable Frequency Three Phase Induction Motor Drive" World Academy of Science, Engineering and Technology 42 2008

[8] Sumedh Tonapia, Piyush Chopadea, Tanmay Kadama, Junaid Julahaa and Priti Tyagib "Speed Control of AC Motor Using VFD" International Journal of Innovative and Emerging Research in Engineering, Volume 2, Issue 3, 2015 161, Available online at www.ijiere.com, e-ISSN: 2394 – 3343, P-ISSN: 2394 – 5494

[9] Pooja Shinde1, Rupali Burungale2, Pooja Kale3, Purvee Jain4 "Speed Control of Induction Motor by Using Variable Frequency Drive" Rupali Burungale et al Int. Journal of Engineering Research and Applications www.ijera.com , ISSN : 2248-9622, Vol. 4, Issue 4(Version 8), April 2014, pp.35-37.

[10] Yogita Y. Garud1, Sayali R. Gole2, Rutuja T. Jadhav3, Seema U. Deoghare4 "A Study on Variable Frequency Drive and It's Applications" ISSN(Online) : 2319-8753, ISSN (Print) : 2347-6710, International Journal of Innovative Research in Science, Engineering and Technology (An ISO 3297: 2007 Certified Organization) Vol. 5, Issue 3, March 2016 Copyright to IJIRSET DOI:10.15680/IJIRSET.2016.0503056 3079

[11] Mr.Ankur P. Desai, Mr.Rakesh.J.Motiyani, Dr.Ajitsinh R.Chudasama "Energy Conservation Using Variable Frequency Drive in Pumping Application" INTERNATIONAL JOURNAL OF ENGINEERING DEVELOPMENT AND RESEARCH I IJEDR(www.ijedr.org) (Two Day National Conference (RTEECE-2014) -17th ,18th January 2014), Energy Conservation Using Variable Frequency Drive in Pumping Application I ISSN: 2321-9939

[12] Tamal Aditya "Research to study Variable Frequency Drive and its Energy Savings" International Journal of Science and Research (IJSR), India Online ISSN: 2319-7064, Volume 2 Issue 6, June 2013, www.ijsr.net

[13] Krupa Gandhi, K.L.Mokariya, Deepa Karvat "Simulation of PWM inverter for VFD application Using MATLAB", International Journal of Engineering Research and Development e-ISSN: 2278-067X, p-ISSN: 2278-800X, www.ijerd.com Volume 10, Issue 4 (April 2014), PP.94-103

[14] Sharda Patwa1 & Hemant Amhia2 "Analysis and Control Of Harmonics Of Three Phase Induction Machine By Space Vector Decomposition Using Matlab Simulation" International Journal of Electrical and Electronics Engineering (IJEEE) ISSN (PRINT): 2231 – 5284, Vol-2, Iss-1, 2012

[15] Sudeep Pyakuryal, Mohammad Matin, "Filter Design for AC to DC Converter" International Refereed Journal of Engineering and Science (IRJES),ISSN (Online) 2319-183X, (Print) 2319-1821, Volume 2, Issue 6 (June 2013), PP. 42-49, www.irjes.com

[16] Amit P. Wankhade[1], Prof. C. Veeresh[2] "Analysis of Advanced Techniques to Eliminate Harmonics in AC Drives" International Journal of Engineering Research and General Science, Volume 2, Issue 5, August – September 2014, ISSN 2091-2730

[17] Poonam Kaur1, Ritula Thakur2 "Harmonic Analysis of VFD's" International Journal Of Modern Engineering Research (IJMER) I IJMER I ISSN: 2249–6645 I www.ijmer.com I Vol. 4 I Iss. 5I May. 2014 I 55 I

[18] Amit Kumar Sharma, Tushar Deep Saxena, Shirazul Islam, Arunkumar Yadav, "Performance Analysis of Three Phase PWM Voltage Source Inverter Fed Three Phase Induction Motor Drive", IJAEEE, ISSN : 2278-8948, Volume-2, 2013.

Probe-fed semi circular microstrip antenna vis-à-vis circular microstrip antenna: a necessary revisit

S K Ghosh[1,4], S K Varshney[2], S Chakraborty[3], L L K Singh[4], S Chattopadhyay[4]

[1] Depart of ECE, Siliguri Institute of Technology, Siliguri, India
[2] Depart of ECE, Sir J C Bose School of Engineering, Mankundu, West Bengal
[3] Device Technology Group, CSIR-Central Electronics Engineering Research Institute, Plani, India.
[4] Depart of ECE, Mizoram University, Aizawl, India.

E-mail: piyalirekha@yahoo.com, s.c.in@ieee.org

Abstract. Microstrip patch antenna of semicircular geometry has been investigated in view of miniaturization of conventional circular geometry. The precise operating frequency of the semicircular microstrip patch antenna is the most significant parameter to be determined in order to design such antenna system to achieve the optimum performance. In the present investigation an improved formulation is presented for accurate determination of the resonant frequency of semicircular patch. Also, the radiation property of such patch is thoroughly investigated. Through comparisons are documented amongst the circular and semicircular patches. It is revealed that, the semicircular patch offers more better radiation performance compared to circular.

1. Introduction

The circular geometry of the microstrip antenna is the most common geometry which has been studied, analyzed and implemented for over last three decades[1]-[4]. In the present era, the scientists and engineers are looking for tiny devices, which can be implemented in miniaturized wireless communication equipment without hampering the performance. In that scenario, semicircular patch antenna is a good choice which can produce such typical qualities like tininess, light weight and compatibility with MMIC along with acceptable radiation performance. In fact, semicircular patch is very advantageous where the space is the key factor to organize a patch with conventional geometry. In fact, 50% of patch area reduction can be achieved using semicircular patch for a particular resonant frequency compared to conventional circular patch geometry. Hence, the accurate determination of mode as well as its radiation performance of a semicircular patch is very important in this scenario.

The electric field beneath the patch and eigen functions of circular sector antennas were presented in [5]. Generalized transmission line modeling of sector antenna has been found in [6]. But all these papers indicated above fail to predict resonant frequencies of sector antenna accurately. In fact, in all of those investigations, the concept of fringing is not
considered. Nevertheless the issue of fringing can not be avoided for accurate computation of dominant mode and its resonant frequency. The effect of fringing has been considered in [7]-[8] for computation of resonant frequency of sector antenna. Still they fail to predict the accurate resonant

frequency. In fact, in these reports, the dominant mode of the sector antenna is not determined properly.

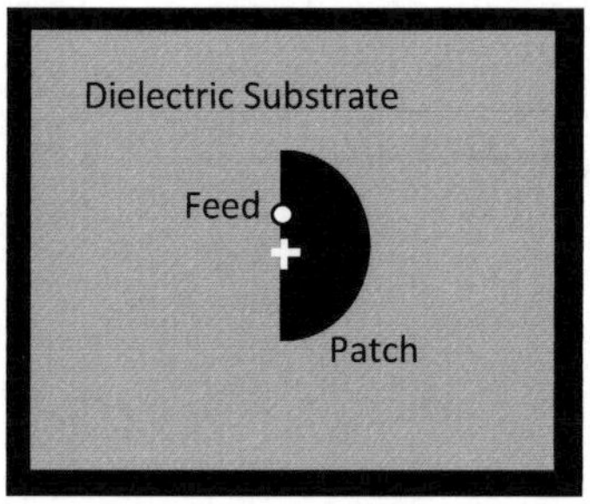

Figure 1. Schematic representation of semicircular patch (Top View).

The radiation characteristic of semicircular patch is also an important issue and it is least investigated topic in last two decades. On the contrary, the radiation property of circular patch is widely investigated. Plenty of works [9]-[10] have been reported to improve the radiation pattern of a conventional circular patch. In that scenario, hardly some investigation has been reported on radiation characteristic of semicircular patch.

Lo *et. al.*[11], and Richards *et. al.* [5] have analyzed the radiation characteristics of semicircular patch for the first time. However, in these reports, clear understanding of the radiation pattern of semicircular patch at its lowest order dominant mode is not apparent. In fact, in the article [5], the radiation patterns for two different modes are portrayed and comparison between them is discussed in terms of their radiation pattern. One very recent report [12] shows the modal analysis of circular sector patch. However, from [12], it is not apparent that whether the semicircular patch can be a best alternative of circular patch. Moreover, the investigation [12] does not deal with the radiation characteristics of the circular sector patch for the dominant mode.

In order to improve the shortcomings of the earlier studies, and also to offer the readers a clear understanding of the dominant mode of semicircular patch, its frequency and its radiation characteristic at its lowest order dominant mode, we have investigated both the circular and semicircular patches thoroughly. The mathematical derivations to determine the accurate dominant mode frequency are based on modified cavity model along with the accurate estimation of the fringing electric fields and resulting effective antenna dimension. Better radiation performance is revealed for semicircular patch compared to that of circular patch. Therefore, it is concluded that a patch of semicircular geometry with 50% reduced patch area becomes the most efficient and suitable alternative to a patch of conventional circular geometry

2. Theory

A coaxially fed CSMA having radial dimension a, printed on a PTFE substrate (ε_r) with thickness h_1 is shown in Fig. 1. The antenna is fed at ρ_c from centre as shown in the figure using a coaxial probe.

In this present model, the top sectorial patch and ground plane act as PEC while three side walls are PMC. Hence, we may write the vanishing boundary condition of magnetic fields at PMC walls.

Thus, ρ component of magnetic field,

$H_\phi = 0$ at $\rho = a$ for $\phi = 0$ to π and $z = 0$ to h

$H_\rho = 0$ at $\phi = 0$ and $\phi = \pi$ for $\rho = 0$ to a and $z = 0$ to h

As we know form [12];

$$A_z = B_{mnp} J_m(k_\rho \rho)[A\cos(m\phi) + B\sin(m\phi)][\cos(k_z z)]$$

$$H_\phi = -\frac{1}{\mu}\frac{1}{\rho}\frac{\partial A_Z}{\partial \rho} \quad (1)$$

$$H_\phi(\rho = a) = \frac{1}{\mu}\frac{1}{\rho} B_{mnp} J'_m(k_\rho a)[A_2 \cos m\phi + B_2 \sin m\phi][\cos(k_z z)] = 0 \quad (1a)$$

It is possible if

$$k_\rho = \chi'_{mn}/a$$

where, $\chi'_{mn} = \chi'_{11} = 1.84$

where, χ'_{mn} is the n th zero of derivative of Bessel function of order m. Physically, m is the number of circumferential variation and n is the number of radial variation. From the chart of derivative of Bessel function it is found that the lowest root is 1.84 and it is happened when $m = n = 1$. Therefore, the semicircular patch is having the same lowest order dominant mode (i.e TM_{11}) like conventional circular patch.

and,

$K_Z = \frac{p\pi}{h}$ where $p = 1,2,3$ as usual like conventional circular patch derived in [13].

Hence, $K^2 = K_\rho^2 + K_Z^2$

$$f_r = \frac{1}{2\pi\sqrt{\mu\varepsilon}}\sqrt{\left(\frac{\chi'_{mn}}{a}\right)^2 + \left(\frac{p\pi}{h}\right)^2}$$

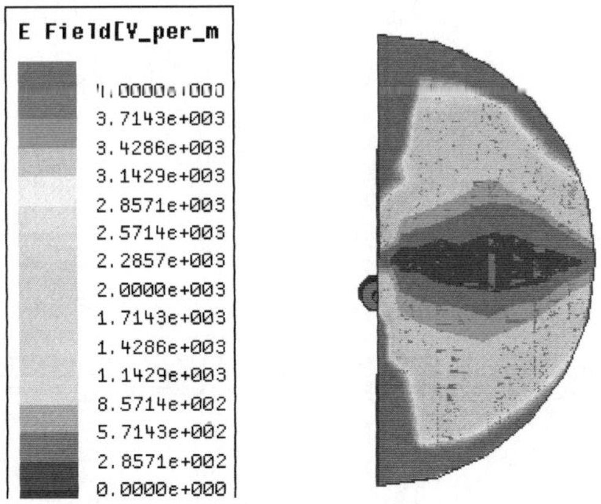

Figure 2. Magnitude of electric field on patch surface to explain the number of field variation in radial and circumferential direction

IOP Conf. Series: Materials Science and Engineering **331** (2018) 012026 doi:10.1088/1757-899X/331/1/012026

Hence, the zeroth order resonance frequency f_r is $f_r = \frac{\chi'_{mn} c}{2\pi a \sqrt{\varepsilon_r}} = \frac{1.84c}{2\pi a \sqrt{\varepsilon_r}}$ where c is the free space velocity.

Now to compute the resonant frequency accurately, one must incorporate the concept of effective dielectric constant and the effective dimension. Therefore, the final equation becomes

$$f_r = \frac{1.84c}{2\pi a_{eff} \sqrt{\varepsilon_{reff}}}$$

where, a_{eff} and ε_{reff} can be computed from [14].

Thus, it is interesting to note that, though the geometry of the patch is different, dominant mode is similar for circular and semicircular patch. Therefore for a particular operating frequency, one need not design the circular patch. Instead, a semicircular patch with 50% reduced patch area can be designed.

The magnitude of electric field over the patch is presented in Fig. 2. It is clear from the figure that, the number of field variation either in circumferential or in radial direction is 1. This confirms that the lowest order mode of semicircular patch is same as circular patch and it is TM_{11}.

3. Results and discussions

The computed results using the present theory are compared with simulations and have been presented in Table-I. Comparison has been made between the lowest order dominant mode resonant frequency of circular and semicircular patches for wide variety of parameters. Computed results are in good agreement with the simulation results. The table shows that, the resonant frequencies are identical for both the circular and semicircular patches for a wide variety of parameters in each case.

Table-I
Comparison of lowest order dominant mode frequency of circular and semicircular patch in terms of simulation and present theory.

Patch radius a in mm	ε_r	h_1 in mm	Lowest order TM_{11} mode frequency at GHz		
			Circular patch [sim]	Semicircular Patch [sim]	Semicircular Patch [Computed] [our]
11.5	2.65	1.5875	4.42	4.40	4.41
10.7	2.65	1.5875	4.72	4.70	4.68
9.6	2.65	1.5875	5.22	5.20	5.24
8.2	2.65	1.5875	6.07	6.06	6.10
7.4	2.65	1.5875	6.63	6.61	6.70
50.0	2.32	1.59	1.128	1.129	1.128
50.0	2.32	1.59	1.286	1.284	1.285
50.0	2.32	1.59	1.350	1.345	1.350
3.0	2.20	0.508	18.54	18.45	18.50
3.7	2.20	0.508	15.88	15.86	15.80

The reflection coefficient profile for semicircular and circular patch of radius a =30 mm on PTFE substrate with h = 1.575 mm, ε_r =2.33 is presented in Fig. 3. It shows that the both the structures resonate at same frequency i.e 1.82 GHz and their profiles are also identical. Nevertheless, semicircular patch offers better matching at its resonance compared to circular patch as is evident from

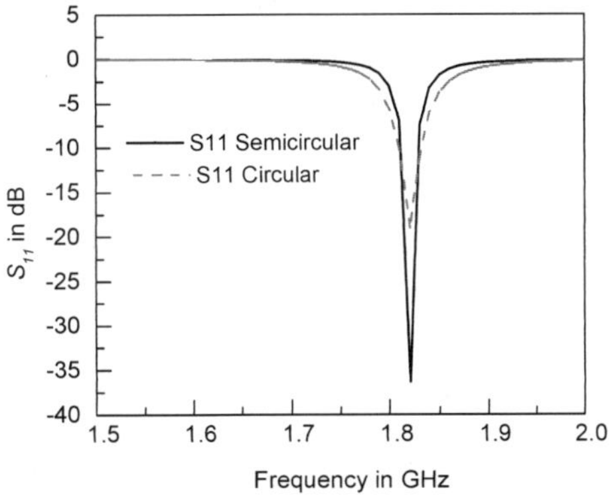

Figure 3. Reflection coefficient profile for circular and semicircular patches.

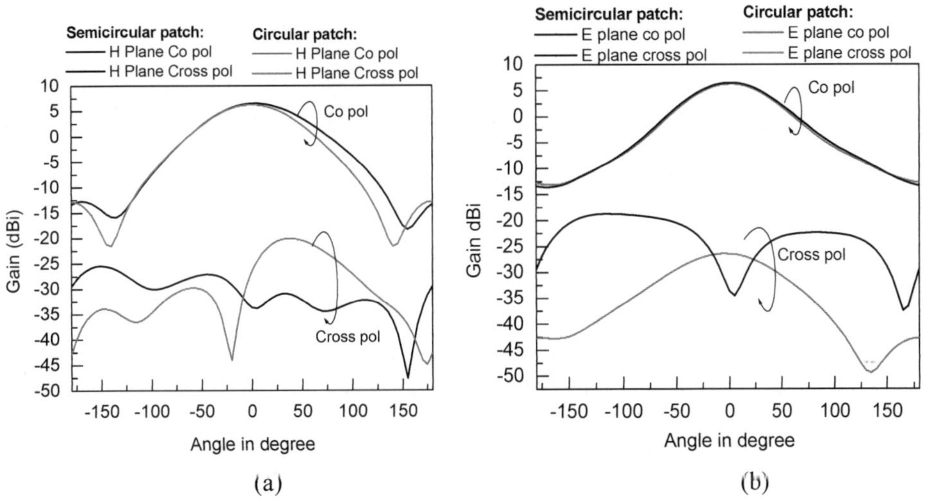

Figure 4. H and E plane radiation patterns for circular and semicircular patches at dominant mode frequency. (a) H plane, (b) E plane.

the figure. Fig. 4. show the comparison between radiation properties of circular and semicircular patch separately in E and H planes. It is found that the gain of both the patches is around 6.4 dBi. However, it may be noted that, the cross polarization level is less for semicircular patch compared to circular patch in their H plane. The isolation between co-polarization to cross polarization radiation for semicircular patch is 32 dB while the same for circular patch is 26 dB. Furthermore, the H plane co-polarized radiation beam is broader in case of semicircular patch compared to that of circular patch. The 3 dB beam width of semicircular patch is around 107^0 in its H plane while the same for circular patch is only 80^0. Therefore, around 34% of improvement in broadening the H plane beam width is evident from semicircular patch compared to that of circular patch. On the contrary, the cross polarization radiation is less in E plane for circular patch compared to semicircular patch. Around 32 dB of Co-Cross polarization isolation is revealed for circular patch in E plane while the same for

semicircular patch is 26 dB. However, the beam of semicircular patch is slightly broader for semicircular patch than circular patch in its E plane.
Therefore, it may be concluded that, the semicircular patch produces same radiation profile in its dominant mode in comparison to circular patch. But it may be noted that, the semicircular patch have more better radiation performance particularly in H plane and partially in E plane compared to the circular patch as discussed above. Nevertheless, as it has 50% reduced physical area, this semicircular patch can be mounted on tiny device. Therefore, the semicircular patch is the most suitable alternative of circular patch for the scientific community looking for tiny antenna with improved radiation performance.

4. Conclusions
Semicircular patch is proposed to be an efficient and most suitable alternative of circular sectorial microstrip antenna for miniaturization and to obtain more improved radiation performance. This is obviously be required in the modern wireless era where, every devices are budding towards tininess and multi-functionality. Therefore, the present investigation is surely be helpful for scientist and practicing engineers to use such semicircular microstrip antenna on tiny wireless equipment.

5. Acknowledgement
Authors would like to thank Prof. Debatosh Guha of Calcutta University, India for his support.
S. Chakraborty would like to thank Director, CSIR-CEERI, Pilani and Dr. S. K. Ghosh, Principal Scientist, CSIR-CEERI, Pilani for always supporting research endeavors.
All of the co-authors express a deep sense of gratitude and thanks to the corresponding institutes for their kind supports and encouragement.

References
[1] Garg R, Bhartia P, Bhal I and Ittipibun A 2001 Microstrip Antenna Design Handbook *Artech House*, Norwood, 2001
[2] Waterhouse R 1995 Small Microstrip Patch Antenna *Electronic Letters*, **31**(8), 604-605.
[3] Lee S, Woo J, Ryu M and Shin H 2002 Corrugated Circular Microstrip Antenna for miniaturization *Electronic Letters*, **38** (6), 262-263.
[4] Guha D 2001 Resonant frequency of circular microstrip antennas with and without air gaps, *IEEE Trans. Antennas Propagat*, 49, 55- 59.
[5] Richards W F , Ou J D, and Long S A 1984, A Theoretical and Experimental Investigation of Annular, Annular Sector, and Circular Sector Microstrip Antennas *IEEE Trans. On Antennas and Propagation* **32**, 864-867.
[6] Bhattacharya A K, and Garg R 1986, Analysis of Annular Sector and Circular Sector Microstrip Patch Antennas *Electromagnetics* **6**, 229-242.
[7] Tiwari V K, Kimothi A, Bhatnagar D, Saini J S, Saxsena V K and Kumar P 2006 Theoretical analysis on circular sector microstrip antennas *Indian Journal of Radio & Space Physics*, **35** , 133-138.
[8] Tiwari V K, Kimothi A, Bhatnagar D, Saini J S, Saxsena V K and Kumar P 2006 Theoretical and experimental investigations of circular sector microstrip antennas *Indian Journal of Radio & Space Physics*, **35** , 206-211.
[9] Guha D, Kumar C and Pal S 2005 Improved Cross Polarization Characteristics of Circular Microstrip Antenna employing Arc Shaped DGS *IEEE Antennas and Wireless Propagat. Lett.*
[10] Duan Z S and Qu S B 2009 Wide bandwidth Broad Beam Width Microstrip Patch Antenna *IET Electronic Letters.*

[11] Lo T T, Solomon D, Richards W F 1979 Theory and Experiments on Microstrip Antennas *IEEE Trans. on Ant. Propagat.*, **27**, 137-145.

[12] Ghosh S K, Chakraborty S, Singh L L K and Chattopadhyay S 2017 Modal Analysis of Probe-fed Circular Sector Microstrip Antenna with and without Variable Air Gap: Investigation with Modified Cavity Model *International Journal of RF and Microwave Computer Aided Engineering*, Wiley, USA, DOI: 10.1002/mmce.21172.

[13] Balanis C A 2001, Antenna Theory: Analysis and Design, 2nd Ed., *John Wiley & Sons*.

[14] Chattopadhyay S, Biswas M, Siddiqui J Y and Guha D 2009 Input Impedance of Probe-Fed Rectangular Microstrip Antennas with Variable Air-Gap and Varying Aspect Ratio *IEE Microwaves, Antennas & Propagation* **3**(8), 1151-1156.

IOP Conf. Series: Materials Science and Engineering **331** (2018) 012027 doi:10.1088/1757-899X/331/1/012027

Real Time Assessment of Potable Water Quality in Distribution Network based on Low Cost Multi-Sensor Array

Jyotirmoy Bhardwaj; Karunesh K Gupta; Punit Khatri

Department of Electrical and Electronics Engineering, Birla Institute of Technology & Science, Pilani

kgupta@pilani.bits-pilani.ac.in

Abstract. New concepts and techniques are replacing traditional methods of water quality parameters measurement systems. This paper proposed a new way of potable water quality assessment in distribution network using Multi Sensor Array (MSA). Extensive research suggests that following parameters i.e. pH, Dissolved Oxygen (D.O.), Conductivity, Oxygen Reduction Potential (ORP), Temperature and Salinity are most suitable to detect overall quality of potable water. Commonly MSA is not an integrated sensor array on some substrate, but rather comprises a set of individual sensors measuring simultaneously different water parameters all together. Based on research, a MSA has been developed followed by signal conditioning unit and finally, an algorithm for easy user interfacing. A dedicated part of this paper also discusses the platform design and significant results. The Objective of this proposed research is to provide simple, efficient, cost effective and socially acceptable means to detect and analyse water bodies regularly and automatically.

1. Introduction

Water quality detection is paramount requirement before consumption by all human activities ranging from potable purpose to industrial and agriculture process. Water quality can be easily monitored in treatment plants and pumping stations. Many parameters, such as pH, turbidity, total organic carbon, chlorine, conductivity and others are measured continuously to monitor the effectiveness of the treatment process. However, once the treated water enters the distribution network, detailed water quality monitoring and analysis takes lots of time and money because sampling becomes a difficult task. Therefore, two dimensional approach has been followed in this paper. Firstly, it discusses the requirements of integrated multi sensor array in water distribution network. Secondly, it presents the design and development steps of low cost autonomous Integrated Multi sensor array that can be used by companies and consumers. Proposed MSA will detect whole spectrum of water quality parameters in the distribution network. Statistics show that 20-60% of water contamination incidents are related to events in the water distribution network, for example sewage leakage, contaminated soil etc. Additionally, it is widely accepted that distribution systems are quite vulnerable to contaminations. Generally, water sensor is an electronic device that is designed to detect the presence of contamination and provide an alert in time [1]. Hall et. al demonstrated changes in water-quality parameters by measuring variations in important parameters using real-time sensors instead of targeting contamination. Empirical evidences suggested that water quality parameters, such as pH, conductivity, free chlorine and dissolved oxygen are sensitive indicators of nicotine, arsenic trioxide, and Escherichia coli [2]. Therefore, instead of direct detection of contaminants, monitoring these parameters are more feasible. Several experimental studies [3], [4] indicate the need for continuous on-line water quality monitoring with efficient data transmission. The conventional water quality methods fail to satisfy this requirement due to higher labor and operating cost. The Multi Sensor Probes shall continuously measure the water quality parameters and thereafter integrated with Arduino MEGA 2560 (a microcontroller) followed by algorithm for easy user interfacing.

Research on water quality monitoring has started in the international arena with the advent of glass electrodes in the early 1930s. However, research was based on detection of single water quality parameter, that cannot effectively judge the overall quality of water. Therefore, it was proposed to detect multiple parameters simultaneously in the last decade of the 20th century. Further research to integrate all the different parameter detection techniques in a single system is difficult due to the varying nature of water parameters. Some researchers proposed the technique for integration of six different parameters, i.e. pH, dissolved oxygen, temperature, conductivity, and water level in a single system. This approach is highly sensitive, offers broad contamination coverage with low cost [5]. However, due to a wide variety of other different water parameters, integration of the above mentioned techniques does not provide an overall solution for water quality parameters detection. O'Flyrm et al. [6] designed a multi- sensor using ZigBee standards for communication and the system accommodated six Plug-and-Play sensors that measure temperature, pH, phosphate, dissolved oxygen, con-ductility, turbidity and water levels based on Tyndall mote. Atkinson et al. (2006) developed a sensor system (YSI Sonde) to simultaneously measure conductivity, salinity, temp, depth, pH, dissolved oxygen, turbidity, chlorophyll and blue-green algae [7]. Dinh et al. used WSN for measuring salinity and underground water level. They used Fleck platform (uses Atmel processors) for sensor node deployment [8]. Chow et al. (2008) proposed Multiple parameter measurement system, applicable for wastewater and potable water that can monitor water quality online, but despite of this, color measurement of this device was not highly sensitive [9]. Real-time Monitoring of the Water Distribution System project in Singapore proposed an end-to-end integrated hardware and software system for monitoring, analysing, and modelling urban water distribution systems in real-time in collaboration with the water utility of Singapore (Public Utilities Board - PUB) [10]. Nasirudin et al. used PICl6F886 as sensor node and GSM modems for contamination detection and updating the central database respectively. Measurements included temperature, pH, turbidity and DO [11]. Similarly, Van den Broeke et al. (2005) proposed UV scanning with a single probe that can be used to measure multiple parameters, including BOD, TOC, UV-254, NO_3, NO_2, Ozone, H2S, TSS, and turbidity [12]. Earlier attempted methods are expensive and not very user friendly. Therefore, based on earlier research we target to develop low cost user friendly mechanism that can effectively detect over all water quality in distribution network.

2. Multi Sensor Array (MSA)

Water quality parameters monitoring requires on-site measurement of a wide range of different chemical and microbial compounds simultaneously. Water may contain broad, varied spectrum of inorganic as well as organic compounds, therefore the measuring equipment has to be robust, sensitive and be able to detect small changes with broad selectivity. Individual sensor can detect single parameter, but cannot detect the overall quality of water; therefore, the concept of Multi Sensor Array (MSA) arises to detect different water parameters all together [13]. The response of MSA is provided by an array of sensors detecting various parameters of water. The output data are treated subsequently with analysis techniques. Commonly MSA is not an integrated sensor array on some substrate, but rather comprises a set of individual sensors measuring simultaneously different water parameters all together [14]. MSA offers following advantages:

- Ensures timely response in the event of a crisis and enhances emergency planning.
- Lower cost, lower power consumption, and lower maintenance with higher reliability and robustness.
- This technique can provide accurate contamination detection and identification.
- MSA can be fully customised according to user requirements because contaminants of interest may vary depending on the location of source.
- Allows easy tracing of a potential waterborne disease
- Facilitates the integration with real time monitoring to support remote sensing.

Signals coming from MSA are treated using microcontrollers and pattern recognition techniques such as the principal component analysis (PCA), the K-nearest neighbour (KNN) and most popular in recent years, the artificial neural networks (ANN) [15]. In the context of this study, MSA will be associated with a microcontroller Arduino Mega 2560. The main challenge of this study is to integrate different water sensors together and develop reliable and low-cost sensing systems for accurate and continuous water quality monitoring. The system intended to be used as a part of a low cost water sensor network to provide water quality information to consumers, water companies and authorities. The data provided by such system can support complex decisions concerning the quality of potable water. Fig. 1 represents block diagram representation of proposed Multi Sensor Array.

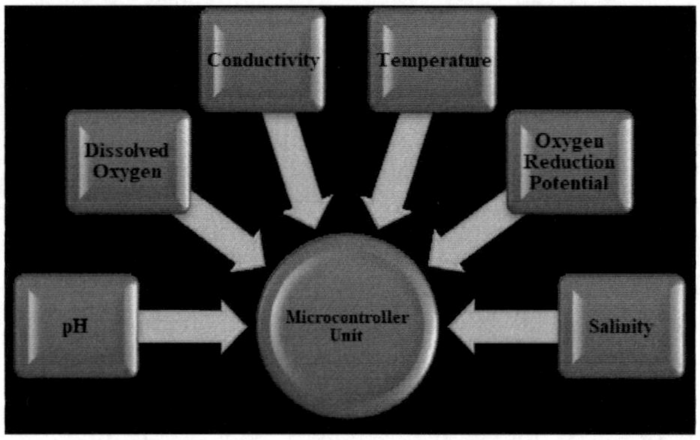

Figure 1. Multi Sensor Array Overview

3. Targeted Parameters

US Environmental Protection Agency has carried out an extensive experimental evaluation of water sensors to assess their performance on several contaminations. The main conclusion was that chemical and biological contaminants generally affect Turbidity, Oxidation Reduction Potential, Electrical Conductivity, pH and detection of variation in above mentioned parameters can significantly detect overall quality of water [16]. For example, Free Chlorine concentration (HOCl) can be approximated based on the ORP, pH and temperature measurements. Therefore, we have selected six parameters based on earlier research and Table 1presents targeted Water Quality Parameters with their acceptable ranges for potable water.

Table 1. Targeted Water Quality Parameters

Sr. number	Parameters	Unit	Range for Potable water
1	**pH**	pH	6.5-8.5
2	**Dissolved Oxygen**	mg/lit	>3 mg/lit
3	**Electrical Conductivity**	uS/cm	0-800
4	**Temperature**	^{o}C	30-35
5	**Oxygen Reduction Potetial**	mV	650-800

4. Results and Discussions

The sensor nodes collect data every second and transmit to Arduino board followed by VPython. Fig 2 presents top view of system prototype, the BNC cord of sensor array is attached to Arduino Mega 2560 board. Primarily, the system has been designed to measure sensor data over a infinity period of time for real time continuous monitoring. Experimental results are presented over X-Y scale. X axis represents time in seconds and Y axis is respective parameters for each result. The potable water sample is collected from Water Laboratory of Birla Institute of Technology and Science. The pH output is as shown if Fig. 3(a). The pH value of sample is 8.8, as confirmed by real time analysis with error of +/- 0.2 value. The dots represent time at which values are taken and for convenience, the first

50 second values have shown. Similarly, Fig. 3(b) represents Dissolved Oxygen with value 9.58 mg/lit with error of +/- 0.2 mg/lit, and Fig. 3(c) Oxygen Reduction Potential with 250 mV and response time is 95% in 1 second. Fig. 3(d) Temperature.

Figure 2. Top view: System Prototype

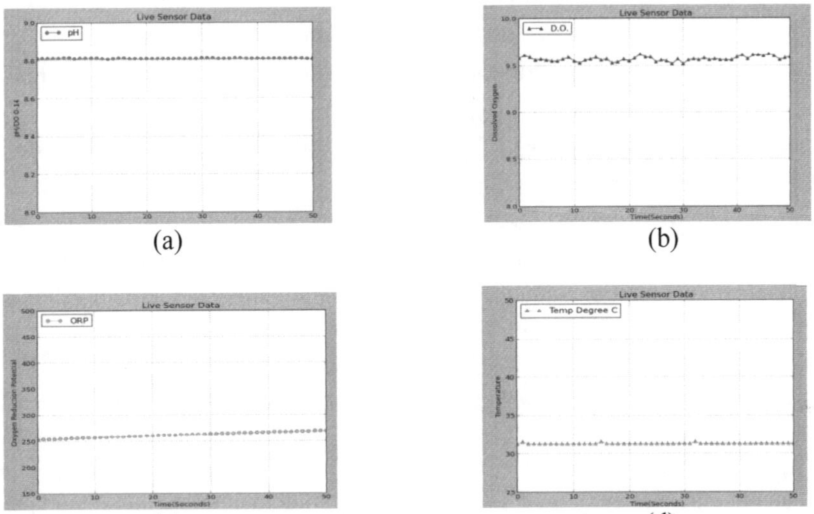

(a) (b) (c) (d)

Figure 3. Real Time Sensor Data (a) pH (b) Dissolved Oxygen (c) Oxygen Reduction Potential (d) Temperature

5. Conclusions and Future work

This paper demonstrates effective low cost mechanism for Real Time Assessment of Potable Water Quality in Distribution Network using MSA. The prototype used Integrated Sensor Array and Arduino Mega 2560 Board with VPython. Calibration can be done by sending command from monitoring stations. This approach demonstrates the effective way to detect water quality in distribution network, environmental protection and tracking pollution sources compared to manual approaches. Moreover, from data analysis acquired from this prototype is user friendly that will assist in better understanding of water distribution network. The MSA can be deployed in user specific places and overall cost of the system is ~1000 USD. The system can be integrated with wireless tools for long distance transmission as well.

References

[1] Sadiq, Rehan, Yehuda Kleiner, and Balvant Rajani. "Aggregative risk analysis for water quality failure in distribution networks." Aqua 53 (2004): 241-261.

[2] Hall, J., Zaffiro, A. D., Marx, R. B., Kefauver, P. C., Krishnan, E. R., Haught, R. C., & Herrmann, J. G. (2007). “On-line water quality parameters as indicators of distribution

system contamination". Journal (American Water Works Association), 66-77.

[3] Aisopou, A., Stoianov, I., & Graham, N. J. (2012). "In-pipe water quality monitoring in water supply systems under steady and unsteady state flow conditions: A quantitative assessment." Water research, 46(1), 235-246.

[4] Panguluri, S., Meiners, G., Hall, J., & Szabo, J. G. (2009). "Distribution system water quality monitoring: Sensor technology evaluation methodology and results". US Environmental Protection Agency, Washington, DC. EPA/600/R-09/076.

[5] Li, Mingli, et al. "A Multi-parameter Integrated Water Quality Sensors System."Computer and Computing Technologies in Agriculture VI. Springer Berlin Heidelberg, 2013. 260-270.

[6] O'Flyrm, B., Ricardo Martinez, John Cleary, Catherine Slater, F. Regan, Dermot Diamond, and H. Murphy. "SmartCoast: a wireless sensor network for water quality monitoring." In Local Computer Networks, 2007. LCN 2007. 32nd IEEE Conference on, pp. 815-816. Ieee, 2007.

[7] Atkinson SF, Mabe JA (2006). Near real-time monitoring and mapping of specific conductivity levels across Lake Texoma, USA. Environ. Monit. Assess. 120: 449-460.

[8] Le Dinh, Tuan, et al. "Design and deployment of a remote robust sensor network: Experiences from an outdoor water quality monitoring network." Local Computer Networks, 2007. LCN 2007. 32nd IEEE Conference on. IEEE, 2007.

[9] Chow, C., Fabris, R., Dixon, M., 2008. Case Studies Using SCAN On-line Monitoring System (Research Report 75) Water Quality Research Australia (Adelaide)

[10] Whittle, Andrew J., Lewis Girod, Ami Preis, Michael Allen, Hock Beng Lim, Mudasser Iqbal, Seshan Srirangarajan, Cheng Fu, Kai Juan Wong, and Daniel Goldsmith. "WATERWISE@ SG: A testbed for continuous monitoring of the water distribution system in singapore." In Proc. 12th Water Distribution Systems Analysis Conference (WDSA). 2010.

[11] Nasirudin, Muhammad Azwan, Ummi Nurulhaiza Za'bah, and O. Sidek. "Fresh water real-time monitoring system based on Wireless Sensor Network and GSM." In Open Systems (ICOS), 2011 IEEE Conference on, pp. 354-357. IEEE, 2011.

[12] Van den Broeke, J., 2005. A Short Evaluation of the Scan Spectroscopy lsser. KIWA Water Research, Nieuwegein.

[13] Rudnitskaya, A., Ehlert, A., Legin, A., Vlasov, Y., & Büttgenbach, S. (2001). "Multisensor system on the basis of an array of non-specific chemical sensors and artificial neural networks for determination of inorganic pollutants in a model groundwater". Talanta, 55(2), 425-431.

[14] Winquist, F., Holmin, S., Krantz-Rülcker, C., Wide, P., & Lundström, I. (2000). "A hybrid electronic tongue." Analytica Chimica Acta, 406(2), 147-157.

[15] Rong, Li, Wang Ping, and Hu Wenlei. "A novel method for wine analysis based on sensor fusion technique." Sensors and Actuators B: Chemical 66.1 (2000): 246-250.

[16] J. Bhardwaj, K. K. Gupta, and R. Gupta. 2015. A review of emerging trends on water quality measurement sensors. In Proceedings of the International Conference on Technologies for Sustainable Development (ICTSD'15). IEEE.

[17] Barrett, Steven F. "Arduino Microcontroller Processing for Everyone!."Synthesis Lectures on Digital Circuits and Systems 8, no. 4 (2013): 1-513.

[18] Scherer, David, Paul Dubois, and Bruce Sherwood. "VPython: 3D interactive scientific graphics for students." Computing in Science & Engineering 5 (2000): 56-62.

Real-Time-Simulation of IEEE-5-Bus Network on OPAL-RT-OP4510 Simulator

Anjali Atul Bhandakkar[1], and Dr. Lini Mathew[2]

[1] PhD. Scholar, Electrical Engineering Department, NITTTR, Chandigarh, India.

[2] Associate Prof., Electrical Engineering Department, NITTTR, Chandigarh, India.

bhandakkar13@gmail.com, lenimathew@yahoo.com

Abstract: The Real-Time Simulator tools have high computing technologies, improved performance. They are widely used for design and improvement of electrical systems. The advancement of the software tools like MATLAB/SIMULINK with its Real-Time Workshop (RTW) and Real-Time Windows Target (RTWT), real-time simulators are used extensively in many engineering fields, such as industry, education, and research institutions. OPAL-RT-OP4510 is a Real-Time Simulator which is used in both industry and academia. In this paper, the real-time simulation of IEEE-5-Bus network is carried out by means of OPAL-RT-OP4510 with CRO and other hardware. The performance of the network is observed with the introduction of fault at various locations. The waveforms of voltage, current, active and reactive power are observed in the MATLAB simulation environment and on the CRO. Also, Load Flow Analysis (LFA) of IEEE-5-Bus network is computed using MATLAB/Simulink power-gui load flow tool.

1. Introduction

The powerful computer with affordable cost has led to the emergence of highly sophisticated simulation software applications that enable simulation of dynamic systems and related controls, and also automatic code generation for implementation in industrial controllers [1]. OPAL-RT-OP4510 is a Real-Time (RT) simulator with 4 cores. RT-LAB system is one of the features of the OPAL-RT-OP4510 Simulator. It has basically two parts. One is host computer and another is RT simulator. Host computer edits Simulink model compiles with RT-LAB and gives user interface. The RT simulator does the real-time model execution. It has REDHAT as an operating system and provides telnet communication with the host [2]. If the execution time, for the simulation of the system is shorter or equal to the selected time step, the simulation is considered to be real-time. If it is greater than its time step size for one or more time-steps, overruns occur and the simulation is considered as non-real-time or offline. In the latter case, either the time-step can be increased or the system model can be simplified to run it in real time. [3].

Power-flow studies are of great importance in planning and designing the future expansion of power systems as well as in determining the best operation of existing systems [4]. Some of the computational tools available for conducting the LFA for a power system network are Educational Simulation Tool (EST), MatPower, Power Analysis Toolbox (PAT), SimPowerSystems (SPS), MatEMTP, and Voltage Stability Toolbox (VST) [5]. Traditionally LFA is computed by means of methods such as Gauss-Siedal Method, Newton-Raphson Method (NRM), Decoupled Method and Fast Decoupled Method. Among these methods, NRM is very popular due to its fast convergence with a less number of iterations and due to this characteristics, it has proved most successful in large power flow studies [6]. MATLAB/Simulink uses NRM by default for load flow computation.

The contents of this paper are, Section II details the IEEE-5-Bus network its data sheet and parameter calculation for Simulink model. Section III gives a procedure for development of a MATLAB / Simulink model compatible to OPAL-RT-OP4510, and LFA using power-gui. Section IV gives results and discussion and Section V gives the conclusions.

2. IEEE-5-Bus Network

The single line diagram of IEEE-5-Bus network is shown in Fig.1. The transmission line parameters, generation, and loads are given in per unit value. The network details are:
Number of lines = 7, Number of buses = 5, Number of generators = 2, and Number of loads = 4.

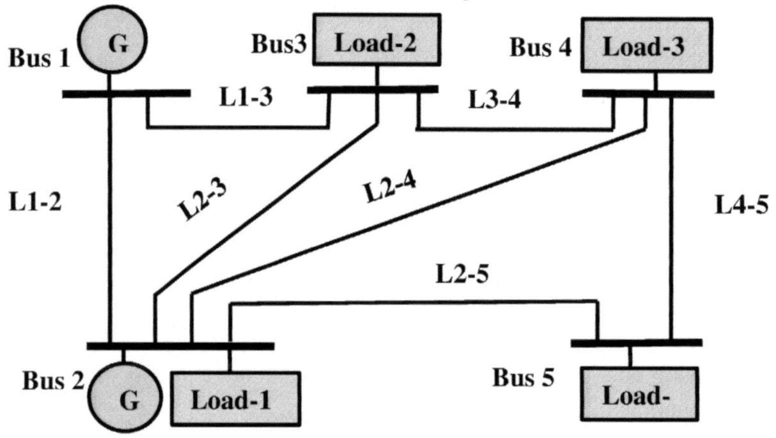

Fig.1. IEEE-5-Bus Network

Transmission line parameters of IEEE-5-Bus are given in per unit value in Table 1.

Table 1. Transmission line parameters of IEEE-5-Bus in per unit value.

From Bus	To Bus	Transmission line Resistance	Transmission line Reactance	Transmission line Conductance	Transmission line Susceptance
1	2	0.02	0.06	0	0.06
1	3	0.08	0.24	0	0.05
2	3	0.06	0.18	0	0.04
2	4	0.06	0.18	0	0.04
2	5	0.04	0.12	0	0.03
3	4	0.01	0.03	0	0.02
4	5	0.08	0.24	0	0.05

By selecting a (suitable) base MVA of 100 MVA (S_b) and base kV of 230 kV (V_b) and using basic per unit concepts of impedance calculation, the values of positive sequence and zero sequence resistance, inductance, capacitance, and transmission line length are calculated and tabulated in Table 2. Length calculation of the transmission line between bus-1 to bus-2 is as follows.

- R=0.02 , X=0.06 , B=0.06 (as per given data-sheet)Length of transmission line $=\left(\frac{\sqrt{X}*\sqrt{B}}{2\pi f}\right)*$ $(Velocity\ of\ LIGHT\ in\ km/sec)$

- Length of transmission line =47 km

Table 2. Actual values of positive sequence and zero sequence parameters of the transmission line.

Bus	Length of Trans. Line km	Positive Seq. R (Ω/km)	Zero Seq. R (Ω/km)	Positive Seq. L (H/km)	Zero Seq. L (H/km)	Positive Seq. C (F/km)	Zero Seq. C (F/km)
1-2	47	0.22	0.66	0.0017	0.0052	$6.3x10^{-9}$	$1.89 x10^{-8}$
1-3	87	0.48	1.45	0.0038	0.0115	$2.87 x10^{-9}$	$8.62 x10^{-9}$
2-3	67	0.46	1.40	0.0037	0.0112	$2.97 x10^{-9}$	$8.91 x10^{-9}$
2-4	67	0.46	1.40	0.0037	0.0112	$2.97 x10^{-9}$	$8.91 x10^{-9}$
2-5	47	0.44	1.32	0.0035	0.0105	$3.15 x10^{-9}$	$9.45 x10^{-9}$
3-4	19	0.27	0.813	0.0021	0.0064	$5.14 x10^{-9}$	$1.54 x10^{-9}$
4-5	87	0.48	1.455	0.0038	0.0115	$2.87 x10^{-9}$	$8.62 x10^{-9}$

Table 3 shows Generation-Load at the bus in per unit values as per the data –sheet. The actual values generation, load are tabulated in Table 4. Also the initial values of nodal bus voltages, their phase angles are given in Table 5. Actual values of voltage are to be used in the MATLAB/Simulink model. Load Flow Bus is to be connected to the buses where voltage and angle are required to be computed by load flow tool in the Simulink model.

Table 3. Generation, load at the bus in per unit values

Bus	P_G	Q_G	Q-Max	Q-Min	P_L	Q_L
1	0	0	5	5	0	0
2	0.4	0	3	-3	0.2	0.1
3	0	0	0	0	0.45	0.15
4	0	0	0	0	0.4	0.05
5	0	0	0	0	0.6	0.1

Table 4. Actual values of generation, load at the bus

Bus	P_G (W)	Q_G (VAR)	Q-Max (VAR)	Q-Min (VAR)	P_L (W)	Q_L (VAR)
1	0	0	$500 x10^6$	$500 x10^6$	0	0
2	$40x10^6$	0	$300 x10^6$	$-300 x10^6$	$20 x10^6$	$10 x10^6$
3	0	0	0	0	$45 x10^6$	$15 x10^6$
4	0	0	0	0	$40 x10^6$	$5 x10^6$
5	0	0	0	0	$60 x10^6$	$10 x10^6$

Table 5. Bus-type, initial voltage, and angle at the IEEE-5-bus network

Bus No.	1	2	3	4	5
Bus Type	Slack	PV	PQ	PQ	PQ
Initial V-magnitude	1.06	1	1	1	1
Initial V-angle	0	0	0	0	0

3. Matlab / Simulink Model Compatible In OPAL-RT-OP4510

The IEEE-5-Bus network has been developed in MATLAB/Simulink environment using the calculated values of transmission line parameters, generation, and load at the respective bus, initial values of voltage, angle and bus type. Real-time simulation of the electric power system is the reproduction of output (voltage/currents) waveforms, with the desired accuracy, and these V/ I wave forms are representative of the behavior of the real power system being modeled. To achieve such a goal, a real-time simulator needs to solve the model equations for one time-step, within the same time as in a real-world clock. Therefore, it produces outputs at discrete time intervals, where the system states are computed at certain discrete times using a fixed time-step [3].

The IEEE-5-Bus network with, breakers, faults are connected in the master subsystem. The control over the variation in breaker position, type of fault is kept in console subsystem. As the network is not too big, it does not require slave subsystem. The block-diagram of subsystems of OPAL-RT-OP4510 with their functionalities are shown in Fig.2.

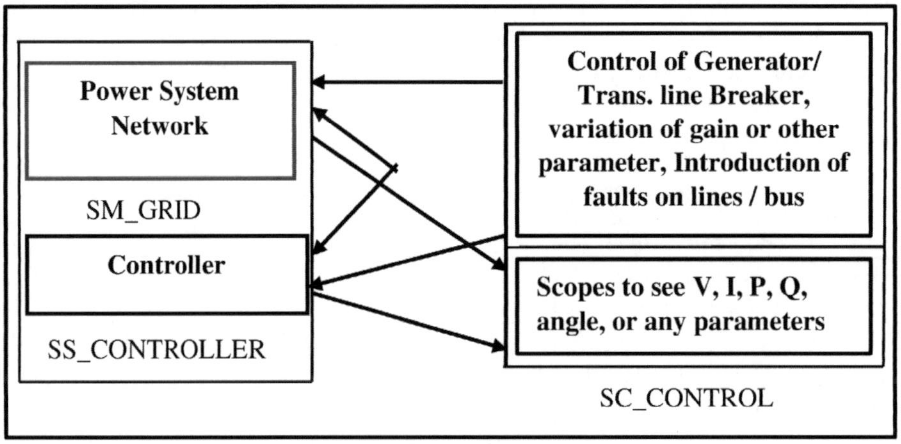

Fig. 2. Functionality of OPAL-RT-OP4510

The master subsystem SM_NET and console subsystem SC_SCOPE connected in top-level OPAL-RT model are shown in Fig.3.

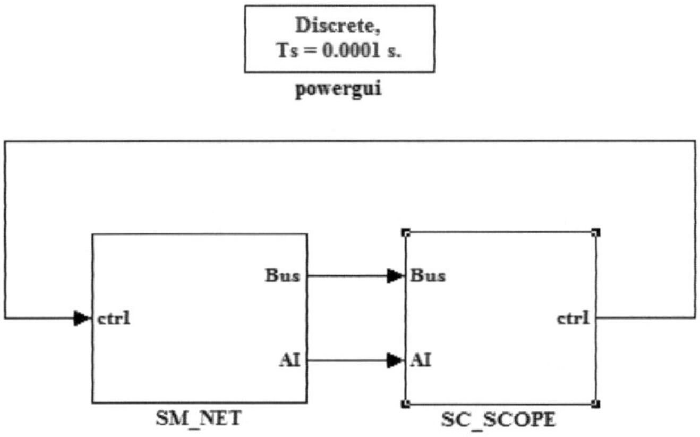

Fig. 3 Top Level Model of OPAL-RT

The details of master subsystem SM_NET are shown in Fig.4. The console subsystem SC_SCOPE and the breaker control system are shown in Fig. 5(a) and Fig. 5(b).

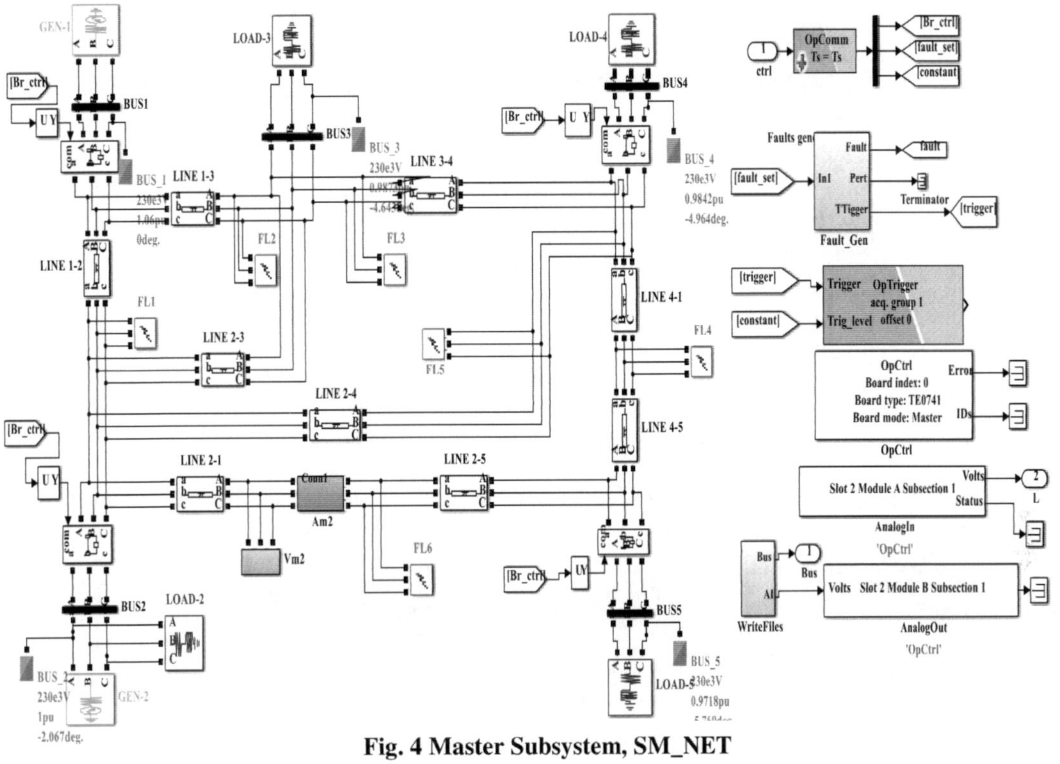

Fig. 4 Master Subsystem, SM_NET

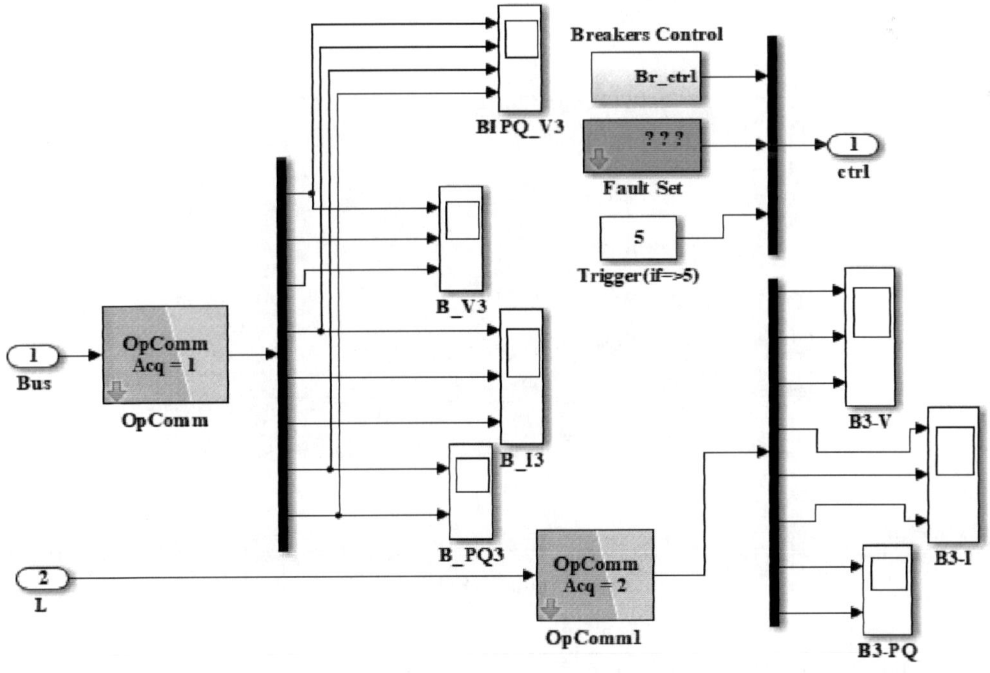

Fig. 5 (a) Console Subsystem SC_SCOPE

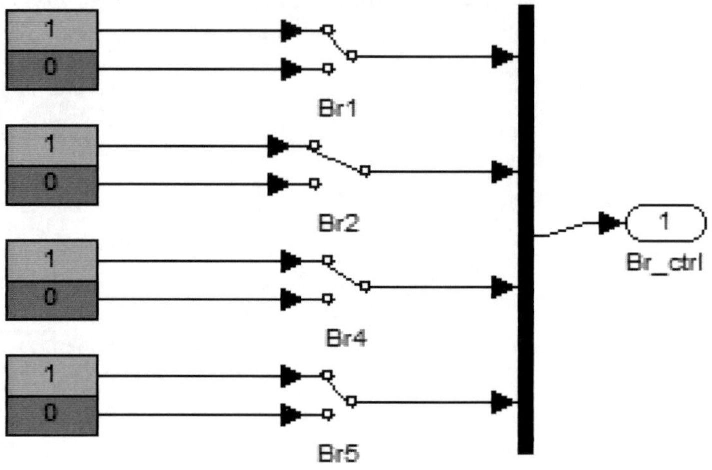

Fig. 5 (b) Breaker Control Subsystem

The photograph of the OPAL-RT-OP4510 Simulator, the outport connector and the whole hardware setup with the laptop (host computer) and CRO are shown in Fig. 6.

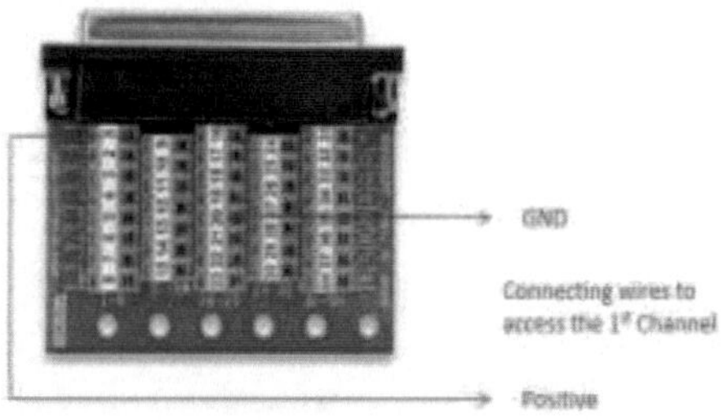

(b)

(a)

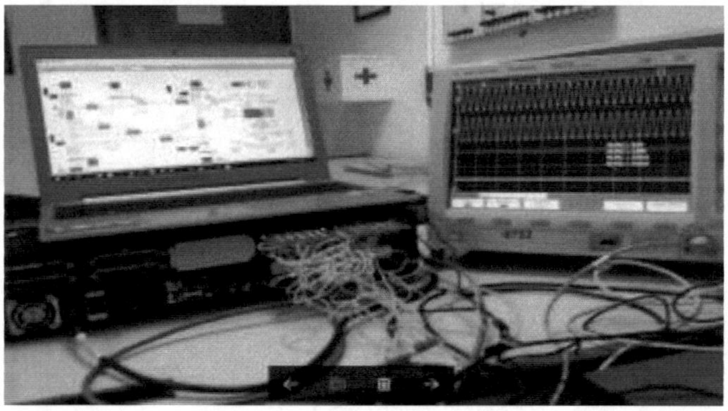

(c)

Fig. 6 Photograph showing (a) OPAL-RT-OP4510 Simulator, (b) I/O connectors, (c) laptop (host computer) and CRO

The Simulink model in real time with the OPAL-RT environment has been built and executed employing the OPAL-RT-OP4510 Simulator. The signals such as voltage, current, active power, reactive power have been observed on CRO

4. Results and Discussion

Results of LFA computed by power-gui and V, I, P, Q waveforms obtained during Real-Time simulation are discussed separately parts in this section.

4.1 LFA of IEEE-5-Bus network

It is computed using power-gui (off-line) Load Flow Tool. The generated report shows that LFA converged in 2 iterations. The Load Flow Bus is required to be connected where ever voltage and angle is to be computed. After performing LF, the magnitude of voltage in per unit and its angle is displayed by the load flow-bus. The significant parameters of LFA obtained from power-gui-Load Flow tab are given in Table 6.

Table 6. Significant Parameters of LFA

Bus	Bus Type	Voltage (pu)	Angle in degree	P (MW)	Q (MVAR)
1	Slack	1.06	0.00	131.43	90.89
2	PQ	1.00	-2.07	20.00	10.00
2	PV	1.00	-2.07	40.00	-61.64
3	PQ	0.99	-4.64	45.00	15.00
4	PQ	0.98	-4.96	40.00	5.00
5	PQ	0.97	-5.77	60.00	10.00

4.2 Real-time-Simulation

Using the outport device, the signals such as voltage (V), current (I), active power (P), reactive power (Q) are taken out from OPAL-RT- OP4510 Simulator. To view them on CRO, gains of appropriate values are used to scale down the magnitudes. The voltage signal has the magnitude of 230 kV, the current has the magnitude of 220 A, active power has 60 MW and reactive power has 10 MVAR, (as per LFA). The gain for voltage signal used in the model is of $1/10^4$, the current signal is 1/100, active and reactive power is of $1/10^7$. The channel-1indicates voltage, channel- 2 current, channel-3 active power, channel-4 reactive power signals.

These signals are fed back to MATLAB environment through import device and observed in the console subsystem. It is found that the magnitudes of signals on CRO and in MATLAB environment are equal. Fig. 7 shows the waveforms of V, I, P, Q, on CRO as well as in the MATLAB environment in the console subsystem during a healthy condition of the network.

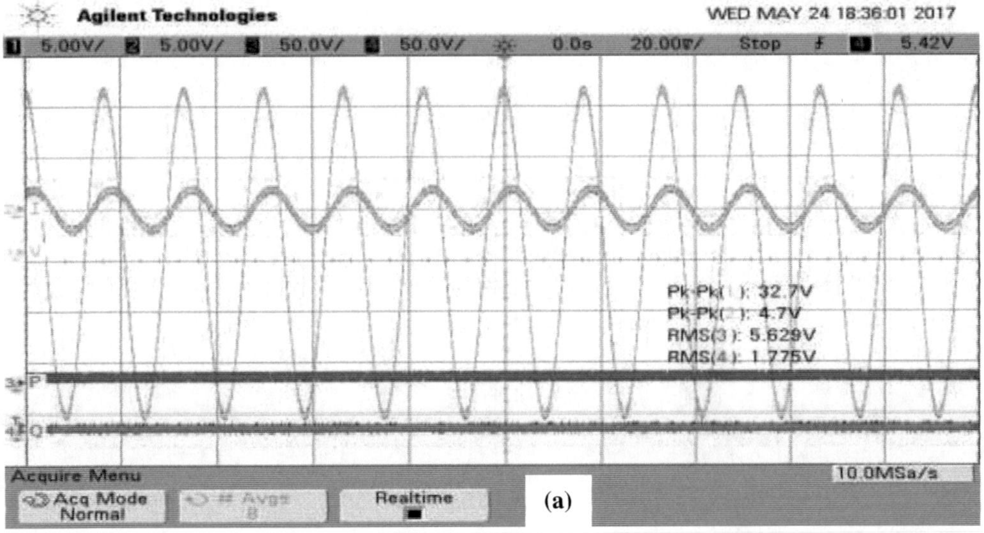

(a)

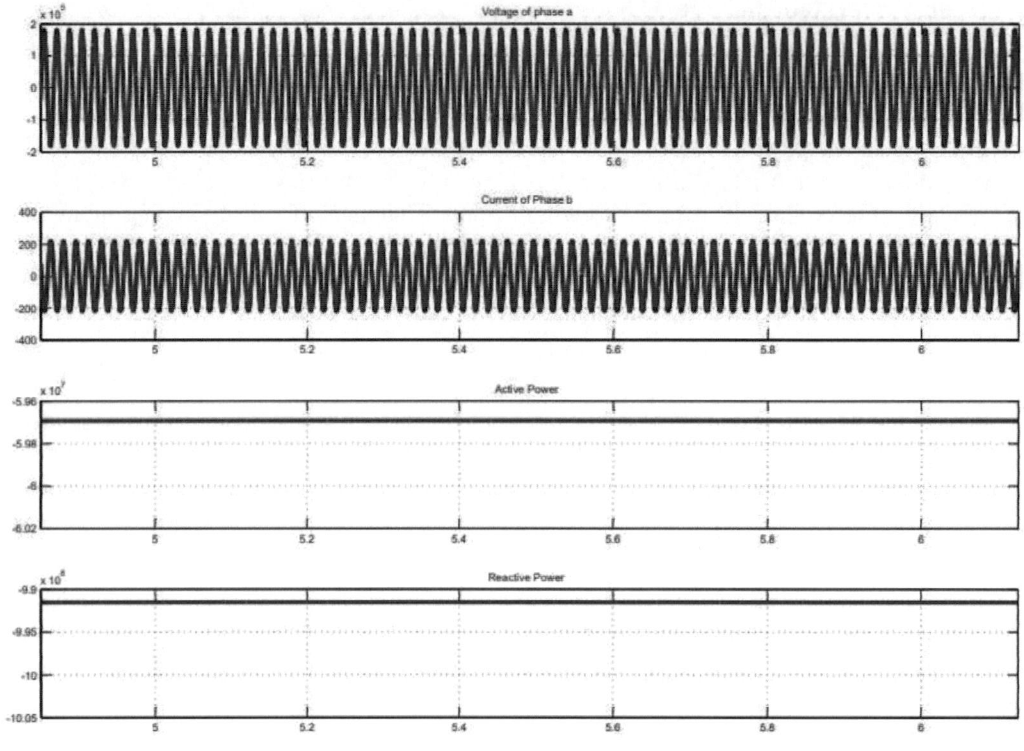

(b)

Fig. 7 Wave forms of V, I, P, Q during Healthy Condition (a) on CRO and (b) on host computer (console)

Fig .8 shows three phase voltages and current of phase-a on CRO which has nearly same value as that of seen in MATLAB environment during healthy condition.

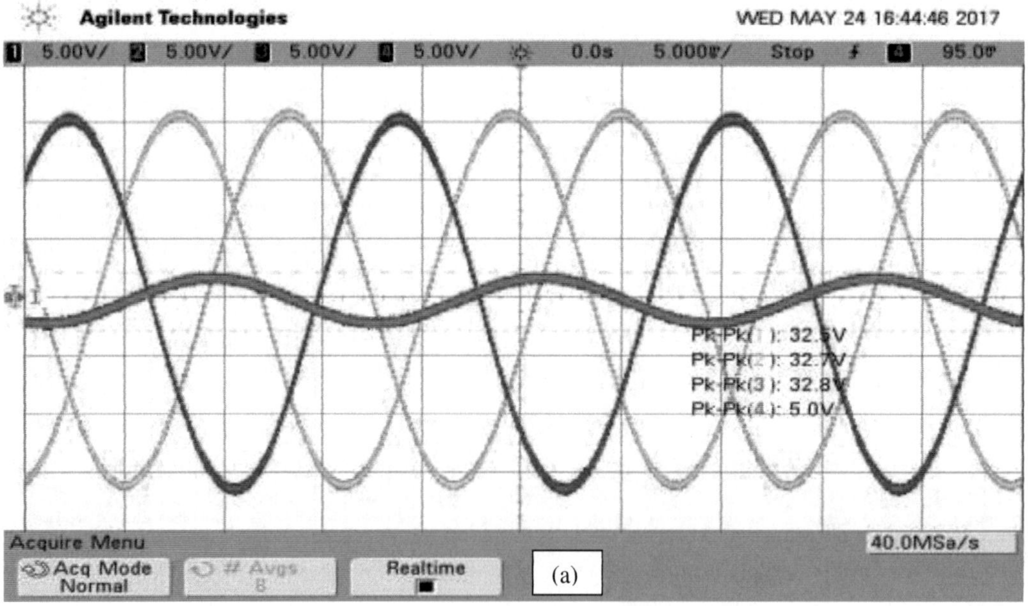

IOP Conf. Series: Materials Science and Engineering **331** (2018) 012028 doi:10.1088/1757-899X/331/1/012028

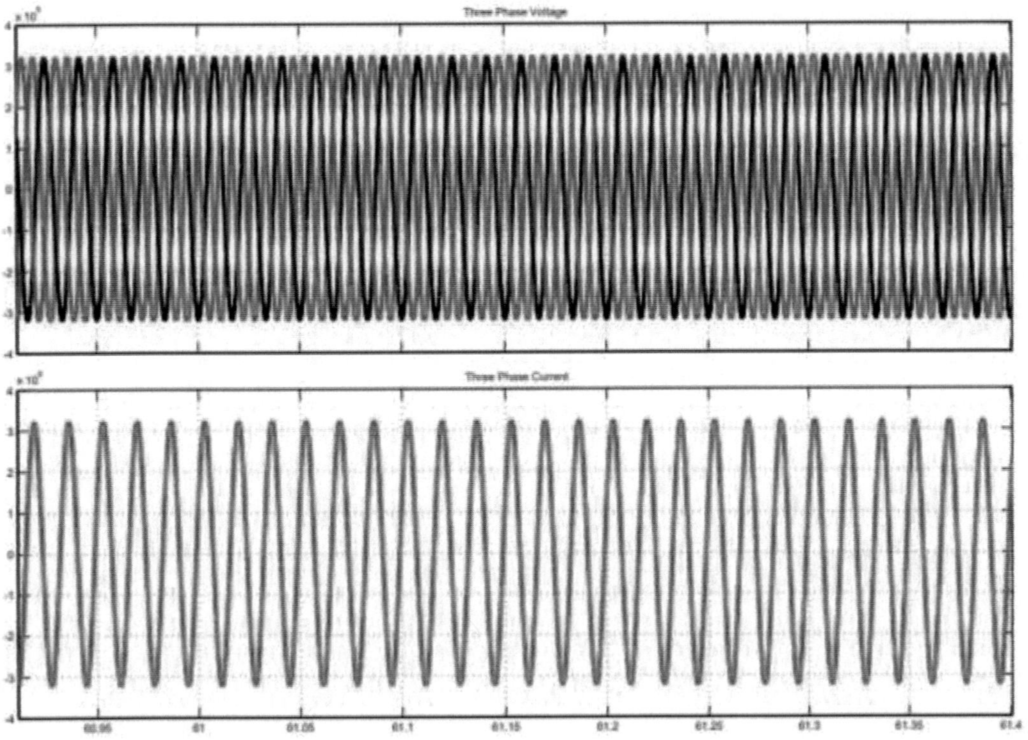

(b)

Fig. 8 Wave forms of Three Phase Voltage and Current during Healthy Condition (a) on CRO and (b) on host computer (console)

The investigator has introduced two types of faults such as (i) L-G, (ii) LLLG into the system and the performance of the IEEE-5-Bus network is analyzed. Fig. 9 shows waveforms of V, I during L-G fault on phase-a, on CRO as well as in the MATLAB environment in the console subsystem

The Fig.9 clearly indicates that during LG fault on phase a, voltage of phase a falls to nearly 300 V and current rises up to 4000 A then becomes 3000 A. The fault start time, its duration of persistence, and its interval of repetition can be varied from the console.

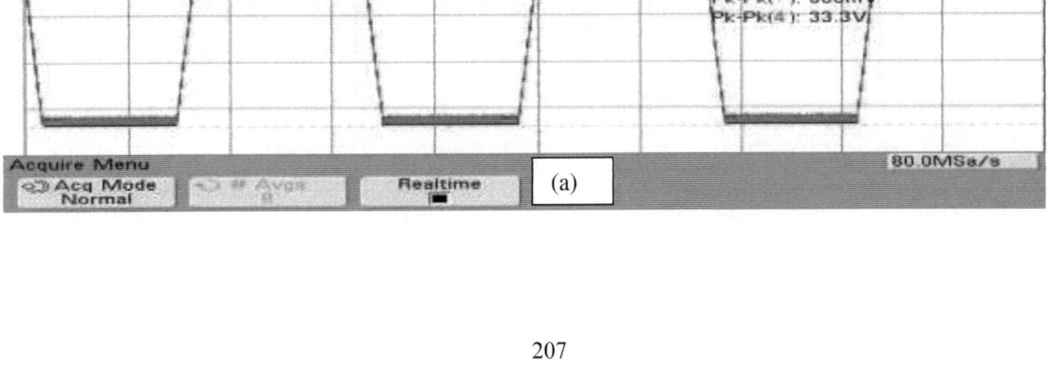

(a)

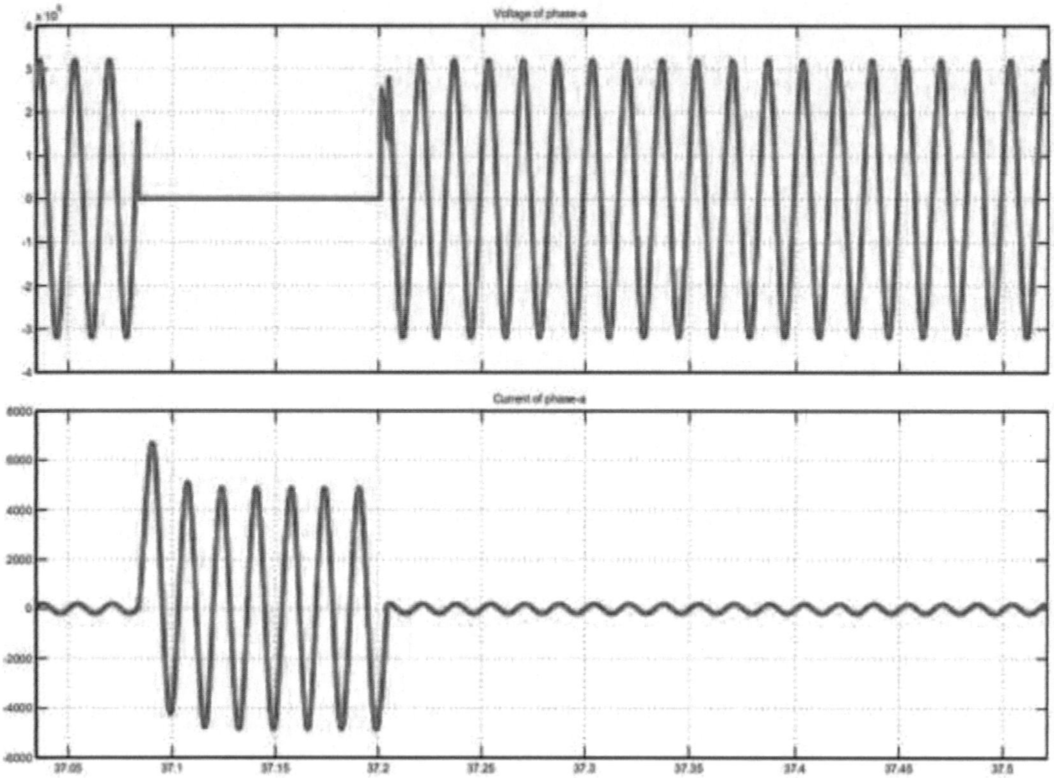

(b)

Fig. 9 Wave forms of V, I during L-G fault on phase-a (a) on CRO and (b) on host computer (console)

Fig.10 shows waveforms of three-phase voltages and three phase currents during LLLG fault on transmission line 1-5. The magnitude of three voltage reduces nearly to zero and current rises to 7000 A then reduces to 5000 A. The same waveforms can also be seen on CRO.

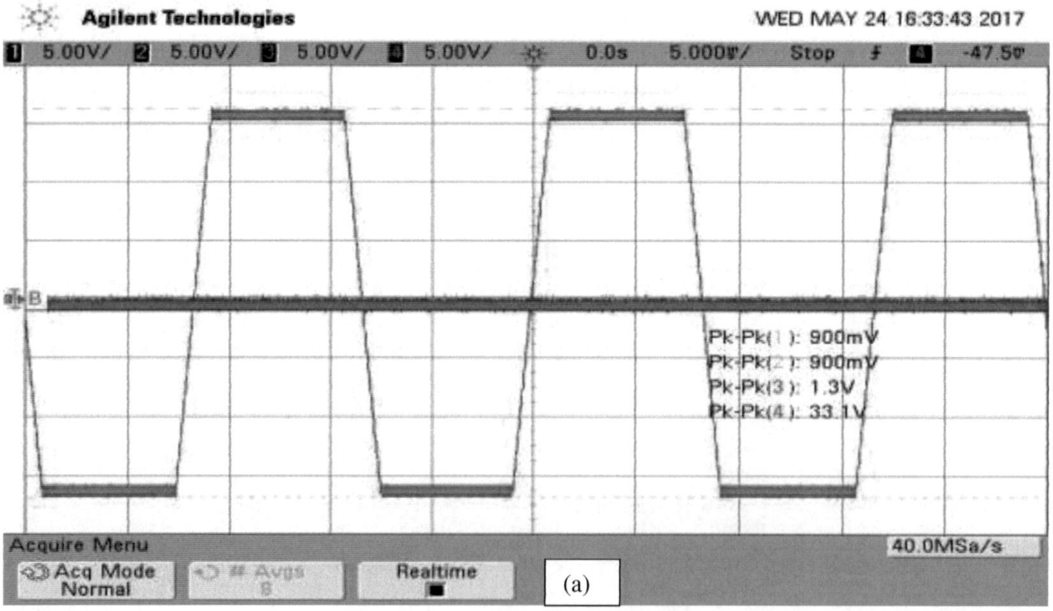

(a)

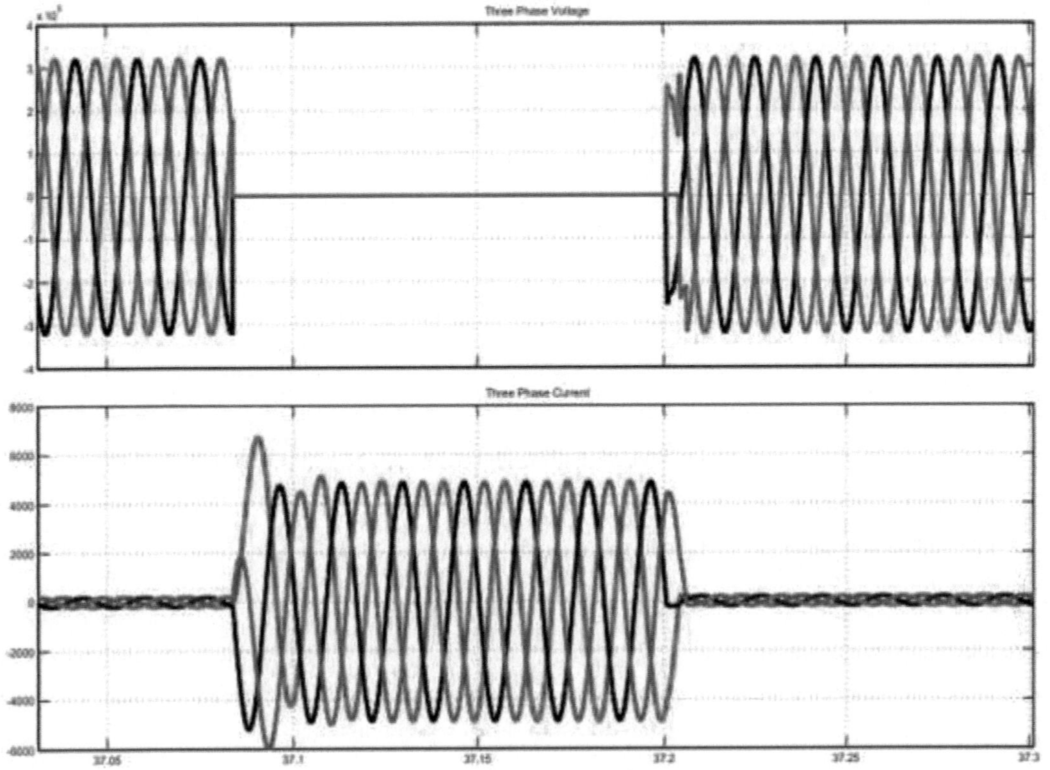

(b)

Fig. 10 Waveforms of V, I during LLLG fault on Line1-5 (a) on CRO and (b) in MATLAB environment

5. Conclusion

By developing IEEE-5-Bus network model in MATLAB/Simulink and run in OPAL-RT-OP4510 following conclusions are drawn.

i) By regrouping the model in the master subsystem and console subsystem, and inserting OPAL-RT blocks, the MATLAB/Simulink model is made compatible to OPAL-RT-OP4510. During a healthy condition of the network, the magnitude of V, I, P, Q saw on CRO, and obtained by LFA are nearly same.

ii) During healthy as well as faulty conditions, the magnitude of the signals in MATLAB environment is similar to that seen on CRO.

iii) Interactions with plant, controller and protection systems, can be easily analyzed and tested during normal and a variety of fault conditions.

iv) The main idea using real-time control is to smoothen the transition from the non-real analysis and simulation to the real-time experiments and implementation. It has applications in large power system network, power electronics, automotive and aerospace. Thus OPAL-RT helps projects to move from imagination to real-time.

v) The future scope is the inclusion of FACTS devices in the network for improvement voltage profiles, and network stability, and load flow analysis of the system with it.

ACKNOWLEDGMENT

The authors would like to thank National Institute of Technical Teachers Training and Research, Chandigarh, India for providing the Real-Time-Simulator, OPAL-RT-OP4510 to carry out this work. The authors would like to express special gratitude and thanks to officials from OPAL-RT.

References:

[1] P. M. Menghal and A. J. Laxmi, "Real-time simulation: Recent progress & challenges," *2012* International Conference on Power, Signals, Controls and Computation, *Thrissur, Kerala, 2012,* pp. 1-6.

[2] OPAL-RT-OP4510 User Manual and RT-LAB Block Library Reference Guide.

[3] M. D. Omar Faruque *et al.*, "Real-Time Simulation Technologies for Power Systems Design, Testing, and Analysis," in *IEEE Power and Energy Technology Systems Journal*, vol. 2, no. 2, pp. 63-73, June 2015.

[4] J. J. Grainger, W. D. Stevenson, "Power System Analysis", McGraw-Hill, International Edition,1994.

[5] R. Anand and V. Balaji, "Power Flow Analysis of Simulink IEEE 57 Bus Test System Model using PSAT", Indian Journal of Science and Technology, Vol 8(23), September 2015

[6] E. Acha, C. R. F.Esquivel, H. A. Perez, C.A. Comacho, "FACTS: Modelling and Simulation in Power network", First. edition, England: John Wiley & Sons, 2004.

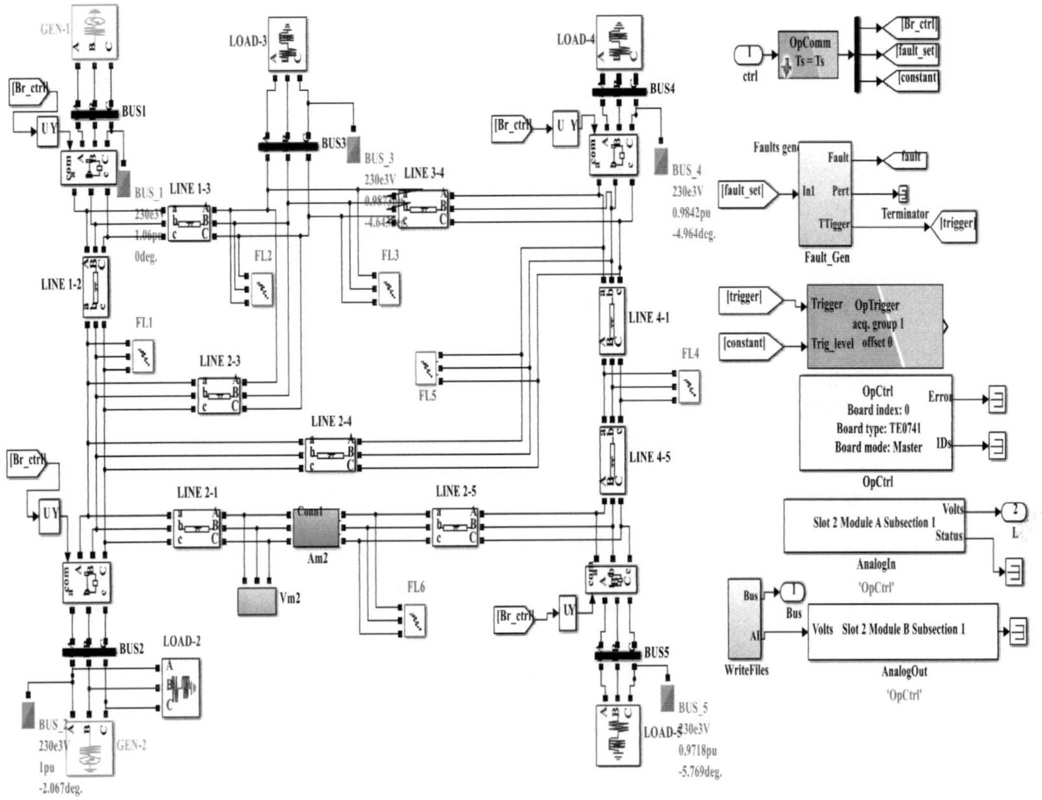
GEN-1
LOAD-3
LOAD-4
BUS1
BUS3
BUS4
BUS_1
BUS_3
BUS_4
230e3V
0.9842pu
-4.964deg.
LINE 1-3
LINE 3-4
LINE 1-2
FL1
FL2
FL3
FL4
FL5
FL6
LINE 2-3
LINE 2-4
LINE 4-1
LINE 4-5
LINE 2-1
LINE 2-5
Am2
Vm2
BUS2
BUS5
LOAD-2
LOAD-5
GEN-2
BUS_2
230e3V
1pu
-2.067deg.
BUS_5
230e3V
0.9718pu
-5.769deg.
ctrl
OpComm
Ts = Ts
[Br_ctrl]
[fault_set]
[constant]
Faults gen
Fault
Pert
TTigger
In1
Fault_Gen
Terminator
[trigger]
[fault]
Trigger
OpTrigger
acq. group 1
Trig_level
offset 0
OpCtrl
Board index: 0
Board type: TE0741
Board mode: Master
Error
IDs
Slot 2 Module A Subsection 1
Volts
Status
AnalogIn
'OpCtrl'
Bus
WriteFiles
Volts Slot 2 Module B Subsection 1
AnalogOut
'OpCtrl'

IOP Conf. Series: Materials Science and Engineering **331** (2018) 012029 doi:10.1088/1757-899X/331/1/012029

Rectangular microstrip antenna with corrugation like defects at radiating edge: A new approach to reduce cross polarization radiation

U A Pawar[1], D Mondal[2], A Nagaraju[3], S Chakraborty[3], L L K Singh[4], S Chattopadhyay[4]

[1]Indian Army Unit, Hisar, India
[2] Radionics Lab, Physics Department, Burdwan University, India
[3] Device Technology Group, CSIR-Central Electronics Engineering Research Institute, Pilani, India
[4] Depart of ECE, Mizoram University, Aizawl, India.

E-mail: piyalirekha@yahoo.com, s.c.in@ieee.org

Abstract. In this paper, single layer, simple and compact RMA, with corrugation like defects at the radiating edge, is studied thoroughly to reduce XP radiation from the patch. Unlike the earlier works reported on defected ground structure integrated patches and defect patch structures , in this work, corrugation like linear defects have been placed at the radiating edges of the patch to reduce cross polarisation radiation. Around 30-40 dB of CP-XP isolation is observed in H-plane with 7% impedance bandwidth and in E-plane also, more than 55 dB CP-XP isolation is found. The proposed structure is very simple to design and easy to fabricate.

1. Introduction

Rectangular microstrip antenna (RMA) is probably the most useful antenna structure for its variety of applications due to its simple design, ease of implementation and low profile. Apart from these desirable properties, a simple RMA suffers from some severe disadvantages such as low gain (around 2-3 dBi) and narrow impedance bandwidth (around 2-3%) when fabricated on a substrate which possesses high dielectric constant. Moreover, a RMA radiates linearly polarized wave along its broadside direction called, co polarized (CP) radiation along with a few degrees of cross polarized (XP) radiation due to spurious and higher order modes [1]. Hence, such antennas suffer from low polarization purity i.e. CP-XP isolation (around 8-10 dB) and improvement of CP-XP isolation is a challenging issue to enhance the radiation characteristics of a RMA without affecting the CP radiation. The XP radiations are much significant in the H plane than in the E plane [2], [3].It is also worthy to mention that polarization purity gets worse at higher frequency as the XP radiation increases with frequency [1], [4].

In [5], a Ψ shaped RMA on two-layer substrate was reported and only 3-10 dB peak CP-XP isolation is achieved over 4-8 GHz. In [6], an E-shaped RMA fabricated on glass epoxy substrate with an finite air gap over the ground plane was reported, and around 191 MHz bandwidth and 18 dB peak CP-XP isolation were found. In [7], a comparative study on the bandwidth and XP characteristics of C shaped, U shaped, E shaped RMAs was discussed.

IOP Conf. Series: Materials Science and Engineering **331** (2018) 012029 doi:10.1088/1757-899X/331/1/012029

Defected ground structure (DGS) is a well-known technique for RMA to improve CP-XP isolation [8]-[11]. It is observed that, the polarization purity of 20-25 dB is achieved from slot DGS [8], bracketed DGS [9] and circular headed dumbbell DGS [10], non-proximal [11] and asymmetric [12] DGS integrated RMA. However, any DGS integrated antenna, in rectangular patch geometry enhances the back radiation from the antenna structure which puts a severe limitation in its use.
Compared to these, defected patch structure (DPS) is a newer technique to suppress XP radiation [13], [14] without enhancing back radiation level from a RMA. It was firstly introduced by Ghosh *et al* for

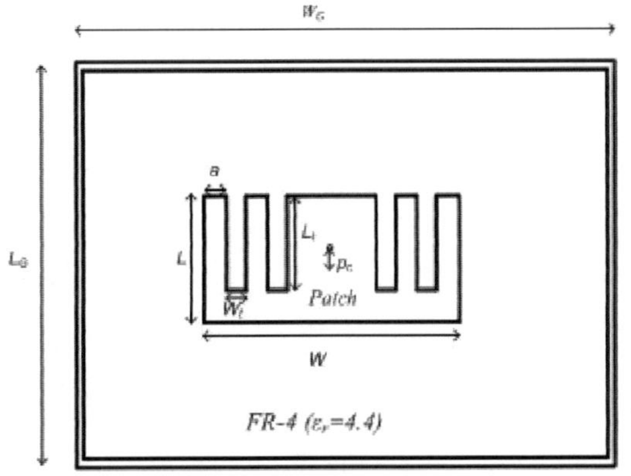

Figure 1. Schematic representation of RMA with corrugation like defects at radiating edge (top view).

circular patch antenna in [13]. Around 25-30 dB of CP-XP ratio is revealed from [13], [14]. In both these work, the defects have been judiciously placed at the non-radiating edges of the patch so that, the electric field distributions of spurious higher order modes particularly that of TM_{02} can be nullified with hampering the electric field distribution of dominant TM_{10} mode.
In this paper, single layer, simple and compact RMA, with corrugation like defects at the radiating edge, is studied thoroughly to reduce XP radiation from the patch (Fig. 1). Unlike the earlier works on DGS and DPS integrated RMAs, the defects have been judiciously placed at the radiating edge of the RMA. Earlier, different research groups reported application of corrugated circular patch [15] for miniaturization and corrugated ground plane [16] for improving CP-XP isolation at φ=45^0 plane and to reduce side lobe levels. Around 18 dB CP-XP isolation was achieved at φ=45^0 plane in [16]. The structures reported in [15], [16] were very much complex to fabricate and XP radiation minimization at principal E and H-planes were not explored in these work.
On the contrary, from the present work around 40dB, 30 dB CP-XP isolation have been revealed at principal E and H plane, respectively, over whole elevation angle (±180^0), whereas, for a RMA the same is around 8-10 dB in the H-plane. The structure resonates around f=7.72 GHz with 7% impedance bandwidth when defect length L_t=4 mm. However, it has also been observed that with a slight increase in the defect length (L_t=5 mm), the structure exhibits dual resonance at f=6.61 GHz and f=7.72 GHz probably due to a weak resonance of TM_{10} mode at the lower side of frequency spectrum. In the latter case, CP-XP isolation are around 11 dB, 30 dB in the principal H plane at f=6.61 GHz and f=7.72 GHz, respectively. The antenna has been designed on FR-4 substrate (ε_r= 4.4) of dimensions 25×25 mm^2. The peak CP gain from the proposed structure is around 5 dBi. The proposed structure is very simple to design and easy to fabricate.

IOP Conf. Series: Materials Science and Engineering **331** (2018) 012029 doi:10.1088/1757-899X/331/1/012029

2. Proposed structures

A thin copper strip of length (L) =8 mm and width (W) = 12 mm, is utilized as patch and it is fabricated on FR-4 substrate with height (h)= 1.575 mm and ε_r= 4.4 of dimensions 25×25 mm^2 (Fig. 1). Here we have studied two antenna structures. In antenna 1, four defects of dimensions (L_t×W_t) = 4×1 mm^2 have been cut on the patch surface at the radiating edge. In antenna 2, the length of each defect has been increased to L_t=5mm but, the width of the slot W_t is kept unaltered. The ground plane dimensions are 25×25 mm^2. The antenna 1 and antenna 2 are fed at p_c=1.4 and 1.6 mm from the center of the patch, respectively, through a co-axial probe. Ground plane dimensions are (L_G×W_G) = 25×25 mm^2.

3. Results and discussions

In this section, the simulated results [17] of a conventional RMA and RMAs with corrugated edge with slot dimensions (L_t×W_t) = 4×1 mm^2 and 5×1 mm^2 have been presented.

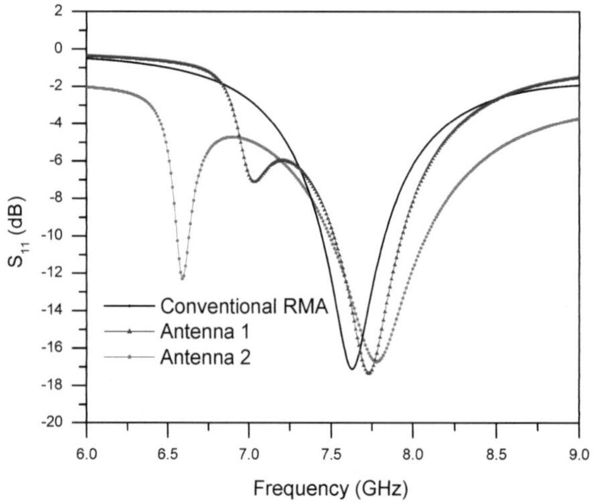

Figure 2. Reflection coefficient profile (S11) for conventional RMA, antenna 1 and antenna 2.

3.1. *Conventional RMA*

Figure. 2 and fig. 3 shows the simulated reflection coefficient (S_{11}) and radiation patterns for a conventional RMA. The conventional RMA resonates at f=7.61 GHz with 4% impedance bandwidth (fig. 2). It can be noted that for the conventional patch, peak CP-XP isolation is nearly 12 dB at around ±50^0 in H-plane whereas, in E plane, CP-XP isolation is around 40 dB (Fig. 3).

3.2. *Antenna 1: RMA with corrugated radiating edge with slot dimensions (L_t×W_t)= 4×1 mm^2*

The antenna 1 resonates at f=7.72 GHz with 4.2% impedance bandwidth (fig. 2). It can be noted that for antenna 1, peak CP-XP isolation isnearly 30 dB at around ±50^0 in H-plane whereas, in E plane, CP-XP isolation is around 50 dB (Fig. 3). it is evident that, 30-40 dB of XP suppression is achieved over whole elevation angle (+180^0 to -180^0) than that of a conventional RMA (fig. 3).

3.3. Antenna 2: RMA with corrugated radiating edge with slot dimensions $(L_t \times W_t) = 5 \times 1\ mm^2$

The antenna 2 exhibits dual resonances at *f*=6.61 GHz and *f*=7.72 GHz with 6% impedance bandwidth (fig. 2). It can be noted that for antenna 2, at *f*=6.61 GHz, peak CP-XP isolation is nearly 12 dB at around $\pm 50^0$ in H-plane whereas, in E plane, CP-XP isolation is around 55 dB (fig. 4). At *f*=7.72 GHz, peak CP-XP isolation isnearly 30 dB at around $\pm 50^0$ in H-plane whereas, in E plane, CP-XP isolation is around 55 dB (Fig.4). At, *f*=7.72 GHz, 30-40 dB of XP suppression is achieved

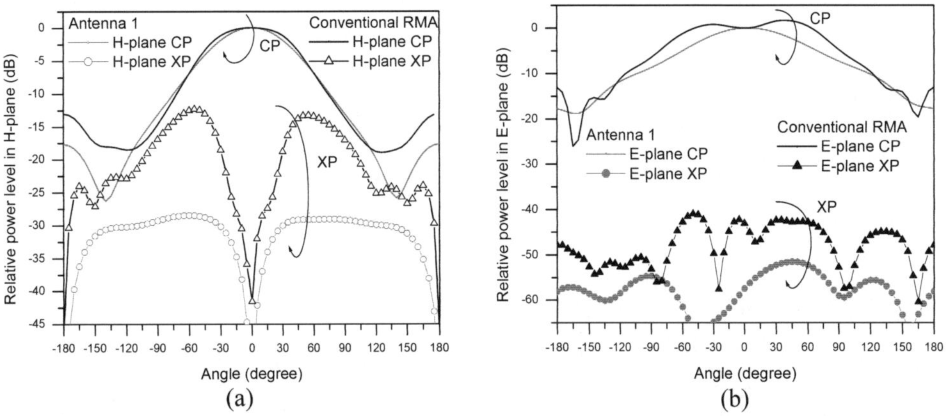

Figure 3. Radiation patterns of conventional RMA and antenna 1 at the centre frequency at: (a) H-plane, (b) E-plane.

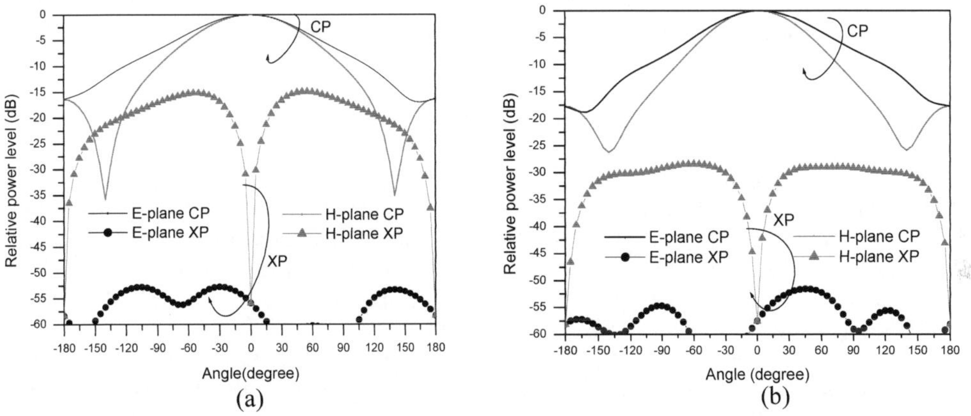

Figure 4. Radiation patterns of antenna 2 at: (a) *f*=6.61 GHz and (b) *f*=7.72 GHz.

over whole elevation angle ($+180^0$ to -180^0) than that of a conventional RMA. It is also observed that the CP radiation patterns are nearly same at *f*=6.61 GHz and *f*=7.72 GHz which indicate excitation of dominant modes at both the frequencies.

4. Conclusions

In this work, single layer, simple and compact RMA, with corrugation like defects at the radiating edge, is studied thoroughly to achieve improved polarization purity over whole elevation angle. To the best of our knowledge, this work happens to be the first report where linear defects have been placed at the radiating edges of the patch to reduce cross polarisation radiation. Around 30-40 dB of CP-XP isolation is observed in H plane with 7% impedance bandwidth in case of antenna 1 and in E plane also, more than 55 dB CP-XP isolation is found. Also, in case of antenna 2, dual resonances have been found due to an increase in defect length. The proposed structure is simple and easy to fabricate. It will surely find applications in array antenna design and establishing wireless links where, improved polarization purity is required over wide angular range.

5. References

[1] Lee K F, Luk K M and Tam P Y 1992 Cross-polarization characteristics of circular patch antennas *Electron. Lett.* **28** (6) 587–89

[2] Garg R, Bhartia P, Bahl I and Ittipiboon A 2001 Microstrip Antenna Design Handbook *Artech House*

[3] Poddar R, Chakraborty S and Chattopadhyay S 2015 Improved Cross Polarization and Broad Impedance Bandwidth from Simple Single Element Shorted Rectangular Microstrip Patch: Theory and Experiment Rectangular microstrip antenna on slot type defected ground for reduced cross polarized radiation *Frequenz.* **70** (1-2) 1-9

[4] Li P, Lai H W, Luk K M and Lau K L 2004 A wideband patch antenna with cross-polarization suppression *IEEE Antennas and Wireless Propagation Letters.* **3** 211-14

[5] Sharma S K and Shafai L 2009 Performance of a Novel Ψ-Shape Microstrip Patch Antenna With Wide Bandwidth *IEEE Antennas and Wireless Propagation Letters.* **8** 468-71

[6] Deshmukh A A and Kumar G 2005 Compact broadband E-shaped microstrip antennas *Electronic Letters*, **41**

[7] Bhardwaj S and Samii Y R 2012 C-shaped, E-shaped and U-slotted Patch Antennas: Size, Bandwidth and Cross-Polarization Characterizations *Proc. 6th European Conference on Antennas and Propagation (EUCAP)*

[8] Ghosh A, Ghosh D, Chattopadhyay S and Singh L L K 2015 Rectangular microstrip antenna on slot type defected ground for reduced cross polarized radiation *IEEE Antennas Wireless Propag Lett.* **14** 321–24

[9] Kumar C and Guha D 2014 DGS integrated Rectangular microstrip patch for improved polarization purity with wide impedance bandwidth *IET Micro. Ant. & Prop.,* **8** (8) 589-96

[10] Chakraborty S and Chattopadhyay S 2016 Substrate Fields Modulation with Defected Ground Structure: A Key to Realize High Gain, Wideband Microstrip Antenna with Improved Polarization Purity in Principal and Diagonal Planes *Int. Journal of RF, Microwave and Computer Aided Engineering (Wiley).* **26** (2) 174-81

[11] Kumar C, Pasha I and Guha D 2015 Microstrip Patch with Non Proximal Symmetric Defected Ground Structure for Improved Cross Polarization Properties Over Principal Radiation Planes *IEEE Antennas Wireless Propag Lett.* **14** 1412–14

[12] Kumar C and Guha D 2016 Asymmetric Geometry of Defected Ground Structure for Rectangular Miocrostrip: A New Approach to Reduce its Cross Polarized Fields *IEEE Trans Antennas Propag.* **64** 2503–06

[13] Ghosh A, Ghosh S K, Ghosh D and Chattopadhyay S 2014 Improved polarization purity for circular microstrip antenna with defected patch surface *International Journal of Microwave and Wireless Technologies.* **8** 89-94

[14] Chakraborty S, Ghosh A, Chattopadhyay S and Singh L L K 2015 Improved cross polarized radiation and wide impedance bandwidth from rectangular microstrip antenna with Dumbbell Shaped Defected Patch Surface *IEEE Antennas and Wireless Propagation Letters.* **15** 84-88

[15] Lee S, Woo J, Ryu M and Shin H 2005 Corrugated circular microstrip patch antennas for miniaturization *Electronics Letters.* **38**

[16] Ying Z and Kidal P S 1996 Improvements of dipole, helix, spiral, microstrip patch and aperture

antennas with ground planes by using corrugated soft surfaces *IEE Proceedings - Microwaves Antennas and Propagation.* **143** 244-48

[17] HFSS, High Frequency Structure Simulator, USA

Acknowledgement

Authors would like to thank Prof. Debatosh Guha of Calcutta University, India for his support.

A. Nagaraju and S. Chakraborty would like to thank Director, CSIR-CEERI, Pilani and Dr. S. K. Ghosh, Principal Scientist, CSIR-CEERI, Pilani for always supporting research endeavors.

All of the co-authors convey a deep sense of gratitude and thanks to the corresponding institutes for their kind supports and encouragement.

IOP Conf. Series: Materials Science and Engineering **331** (2018) 012030 doi:10.1088/1757-899X/331/1/012030

RF MEMS Based Tunable Bandpass Filter For X-Band Applications

Mahesh Kumar Chaubey[1] and Avanish Bhadauria[2]
CSIR-Central Electronics Engineering Research Institute, Pilani -333031, India
[2]Email: avanish@ceeri.ernet.in

Abstract. In this paper, we present the design and simulation of RF MEMS based Tunable combline band pass filters for X-band applications at different substrate thicknesses and studied the effect of thickness on tuning. The proposed filters are designed on high resistive silicon substrate of 500μm and 300 thicknesses. The tunability is achieved by using MEMS based varacter replaced with fix capacitor in conventional combline filter. First, the microstrip combline filter is designed at the centre frequency of 9.5 GHz and then tuning is achieved by varying the capacitance in the designed combline filters. The electromagnetic simulation has been carried out using HFSS v15 software based on finite element method (FEM). The tuning of the filter on silicon substrate of 500 μm is achieved by changing the capacitance value from 0.2035 pF to 0.4035 pF in the model in HFSS, which resulted the tuning in the frequency range of 7.85 to 10.35GHz. Insertion loss of design filter is in the range of 1dB within the tuning range. In case of substrate thickness 300 μm the tuning of the filter is achieved by changing the capacitance value from 0.293 pF to 0.403 pF in the model in HFSS, which resulted the tuning in the frequency range of 8.80 to 9.90 GHz. Insertion loss of design filter is in the range of ~1.2dB within the tuning range.

1. INTRODUCTION

Tunable or reconfigurable RF filters make microwave systems adaptable to multiple frequency bands of operation using a single filter, which has tremendous applications in modern communications systems like satellite communication, radar, medical electronics and software defined radio and cognitive radios. Tunable filters can meet the requirement of several filters to have more than one filter response by introducing tuning elements introducing into a filter topology [1]. These tunable filters can be divided into two groups (i) filters with discrete tuning and (ii) filters with continuous tuning. Filter topologies for a discrete tuning can be realized by using PIN diodes or MEMS switches. While the continuous tuning can be achieved by using varactor diodes, MEMS capacitors, ferroelectric materials or ferromagnetic materials are frequently used to obtain a continuous tuning device. Center frequency is the most common filter parameter to reconfigure. Fewer designs reconfigure other parameters, such as the bandwidth or selectivity. When deciding which technology is adequate for a given application, the designer must consider the following issues: cost, power consumption, size, performance and operating frequency.

This paper present and discusses the use of Micro Electro Mechanical Systems (MEMS) varactors as tuning elements on combline filter topology for x-band applications [2,3]. X-band tunable band pass

IOP Conf. Series: Materials Science and Engineering **331** (2018) 012030 doi:10.1088/1757-899X/331/1/012030

filter is widely used by the military applications for beyond line of sight communication. These systems are less prone to interference, rain resilience, and terminal size. X-band has several applications in various radars with continuous wave, pulsed wave, single polarization, dual-polarization, synthetic aperture radar, and phased arrays.

The proposed filters are designed on high resistive silicon substrate of 500μm and 300μm thicknesses. First, the microstrip combline filters are designed at the center frequency of 9.5 GHz and then tuning is achieved by varying the capacitance in the designed combline filter. To study the performance of the filters the simulations are carried out using HFSS v15 software based on finite element method (FEM). For the filter designed on substrate thickness 500μm, the tuning of the filter is achieved by changing the capacitance value from 0.2035 pF to 0.4035 pF, resulted the tuning in the frequency range of 7.85 to 10.35GHz with the insertion loss in the range of 1dB within the tuning range. Similarly, the filter designed on substrate thickness 300μm, the tuning of the filter is achieved by changing the capacitance value from 0.2930 pF to 0.4030 pF, resulted the tuning in the frequency range of 8.80 to 9.90 GHz with the insertion loss in the range of ~1.2 dB within the tuning range. The First to realize tunable filter the combline filter topology is used to realize the and designed at center frequency of 9.5 GHz. The tuning is achieved by varying the capacitances in the designed combline filter.

The typical combline filter is shown in Figure 1. The microstrip resonators in this type of filter based on quasi transverse electromagnetic (TEM)-mode transmission lines that are shorted at ground through via holes at one end and terminated to the ground by a lumped capacitor at the another end of each resonator line [4]. Coupling between microstrip resonator lines is achieved in by fringing fields between the resonator lines. The resonator lines have length, $\lambda_o/8$ long at the designed center frequency (f_o). The combline filters topology has been chosen due to following attractive features:

(a) They are compact.
(b) They have strong stop-bands, and the stop-bands above the primary pass-band can be made to be strong.
(c) If desired, they can be designed to have an unusually steep rate of cut-off on the high side of the passband.
*(d) A*dequate coupling can be maintained between resonator elements with sizeable spacing between resonator lines. (This feature means that the proper coupling can be maintained in manufactured filters without unreasonable production tolerance requirements.)
(e) Filters of this type can usually be fabricated without dielectric support materials so that, if desired, dielectric losses can be eliminated.

A five pole comb-line filter topology has been chosen for our study and schematics have been shown in figure 1(a) and (b). The input is feed through the resonator line of length L_1 and output is taken from resonator line length L_7. Filters are designed on a 500 μm and 300μm thick silicon substrate with thick high resistivity.

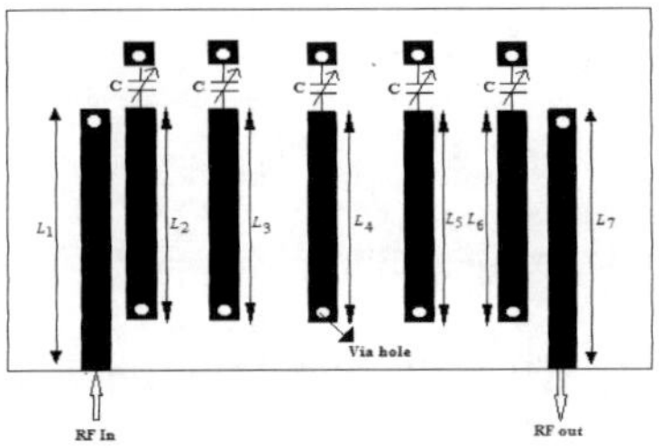

(a)

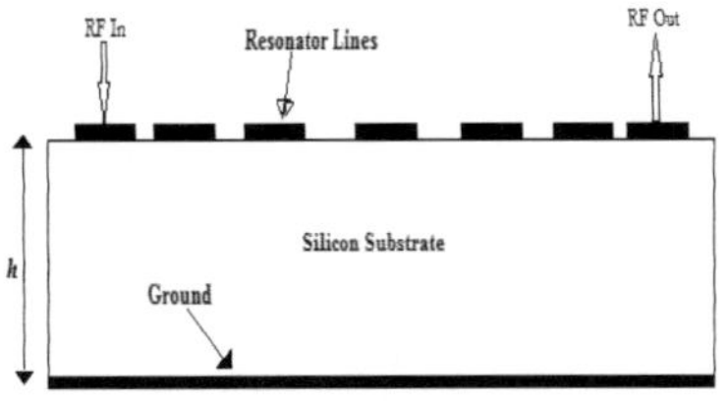

(b)

Figure 1. Combline BPF (a) Top view (b) Side view

The typical MEMS varactor is shown in Figure 2. The capacitance is controlled by varying the gap between the two parallel plates of the capacitor using electrostatic actuation, Where one plate is fixed and another kept suspended for movement under electrostatic actuation due to force of attraction be applied electric field [5]. When a dc potential is applied to between these two electrodes, due to electrostatic actuation, the gap between electrodes reduces; resulting in an increase in the parallel plate capacitance. Thus, by changing the applied potential, the structure can be made to operate as a variable capacitor [6].

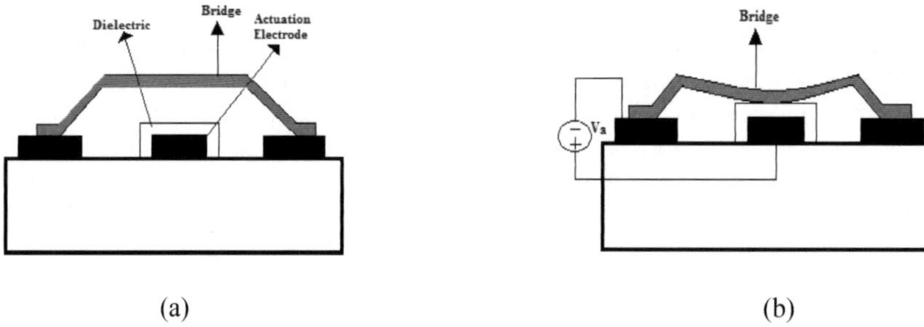

Figure 2. Schematic of MEMS varactor (a) without bias (b) with bias

2. DESIGN AND SIMULATION

(i) Design Methodology

The combline bandpass filter is comprised of an array of coupled resonators as shown in as shown in Figure 3. The resonators consist of line elements 1 to n, which are short-circuited at one end, with a lumped capacitance C_{Li} loaded between the other end of each resonator line element and ground. The input and output of the filter are through coupled-line elements 0 and n + 1, which are not resonators. With the lumped capacitors present, the resonator lines will be less than $\lambda_{go}/4$ long at resonance, where λ_{go} is the guided wavelength in the medium of propagation at the midband frequency of filter [7]. It is interesting that if the capacitors were not present, the resonator lines would be a full $\lambda_{go}/4$ long at resonance, and the filter structure in Figure 1, would have no passband at all when the line elements are constructed from a pure TEM-mode transmission line such as stripline [8]. This is because the magnetic and electric couplings totally cancel each other out in this case. The larger the loading capacitances C_{Li}, the shorter the resonator lines, which results in a more compact filter structure with a wider stopband between the first passband (desired) and the second passband (unwanted). The lumped capacitors may offer a convenient means for filter tuning, which may be required particularly for narrow-band filters.

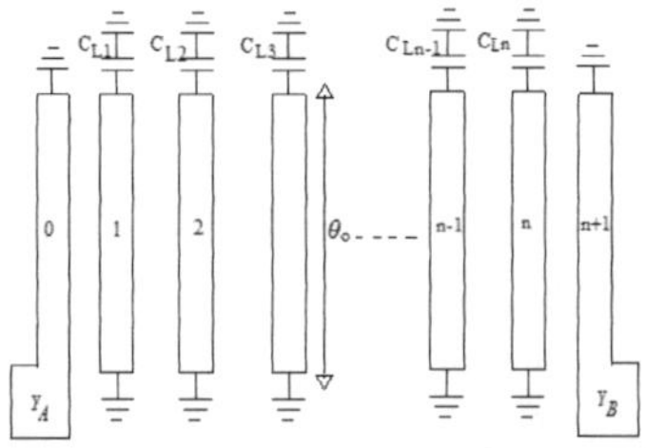

Figure 3. Schematic of n-pole combline filter

Filter can be designed from a chosen ladder type of lowpass prototype with the lowpass parameters g_i for i = 0 to n + 1. A design procedure as described in [9] starts with choosing the resonator susceptance slope parameters b_i

$$\frac{b_i}{Y_A} = \frac{Y_{ai}}{Y_A}\left(\frac{\cot\theta_o+\theta_o\csc^2\theta_o}{2}\right) \qquad \text{for i = 0 to n} \qquad (1)$$

where Y_A is the terminating line admittance, θ_o is the midband electrical length of the resonators, and Y_{ai} is interpreted physically as the admittance of line with the adjacent lines i – 1 and i + 1 grounded. The choice of Y_{ai} fixes the admittance level within the filter and can influence the unloaded quality factors of the resonators. For a specified fractional bandwidth *FBW*, calculate

$$\frac{J_{0,1}}{Y_A} = \sqrt{\frac{FBW\frac{b_1}{Y_A}}{g_o g_1}}, \qquad \frac{J_{n,n+1}}{Y_A} = \sqrt{\frac{FBW\frac{b_n}{Y_A}}{g_n g_{n+1}}} \tag{2}$$

$$\frac{J_{i,i+1}}{Y_A} = FBW\sqrt{\frac{\left(\frac{b_i}{Y_A}\right)\left(\frac{b_{i+1}}{Y_A}\right)}{g_i g_{i+1}}} \qquad \text{for i = 0 to n-1} \tag{3}$$

where, the lowpass element values g_i are given for a normalized cut-off frequency Ω_c=*1*. The design procedure leads to the determination of the lumped capacitances as well as the self- and mutual capacitances per unit length of the distributed line elements.

The lumped capacitances C_{Li} are:

$$C_{Li} = Y_A\left(\frac{Y_{ai}}{Y_A}\right)\frac{cot\theta_o}{\omega_o} \qquad \text{or i = 0 to n} \tag{4}$$

where, ω_o is the angular frequency at the midband.

The self-capacitances Ci are given by

$$\frac{C_o}{\varepsilon_o\sqrt{\varepsilon_{re}}} = \eta_o Y_A\left(1-\frac{J_{0,1}}{Y_A}\right), \qquad \frac{C_{n+1}}{\varepsilon_o\sqrt{\varepsilon_{re}}} = \eta_o Y_A\left(1-\frac{J_{n,n+1}}{Y_A}\right) \tag{5}$$

$$\frac{C_o}{\varepsilon_o\sqrt{\varepsilon_{re}}} = \eta_o Y_A\left(\frac{Y_{a1}}{Y_A}-1+\left(\frac{J_{0,1}}{Y_A}\right)^2-\frac{J_{0,1}}{Y_A}\tan\theta_o\right)+\frac{C_o}{\varepsilon_o\sqrt{\varepsilon_{re}}} \tag{6}$$

$$\frac{C_n}{\varepsilon_o\sqrt{\varepsilon_{re}}} = \eta_o Y_A\left(\frac{Y_{an}}{Y_A}-1+\left(\frac{J_{n,n+1}}{Y_A}\right)^2-\frac{J_{n-1,n}}{Y_A}\tan\theta_o\right)+\frac{C_{n+1}}{\varepsilon_o\sqrt{\varepsilon_{re}}} \tag{7}$$

$$\frac{C_i}{\varepsilon_o\sqrt{\varepsilon_{re}}} = \eta_o Y_A\left(\frac{Y_{ai}}{Y_A}-1+\frac{J_{i-1,i}}{Y_A}\tan\theta_o-\frac{J_{i,i+1}}{Y_A}\tan\theta_o\right)+\frac{C_{n+1}}{\varepsilon_o\sqrt{\varepsilon_{re}}} \qquad \text{for i = 2 to n-2} \tag{8}$$

And the mutual capacitances $C_{i.i+1}$ are

$$\frac{C_{0,1}}{\varepsilon_o\sqrt{\varepsilon_{re}}} = \eta_o Y_A - \frac{C_o}{\varepsilon_o\sqrt{\varepsilon_{re}}}, \qquad \frac{C_{n,n+1}}{\varepsilon_o\sqrt{\varepsilon_{re}}} = \eta_o Y_A - \frac{C_{n+1}}{\varepsilon_o\sqrt{\varepsilon_{re}}} \tag{9}$$

$$\frac{C_{0,1}}{\varepsilon_o\sqrt{\varepsilon_{re}}} = \eta_o Y_A\frac{J_{i,i+1}}{Y_A}\tan\theta_o \tag{10}$$

Note that in (5 to 8) and (9 to 10) the self- and mutual capacitances are normalized with the free space permittivity ε_o and scaled by $\sqrt{\varepsilon_{re}}$ with ε_{re} denoting the relative effective dielectric constant constant of the line element and η_o = 120π ohm the wave impedance in the free space.

Using the above steps, we have designed the 5 pole microstrip combline filter at the center frequency of 9.5GHz, using high resistivity silicon substrates of 500μm and 300μm thickness. We have used

Chebyshev low pass prototype element having pass band ripple of 0.1dB. the low pass prototype parameter are:

Table1. Low pass prototype parameter

g_0	g_1	g_2	g_3	g_4	g_5	g_6
1.0	*1.1468*	*1.3712*	*1.9750*	*1.3712*	*1.1468*	*1.0*

Using above parameters, the microstrip resonators are calculated on different substrates and further optimized which are shown given in following tables.

Table 2. Design Parameter of Combline Filter on silicon substrates of 500μm thickness

Initial design Parameter			*Optimized Parameter*		
Length (mm)	*Width (mm)*	*Spacing (mm)*	*Length (mm)*	*Width (mm)*	*Spacing (mm)*
$L_1 = 2.8273$	W_1=0.39492	S_o= 0.0523	$L_1 = 2.58938$	W_1=0.39492	S_o= 0.055
$L_2 = 1.4186$	W_2=0.39492	$S_{2,3}$ = 0.40494	$L_2 = 1.27969$	W_2=0.39492	$S_{2,3}$ = 0.53
$L_3 = 1.4186$	W_3=0.39492	$S_{3,4}$= 0.53503	$L_3 = 1.34969$	W_3=0.39492	$S_{3,4}$= 0.73
$L_4 = 1.4186$	W_4=0.39492		$L_4 = 1.34969$	W_4=0.39492	
$L_5 = 1.4186$	W_5=0.39492		$L_5 = 1.34969$	W_5=0.39492	
$L_6 = 1.4186$	W_6=0.39492		$L_6 = 1.27969$	W_6=0.39492	
$L_7 = 2.8273$	W_7 =0.39492		$L_7 = 2.58938$	W_7 =0.39492	

Table 3. Design Parameter of Combline Filter on silicon substrates of 300μm thickness

Initial design Parameter			*Optimized Parameter*		
Length (mm)	*Width (mm)*	*Spacing (mm)*	*Length (mm)*	*Width (mm)*	*Spacing (mm)*
$L_1 = 2.8677$	W_1=0.395	S_o= 0.0273	$L_1 = 3.0555$	W_1=0.395	S_o= 0.0346
$L_2 = 1.434$	W_2=0.395	$S_{2,3}$ = 0.233	$L_2 = 1.4456$	W_2=0.395	$S_{2,3}$ = 0.412
$L_3 = 1.434$	W_3=0.395	$S_{3,4}$= 0.3066	$L_3 = 1.5056$	W_3=0.395	$S_{3,4}$= 0.603
$L_4 = 1.434$	W_4=0.395		$L_4 = 1.5056$	W_4=0.395	
$L_5 = 1.434$	W_5=0.395		$L_5 = 1.5056$	W_5=0.395	
$L_6 = 1.434$	W_6=0.395		$L_6 = 1.4456$	W_6=0.395	
$L_7 = 2.8677$	W_7 =0.395		$L_7 = 3.0555$	W_7 =0.395	

The terminating capacitance value has been chosen uniform and the capacitance value for each resonators is C = 0.335063pF. The spacing between the input feed line and first resonator isS_o, while resonator *2-3* and *3-4* are $S_{2,3}$ and $S_{3,4}$ respectively. Similarly for 300μm Using microstrip design formula calculated value and optimized value of the filter are given in tabular form. The capacitance value the resonators are *C = 0.333pF.*

(ii) Electromagnetic Simulation

To study the transmission behaviour of proposed filters, we have carried out finite element method (FEM) based simulations on a proposed structure using commercial software ANSOFT HFSS v15. In FEM the unknown fields are discretized using a finite element mesh; typically triangular elements are for surface meshes and tetrahedrons for volumetric meshes [10]. Both these geometries are chosen as with these geometries two dimensional and three dimensional regions respectively can be meshed [11]. To design the filter we have chosen the dielectric constant of the silicon ε_r = 11.9, the aluminium metal of bulk conductivity 3×10^7 S/m is used for resonator lines and ground plane The microstrip is shorted to ground by via holes of radius of 0.1mm for 500μm and 300μm thick high resistivity silicon substrate. These via holes are filled with copper metal. To carry out simulation in HFSS, we have used driven modal solution type. The wave ports were assigned for port excitation. An air box is created to cover the device and to provide the appropriate boundary conditions. The length of input lines and output lines are larger than the remaining resonator lines. In our simulation the capacitance are connected through the lumped ports.

We have tuned the pass band of combline band pass filter by changing the capacitance values. We have also studied the group delay of the filter with different capacitance. The group delay can be defined as the rate of change of transmission phase angle with respect to frequency. Mathematically group delay can be expressed as

$$\tau_g = \frac{\Delta\varphi}{\Delta\omega} \tag{11}$$

where $\Delta\varphi$ is the change in phase of $|S_{21}|$ in radians and $\Delta\omega$ is the change in frequency.

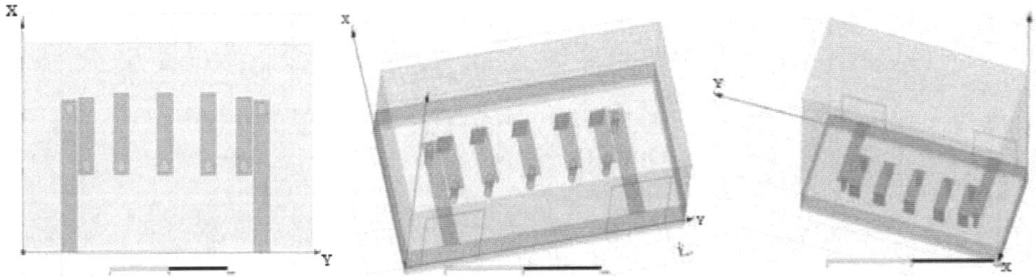

(a) (b) (c)

Figure 4. Combline band pass filter model in HFSS (a)Top view (b) Side view (c) Simple view

3. RESULTS AND DISCUSSION

The simulated transmission behavior of proposed filters are shown in Figures 5, 6, and 7. Figure 5 shows the band pass behavior$|S_{11}|$ and $|S_{21}|$ of combline filters designed at 9.5 GHz at different substrates thicknesses. In the Figure 5, at center frequency 9.5GHz the insertion loss is 0.9579dB for the substrate thickness of 500μm while 1.2dB for substrate thickness of 300μm.

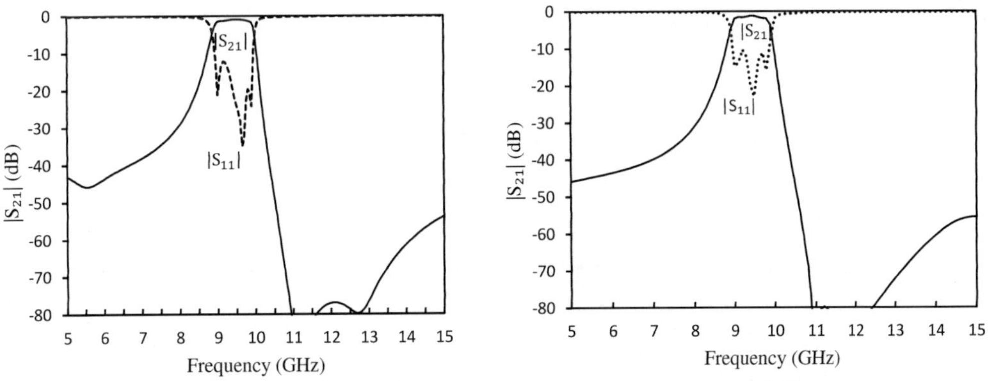

(a) (b)

Figure 5. Simulated S-parameters of combline filter at center frequency of 9.5 GHz
(a) For 500μm (b)For 300μm

Figures 6(a) and 7(a) show the $|S_{21}|$ of the combline band pass filter by changing the value of lumped capacitance. The simulation results show that frequency band can be tuned by varying the capacitance of MEMS varactor with keeping bandwidth constant. Figure 6(b) and 7(b) show the group delays in during tuning by varying the capacitance.

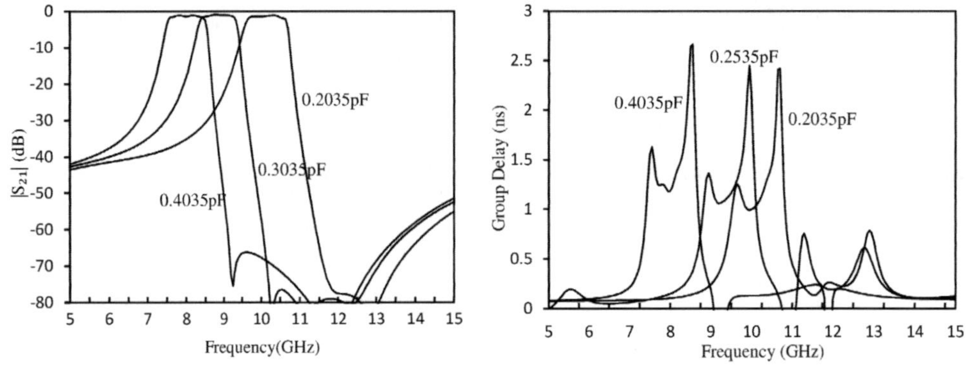

(a) (b)

Figure 6. Plots (a) $|S_{21}|$ and (b) Group Delay of MEMS Tunable filter on substrate thickness 500μm

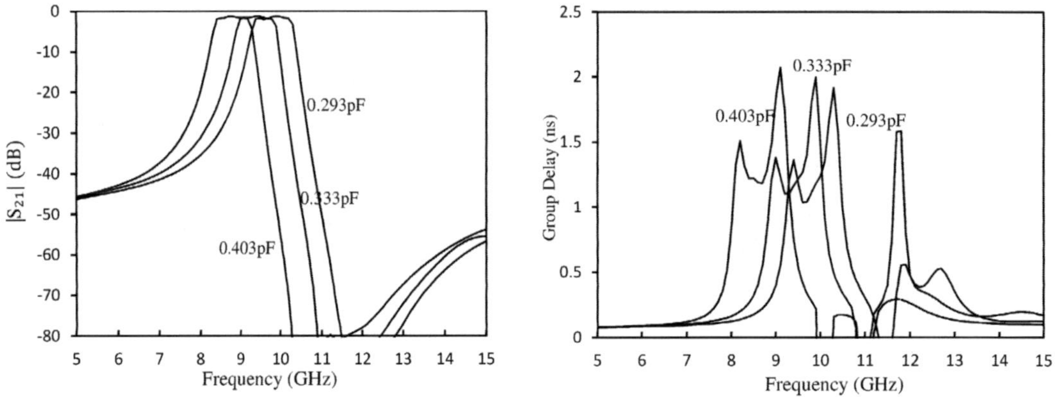

(a) (b)

Figure 7. Plots (a) $|S_{21}|$ and (b) Group Delay of MEMS Tunable filter on substrate thickness 300μm

4. CONCLUSION

We have designed and simulated the tunable band pass filter using microstrip combline filter topology. The proposed filters were designed and its transmission behaviour studied on two different substrate thicknesses of high resistivity silicon. The five-pole combline filter with insertion loss 0.9579dB and 1.2dB have been designed at the center frequency of 9.5GHz on 500μm and 300μm thick silicon substrates respectively. It has been seen that insertion loss is better and tuning range is wider for filter designed on 500μm thick silicon substrate while in case of 300μm thick silicon substrates we have obtained lesser tuning range and higher insertion loss. RF MEMS filter show great potential for weight, power consumption, and size reduction. The group delay is lesser in case of 300μm thick silicon substrates as compared to 500μm thick silicon substrates. Therefore, to design a tunable filter the substrate thickness also plays an important role.

ACKNOWLEDGEMENT

Authors are grateful for the Director, CSIR CEERI Pilani for his support and encouragement.

REFERENCES

[1] J. Brank, J. Yao, and M. Eberly et al., "*RF MEMS-based tunable filters*", Int JRF and Microwave CAE., vol. 11.276-284,.

[2] R. Borwick, P. Stupar, and J. DeNatale. 2003, "*Variable MEMS capacitors implemented into RF Filter system",* IEEE Trans. Microwave Theory Tech., vol.51, pp. 315-319.

[3] S.caspi and L. Adelman, April 1988, "*Design of combline and interdigital Filters with tapped-line input*", IEEE Transaction, MTT-36, 759-763.

[4] Byung-Wook Kim, Sang-Won Yun, April 2004, "*Varactor-Tuned Combline Bandpass Filter Using Step Impedance Microstrip",* IEEE Transactions on Microwave Theory and Techniques, vol. 52, No.4,1279-1283.

[5] S. C. Saha, U. Hanke, H. Segberg, T. A. Fjeldly, T Saether, 2011, "*Tunable Band-Pass Filter Using RF MEMS Capacitance And Transmission Line*", Progress In Electromagnetics Research C, Vol.23, 233-247.

[6] Zou, J., C. Liu, J. Schutt-Aine, J. Chen, and S. Kang, 2000, "*Development of a wide tuning range MEMS tunable capacitor for wireless communication systems*," International Electron. Devices Meet. Tech. Digest, 403-406.

[7] Xu-Guang Wang, Young-Ho Cho, Sang-Won Yun, 2012, "*A Tunable Combline Bandpass Filter Loaded With Series Resonator*", IEEE Transactions on Microwave Theory and Techniques, 1569 - 1576.

[8] Jia-Sheng Hong, "*Microstrip Filters for RF/Microwave Applications*", Second Edition, published by John Wiley & Sons, Inc., New Jersey.

[9] G. L. Matthaei, L. Young, and E. M. T. Jones, 1980, "*Microwave Filters, Impedance-Matching Networks, and Coupling Structures".* Norwood, MA: Artech House.

[10] Q. Xiang, Q. Feng, X. Huang, and D. Jia, Mar. 2013, "*Electrical tunable microstrip LC bandpass filters with constant bandwidth*," IEEE Trans. Microw. Theory Techn., vol. 61, no. 3, pp. 1124–1130.

[11] C.-W. Tang, C.-T. Tseng, and S.-C. Chang, Nov. 2014, "*Design of the compact tunable filter with modified coupled lines*," IEEE Trans. Compon., Packag., Manuf. Technol.,vol.4, no.11, pp. 1815–182.

IOP Conf. Series: Materials Science and Engineering **331** (2018) 012031 doi:10.1088/1757-899X/331/1/012031

Simulating 1.30 μm Optical Gain in Type-I InGaAsN/GaAs Nano-Heterostructure

Garima Bhardwaj[1], Sandhya K.[2], Pyare Lal[2], P. A. Alvi[2*]

[1]Department of Electronics, Banasthali University, Banasthali-304022, Rajasthan (INDIA)
[2]Department of Physics, Banasthali University, Banasthali-304022, Rajasthan (INDIA)

E-mail corresponding author: drpaalvi@gmail.com

Abstract. This paper reports theoretical simulation of optical gain achieved at 1.30 μm in the type-I InGaAsN/GaAs nano-scale heterostructure having active region of width ~ 40 Å. In addition, the modal gain, behaviour of differential gain and anti-guiding factor of the heterostructure is also simulated and reported. From simulation results, it is concluded that the proposed InGaAsN/GaAs heterostructure can produce the maximum optical gain ~2100/cm at 1.30 μm.

1. Introduction

The III-nitride based semiconductors, for examples: GaN, InN, AlN and their alloys, have been found to provide unique electrical, electronic, optical and mechanical properties which can be utilized for designing of future photonic and optoelectronic devices [1-3]. Since these semiconductors have wide band gap ranging from 1.9 eV to 6.2 eV, and therefore these materials can be utilized in the fabrication of optoelectronic devices operating in visible-to-ultraviolet regime. Examples are green, blue lasers and photo detectors. To date, III-nitrides based photodetectors operating in ultraviolet regime have exhibited promising output. In addition, III-nitrides based LEDs have been commercialized due to emission of super bright blue and green light. Moreover, these materials have some exceptional physical properties which make the devices ready for applications beyond the data storage and imaging [4]. These materials are also useful in high temperature electronics and space applications because these are physically and chemically strong. This property makes them ideal for operation in cruel environment.

For the application point of view, the III-nitride materials based devices are more beneficial in comparison to conventional semiconducting materials. For example, III-nitride materials based UV photodetectors are much sensitive to ultraviolet radiations while insensitive to infrared radiation. Thus, these devices are applicable in order to detect the ultraviolet radiations existing in the infrared and visible background [5]. Due to this, the III-nitride devices can be utilized in engine and furnace monitoring for automotive, flame detection, undersea communication, petroleum industry, and space-to-space communication, and in chemical analysis systems. One of the major advantages is that the III-nitride devices can be expected to work without optical filters due to their low dark currents and theoretical intrinsic solar blindness and, obviously, this behavior of these devices reduce the launch weight for space significantly.

IOP Conf. Series: Materials Science and Engineering **331** (2018) 012031 doi:10.1088/1757-899X/331/1/012031

2. Device Geometry and Theory

The geometry of device is very simple and consists of type-I double heterojunctions. The active region of the device is made of InGaAsN material, which is sandwiched between two barriers followed by claddings. The barriers and claddings are considered of GaAsN (with dilute N) material. The width of active region (quantum well region) is taken as 40 Å; while barriers are of width ~ 60 Å. The region behind the chosen of such size of active region is to have quantum confinement of the envelope function (or wave function) associated with the quantum well. The entire device is supposed to grow on GaAs substrate. This substrate is selected to have lattice matched condition of quantum well with the substrate in order to avoid unnecessary strain in the device. The unnecessary strain produced in the device may degrade the device performance.

The main object of this paper is to simulate the optical and modal gain. The optical gain for the heterostructure can be given as [6, 7]:

$$G(\hbar\omega) = \frac{2.\pi e^2}{nc\epsilon\omega Lm^2} \sum_{\sigma=U,L} \sum_{n,m} \int \left|\left(\hat{e}.\ M_{nm}^{\eta\sigma}(k_t)\right)\right|^2 \times \frac{\left(f_n^c(k_t) - f_{\sigma m}^v(k_t)\right)\left(\frac{\gamma}{\pi}\right)}{\left(E_{\eta,\sigma nm}^{c,v}(k_t) - \omega\hbar\right)^2 + \gamma^2} \frac{k_t dk_t}{2\pi}$$

Where $f_n^c(k_t)$ and $f_{\sigma m}^v(k_t)$ are quasi Fermi levels associated with conduction and valence band of the heterostructure respectively and can be given as:

$$f_n^c(k_t) = \frac{1}{1+\exp\left(\frac{E_n^c(k_t)-F_c}{K_B T}\right)} \quad \text{and} \quad f_{\sigma m}^v(k_t) = \frac{1}{1+\exp\left(\frac{E_{\sigma m}^v(k_t)-F_v}{K_B T}\right)}$$

The details of symbols used in the above expressions can be seen in reference [7, 8].

3. Results and Discussion

Prior to optimize the optical gain of any heterostructure, it is essential to study the behavior of quasi Fermi levels associated with the conduction and valence bands of the heterostructure. Under the non equilibrium cases, these quasi Fermi levels play a very important role in specifying the carrier concentrations. The separation between quasi Fermi energy levels plays a critical role in understanding the optical behavior in the semiconducting heterostructure because of its relationship with the density of injected carriers and the intensity of optical gain [9]. Therefore, for the InGaAsN/GaAs nano heterostructure, the plot of quasi Fermi levels associated with the conduction and valence bands is shown in figure 1. In Figure 1, the behavior of quasi Fermi levels for electrons in the conduction and holes in the valence bands in the left (black) and right (blue) y-axes, respectively, is shown for $In_{0.29}Ga_{0.71}As_{0.99}N_{0.01}$/GaAs Straddled Heterostructures. This figure predicts that the quasi Fermi energies for both the conduction and valence bands are increased as increase in carrier density.

The optical gain is the fundamental characteristic of the quantum well heterostructures based lasers. The optical gain which depends on the confinement factor, GRIN (graded refractive index) layers and number of QWs (quantum wells) is termed as modal gain. For the InGaAsN/GaAs nano heterostructure, the behaviors of optical gain and modal gain with lasing wavelength are plotted in figure 2. In figure 2, the left (black) y-axis is chosen for optical gain; while the right (blue) y-axes is for modal gain. From figure 2, it is clear that the maximum optical of the heterostructure is found ~ 2100/cm at the wavelength of 1.30 μm. In the both spectra of optical gain and modal gain, two peaks are obtained. These two different peaks are due to different optical transitions which occur between different conduction and valence sub bands of the quantum well region of the heterostructure.

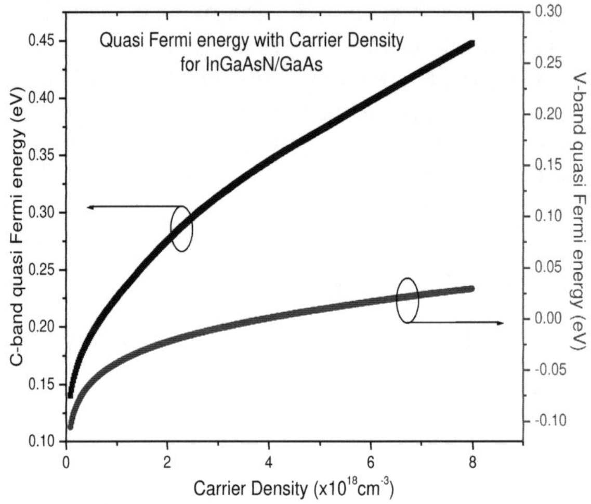

Figure 1. Conduction and valence bands quasi Fermi levels as a function of Carrier density for $In_{0.29}Ga_{0.71}As_{0.99}N_{0.01}$/GaAs Straddled Heterostructure.

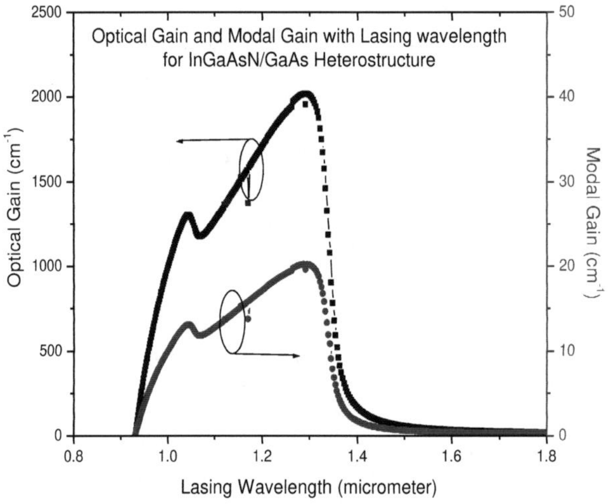

Figure 2. Optical gain and Modal gain with Lasing wavelength for $In_{0.29}Ga_{0.71}As_{0.99}N_{0.01}$/GaAs Straddled Heterostructure.

The differential gain for the $In_{0.29}Ga_{0.71}As_{0.99}N_{0.01}$/GaAs straddled heterostructure is also simulated and shown in figure 3. The differential gain is defined as first derivative of optical gain with respect to the carrier density. It is shown on the right (blue) y-axis. In addition, an interesting feature of the heterostructure, the anti-guiding factor is also calculated. The differential gain and anti-guiding factor are substantial parameters in the dynamics performance of quantum well heterostructures based lasers [10]. The anti-guiding factor plays a very crucial role in deciding the optical gain of lasing heterostructure and can be defined in terms of refractive index change and differential gain. Figure 3 shows the proportional and reciprocal behaviours of anti-guiding factor and differential gain, respectively as a function of carrier density.

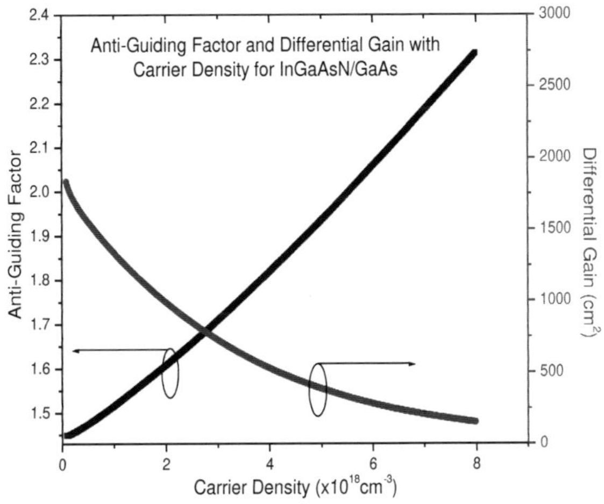

Figure 3. Anti-Guiding factor and Differential gain with Carrier density for $In_{0.29}Ga_{0.71}As_{0.99}N_{0.01}/GaAs$ Straddled Heterostructure.

Conclusion

The gain spectra of type-I InGaAsN/GaAs nano-scale heterostructure has been simulated and reported successfully. Apart from this, the behavior of the modal gain, differential gain and anti-guiding factor of the heterostructure is also simulated and reported. According to the results, the InGaAsN/GaAs heterostructure can produce the maximum optical gain ~2100/cm at 1.30 μm.

Acknowledgement

Authors are thankful to "Banasthali Center for Research and Education in Basic Sciences" under CURIE programme supported by the DST, Government of India, New Delhi.

References

[1] Alvi P A, Gupta S, Shrama M, Jha S, Rahman F, *Physica E: Low-dimensional systems and Nanostructures*, 2011, **44**, pp. 49-55.
[2] Alvi P A, *Advanced Science Letters,* 2012, Vol. 5, No. 17, pp. 101-107.
[3] Gupta S, Rahman F, Siddiqui M J, Alvi P A, *Physica B: Condensed Matter,* 2013, Vol. 411, pp. 40-47.
[4] Park Y S, *Optoelectronics Review*, 2001, 9(2), 117-124.
[5] M. Razeghi and A. Rogalski, *J. Appl. Phys.,* 1996, **79,** 7433.
[6] Alvi P A, *Phys. Status Solidi B,* 2017, **254**, No. 5, p. 1600571.
[7] Shun Lien Chuang, "Physics of optoelectronic devices." John Wiley & Sons, 1995.
[8] Alvi P A, *Materials Science in Semiconductor Processing*, 2015, Vol. 31, pp. 106-115.
[9] Wu Lin-Zhang, Tian Wei and Gao Feng, *Chinese Physics Letters,* 2004, Vol. 21, No.7, pp. 1359-1361.
[10] Y. Arakawa and A. Yariv, *IEEE J. Quantum Electron,* 1985, Vol. 21, pp. 1666-1674.

Various Indices for Diagnosis of Air-gap Eccentricity Fault in Induction Motor-A Review

Nikhil, ***Dr. Lini Mathew, Amandeep Sharma***

Department of Electrical Engineering, NITTTR, Chandigarh

nikhilinsan@gmail.com

Abstract. From the past few years, research has gained an ardent pace in the field of fault detection and diagnosis in induction motors. In the current scenario, software is being introduced with diagnostic features to improve stability and reliability in fault diagnostic techniques. Human involvement in decision making for fault detection is slowly being replaced by Artificial Intelligence techniques. In this paper, a brief introduction of eccentricity fault is presented along with their causes and effects on the health of induction motors. Various indices used to detect eccentricity are being introduced along with their boundary conditions and their future scope of research. At last, merits and demerits of all indices are discussed and a comparison is made between them.

1. Introduction

Because of its simple construction and cheap maintenance and repair, the induction motor is the most commonly used motor in the industry. Since induction motors are used very frequently, their protection and fault diagnosis schemes were developed along with their invention [1]. At that time schemes like overvoltage, ground fault and overcurrent protection were only applied. But now day's scenario has been completely changed. Now, induction motors are exclusively used for severe process dynamics and control process and any fault in the motor will result in shut down of the whole process. So, in order to prevent huge financial loss, it is mandatory to have a trustworthy fault detection and diagnosis system for early detection of incipient fault.

There are basically electrical and mechanical faults in induction motors, of which 60% are mechanical faults and 80% of mechanical faults results in air-gap eccentricity [2], [3]. There are basically three types of eccentricity i.e. static, dynamic or mixed eccentricity.

2. Air gap eccentricity fault

There are basically three axes i.e. rotor symmetry axis O_r, stator symmetry axis O_s and rotor rotation axis O_w in an induction motor which coincides with each other in ideal condition. If eccentricity exists then there is a dislocation of any of the axis from the other two axes or dislocation of all the axes [4]. Eccentricity results in the unequal air-gap length along the stator to rotor circumference. This unequal air-gap length may either be static or dynamic depending upon the type of eccentricity. If stator symmetry axis O_s separates from the other two axes, the minimum/maximum air gap length is static and is called as static eccentricity. If rotor symmetry axis O_r separates from the other two axes and the minimum/maximum air gap length is dynamically moving then it is dynamic eccentricity. If all the three axes separate then it is mixed eccentricity. Three types of eccentricity are shown in (Fig. 1).

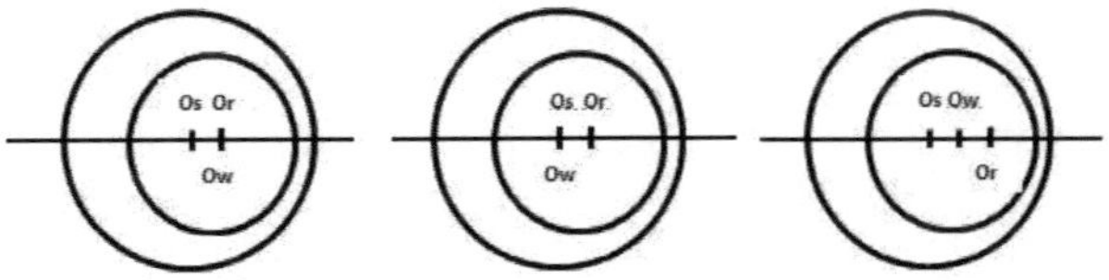

Fig.1. Different eccentricity types: (a) static, (b) dynamic and (c) mixed

The reasons which cause eccentricity are described as follow [5]–[7]:

i. Operation of the motor at critical speed.
ii. Wrong positioning of rotor and stator at the time of manufacturing.
iii. Mechanical resonance at critical load.
iv. Oval inner cross section of the stator.
v. Wrong positioning of load axes and the rotor shaft.
vi. Wear and tear of ball bearings.

There is a generation of radial Unbalanced Magnetic Pull (UMP) between rotor and stator due to eccentricity which amplifies the degree of eccentricity within the motor. An enormous amount of burden is created on the machine which causes the wear and tear of the ball bearings. Also, the undesirable vibrations are produced in the stator windings due to radial magnetic forces caused by eccentricity. The worst happens when eccentricity causes rotor to stator rub which permanently damages stator core and rotor cage [6], [8].

Here are some adverse effects on the characteristics of the motor due to eccentricity:

i. Power losses increase, due to which efficiency decreases.
ii. The temperature of the windings increases.
iii. Overall average torque decreases.
iv. Variations in speed and torque increases.

Any change from the normal operation of the motor which can be measured can act as fault index for detection and diagnosis of faults. An ideal index can give results in all possible conditions. But sometimes it may happen that we are measuring a change in an index or a physical quantity due to the simultaneous existence of two or more faults. At that time, we require such an index which shows changes due to a fault and remains unchanged due to another fault.

This will help us in separately diagnosing simultaneously existing two or more faults. Different indices require different sensors and transducers for measurement. Therefore a comparison can also be made between indices on the basis of easiness of measurement, how precisely they detect and diagnose the fault, how the index behaves when there are fluctuating load conditions, machine gets magnetically saturated, voltage variations and voltage unbalance condition.

The main aim of this paper is to make a comparison between different fault indices in order to provide the researcher with complete information about which index will be best suited for the prevailing conditions and setup of the experiment.

3. Various indices used for detection of eccentricity

3.1. Normalized splitting severity factor (NSS)

For an induction motor with 4 poles, with the rotating magnetic field's one complete rotation, two stator current cycles are produced. Rotating magnetic field encounters with unequal air-gap with the existence of static eccentricity within the motor. Due to magnetic reluctance, the different amplitudes of current will be produced as compared to the ideal or healthy motor. If park transformation over a complete rotation of stator current is drawn, it can be seen that there is a difference between two current cycles if eccentricity fault exists in the motor. This difference can be used to detect the static eccentricity fault. The above mentioned index will be used for this objective.

If total P samples are received for two subsequent cycles of current and park's current idn and iqn are found for P= 1, 2..n then normalized splitting severity factor will be given by:

$$(\Delta\rho)_{avg} = \frac{(\Delta\rho)_{avg}}{\rho_{avg}} \tag{1}$$

Where the numerator of equation (1) is the difference of park's current for two successive cycles and is given by:

$$(\Delta\rho)_{avg} = \frac{\sum_{P=1}^{\frac{n}{2}}\left(\sqrt{i_{dn}^2+i_{qn}^2}-\sqrt{i_{dn/2+k}^2+i_{qn/2+k}^2}\right)}{P/2} \tag{2}$$

And the denominator of (1) is the park's current's average length given by:

$$\rho_{avg} = \frac{\sum_{P=1}^{n}\sqrt{i_{dn}^2+i_{qn}^2}}{P} \tag{3}$$

Using this technique the static eccentricity will only be eliminated if and only if the degree of eccentricity is above 60% but this method has an edge on others as it is simply measurable [9]-[11].

3.2. Difference of area under Park's vector current for two consecutive current cycles to the average of all the areas (APC) [12]-[14]

For a 4-pole eccentric induction motor, if the magnitude of Park's vector of current is plotted for two subsequent cycles against its angle from zero to 2π in electrical radians, then two different curves are obtained. If the difference of average area of two given curves and the sum of area of two curves is calculated then it can be used as an index for detection and diagnosis of eccentricity fault. Research can be further expanded to detect the dynamic and mixed eccentricity. Also, the effect of the variations in loading conditions and magnetic saturation can be determined.

3.3. High-frequency component of stator current (HFS) [15]–[21]

Harmonic components in the stator current of an eccentric induction motor are given by the following equation:

$$f_e = f_1\left[(kR \pm n_d)\frac{1-s}{P} \pm \vartheta\right] \tag{4}$$

where the power supply's time harmonic order of the motor is ϑ, k is an arbitrary integer, P is the number of pole pairs, s is the slip, f_1 is the fundamental component of supply frequency, R is rotor slot number, n_d is the degree of dynamic eccentricity given by 1, 2, 3,… and is zero for static eccentricity. These harmonic components will only be produced if it is not the third multiple of fundamental frequency i.e. $3(f_1)$ and also there exists a special relation between rotor slots and pole number [17]. There is a well set technique to detect the given harmonics. Firstly the stator current signal should be sampled with a particular frequency which satisfies the nyquist criterion. Now these sampled signals are passed through a specified filter in order to make the amplitude of the proposed harmonics visible. Since the amplitude of these harmonics is very low as compared to fundamental component it is mandatory to apply the filter. Now the required harmonic components are determined by applying Fast

Fourier Transform (FFT) to the filtered signal. This index can be well used for detection of fault as it does not show the changes when there is an unbalance in the voltage. Also it is precise and certain in making a difference between static and dynamic fault.

3.4. Low-frequency components of stator current (LFS) [22]–[27]

There is a drawback of the third index i.e. as it is based on high-frequency components noise signals also interfere highly with the main signal. To detect and to separate these signals is not so easy and requires very expensive hardware and software setup. Therefore such an index is required which is based on low-frequency components. It is found that following harmonic components of stator current exists in a motor with mixed eccentricity:

$$f_e = |f_1 \pm k f_r| \quad (5)$$

where f_1 is the fundamental frequency component of the supply, f_r is the frequency of rotor rotation in rotation per second (rps) and k is any random integer. The fundamental component of supply is surrounded with two harmonic components around it with $k = 1$. These two harmonics can be used as an eccentricity fault index for detection of fault. Unlike HS it is independent of the number of rotor slots and motor's number of poles. But it depends upon the presence of mixed eccentricity within the machine. It has an advantage of comparatively easy computation and measurement. But till now the effect of unbalanced voltage, fluctuating load, magnetic saturation has not been found out and can be considered for further experimentation.

3.5. Stator current's low and high frequency component (SLH) [28]–[33]

Both the low and high-frequency components must be found out and analyzed carefully in order to get more accurate results for mixed eccentricity. In reality low and high-frequency components are interconnected as due to the presence of both dynamic and static eccentricity stator current's low-frequency components given by (5) also create some high-frequency components given by (4). Therefore in order to get more explicit results for mixed eccentricity both high and low frequency must be detected and analyzed. The motor's number of poles and fluctuating load conditions does not effect the given fault index which makes it advantageous. But the effect of magnetic saturation is yet to be found.

3.6. Ratio of summation of low-frequency stator current components and no-load current of motor (RSN)

As stated in the fourth index, if the machine has mixed eccentricity fault then it will generate low-frequency components around the fundamental stator current frequency which are given by:

$$f_e = |f_1 \pm k f_r| \quad (6)$$

as similar to (5). Taking the value as 1 in an induction motor having 4 poles, the resultant frequencies will come out to be approximately near to $0.5f_1$ and $1.5f_1$ named as f_{eh} and f_{el} respectively. If the fault index is given by

$$FI = \frac{f_{el+}f_{eh}}{I_{nl}} \times 100 \quad (7)$$

where FI is fault index and I_{nl} is no load current. While going through [34] it has been found that the value of FI comes out to be 3.8% for a healthy motor. This is due to naturally or inherently existing static and dynamic eccentricity with in the motor. But if there is an introduction of 67% static eccentricity in the machine then the value of FI comes out to be 10.51%. Therefore it can be well used as an index to detect eccentricity fault in the motor. It has some other advantages of not affected by fluctuations in load and unbalance in voltage. But the effect of magnetic saturation of motor should be further investigated.

3.7. Gyration radius of the torque developed (GRT)

The measurement of the developed torque is not an easy task as some special types of sensors are to be applied to the motor case which makes it difficult and expensive. It can be used as an index because stator current can be used as a replacement for it [35], [36]. If sampling is applied over the developed torque signals for a complete cycle and total N samples are gathered then for the first torque difference the time series is given by:

$$\Delta T = \{\Delta\tau(k), \quad k = 2, 3, \ldots N\} \tag{8}$$

where $\Delta\tau(k) = \tau(k) - \tau(k-1)$ and k is the number of torque samples. If a 2D phase space diagram is made with $\Delta\tau(k)$ on the vertical axes and $\Delta\tau(k-1)$ on the horizontal where and is the number of torque samples. If a 2D phase space diagram is made with on the vertical axes and on the horizontal axes, it will show a considerable difference for eccentric motor as compared to healthy induction motor. Therefore it can definitely be used as an eccentricity fault index. On the 2D diagram, for a unit mass at any point, the mass center lies on the coordinates μ_0 and μ_1 for the total diagram. These μ_0 and μ_1 are given by:

$$\mu_0 = \frac{\sum_{k=1}^{N-l} \Delta\tau(k)}{N-l} \tag{9}$$

$$\mu_1 = \frac{\sum_{k=l+1}^{N} \Delta\tau(k)}{N-l} \tag{10}$$

The total distance between the mass center and any point is given by $d(k)$ as:

$$d(k)^2 = [\Delta\tau(k) - \mu_0]^2 + [\,\Delta\tau(k-1) - \mu_1\,]^2 \tag{11}$$

And, finally, the gyration radius is given by:

$$r = \sqrt{\frac{\sum_{k=1+l}^{N} d(k)^2}{N-l}} \tag{12}$$

Around the mass center, there will be less concentration of the 2D diagram if the gyration radius is large which makes it able to be used as one of the index for eccentricity fault diagnosis.

4. Comparison of all the indices

An index is called as perfect if it helps us to identify different types of faults under different operating conditions. But in the real world, it is approximately impossible to have all the qualities in a single index. The reliability and feasibility of an index depends on various factors. It has been found from the literature review that deviations in voltage, levels of loading, unbalanced voltage, and magnetic saturation are some of the factors, which if altered affects the index value. Moreover factors like easiness in making calculations and measurements, precision in diagnosing fault intensity and type are some of the factors which decides the quality of an index. In the given Table 1 and Table 2 a comparison is made between all the indices for the manufacturers and researchers which will help them proceed further in their respective areas. Here the tick sign (✓) shows that the given index is affected, the cross sign (×) indicates its contrast and the asterisk sign (*) indicates that the attitude of the index is not known towards the following weakness and is still under consideration for the future experimental scope of work.

Table 1. The behaviour of different indices corresponding to the following variations

Index number	Magnetic saturation	Voltage deviation	Voltage unbalance	Load level
NSS	*	*	*	*
APC	*	*	*	*
HFS	✓	✓	×	✓
LFS	*	✓	×	✓
SLH	*	✓	×	×
RSN	*	✓	×	×
GRT	×	×	*	*

Table 2. The behaviour of Indices in front of their merits

Index number	Precision and certainty in diagnosis of the fault type	Simplicity of measurement and calculation	Precision and certainty in diagnosis of the fault intensity
NSS	×	✓	×
APC	×	✓	×
HFS	✓	×	×
LFS	×	✓	×
SLH	×	×	×
RSN	×	✓	✓
GRT	✓	✓	✓

From Table 1 we can say that HFS is unaffected by the unbalanced voltage but it is affected by the magnetic saturation of motor, when load fluctuates, when voltage deviates, noise occurs etc. Similarly, GRT also remained unaltered with deviations in voltage and magnetic saturiation. But the effect of voltage unbalance and load level on the index is not known yet.

Similarly from Table 2 we can say that neither it is easy to measure HFS nor it diagnose the intensity of fault precisely. But developed GRT can easily be measured and calculated and both the type and intensity of fault are diagnosed precisely.

5. Conclusion

In this paper first, we have discussed the eccentricity and its different types. Then we discussed the causes, effects and consequences of eccentricity. Then in Section 3 of this paper different indices are mentioned along with nature and behavior they show towards different changes. At last a brief tabular comparison is made between all the indices corresponding to different variations and their relative advantages. It can be said that the gyration radius of the torque developed (GRT) is the best index for detection and diagnosis of eccentricity fault. The researchers can use the results of this paper to choose the appropriate fault index in order to get better results. Moreover, it is beneficial for the manufacturers too. Considering the strength and weakness of indices they can apply the amendments at the manufacturing stage.

References

[1] J. Faiz and M. Ojaghi, "Different indexes for eccentricity faults diagnosis in three-phase squirrel-cage induction motors: A review," Mechatronics, Vol. 19, No. 1, pp. 2–13, 2009.

[2] J. Faiz, B.M.Ebrahimi, B. Akin, and H. A. Toliyat, "Comprehensive Eccentricity Fault Diagnosis in Induction Motors Using Finite Element Method," IEEE Transaction on Magnetics., Vol. 45, No. 3, pp. 1764–1767, 2009.

[3] J. Faiz, B. M. Ebrahimi and M. B. B. Sharifian, "Finite Element Transient Analysis of an On-Load Three-Phase Squirrel-Cage Induction Motor with Static Eccentricity,"Electromagnetics, Vol. 27, No. 4, pp. 207–227, 2007.

[4] J. Faiz and I. Tabatabaei, "Extension of Winding Function Theory for Nonuniform Air Gap in Electric Machinery," IEEE Transactions on Magnetics, Vol. 38, No. 6, pp. 3654–3657, 2002.

[5] K. Yahia, A.J.M.Cardoso, A.Ghoggal, and S.-E. Zouzou,"Induction Motors Air-Gap Eccentricity through the Discrete Wavelet Transform of the Apparent Power Signal under non-stationary Operating Conditions," Electric Power Components and Systems, Vol. 42, No. 7, pp. 682–692, 2014.

[6] A. J. M. Cardoso, M. Drif, "Airgap Eccentricity Fault Diagnosis in Three- Phase Induction Motors by the Complex Apparent Power Signature Analysis," IEEE Transactions on Industrial Applications, Vol. 55, No. 3, pp. 18–22, March 2008.

[7] D. G. Dorrell, W.T. Thomson, and S. Roach, "Analysis of Airgap Flux, Current, and Vibration Signals as a Function of the Combination of Static and Dynamic Airgap Eccentricity in 3-Phase Induction Motors," IEEE Transactions on Industry Applications, Vol. 33, No. 1, pp. 24–34, Feburary 1997.

[8] M. Ahmadi, J. Poshtan, and M. Poshtan, "Static Eccentricity Fault Detection in Induction Motors using Wavelet Packet Decomposition and Gyration Radius," 2013 1st International Conference on Communications, Signal Processing, and their Applications (ICCSPA), pp. 1–5, 2013.

[9] A. J. Marques Cardoso and E. S. Saraiva, "Predicting the Level of Airgap Eccentricity in Operating Three-phase Induction Motors, by Park's Vector Approach," Proceedings of the IEEE industry applications society annual meeting conference, Vol. 1, pp. 132–135, 1992.

[10] J. K. Park, J. Hur, "Detection of Inter-Turn and Dynamic Eccentricity Faults using Stator Current Frequency Pattern in IPM-Type BLDC Motors," IEEE Transactions on Industrial Electronics, Vol. 63, No. 3, pp. 1771–1780, 2015.

[11] T. G. Vilhekar, M. S. Ballal, " Application of Double Park's Vector Approach for Detection of Inter-Turn Fault in Induction Motor" ,International Conference on Condition Assessment Techniques in Electrical Systems (CATCON), 2015.

[12] A. J. M. Cardoso and S. Saraiva, "Computer-Aided Detection of Airgap Eccentricity in Operating Three-phase Induction Motors by Park's Vector Approach," IEEE Transactions on Industry Applications, Vol. 29, No. 5, pp. 897–901, 1993.

[13] J. Zarei, S.Yousefizadeh, "Fault Detection in Induction Motor using Park's Vector Approach and Wavelet Analysis,"International conference on Mechatronics and Control (ICMC), 2014.

[14] I. Y. Onel, M.H. Benbouzid," Induction Motor Bearing Failure Detection and Diagnosis: Park and Concordia Transform Approaches Comparitive Study," IEEE Transaction on Mechatronics, Vol. 13, No. 2, pp. 257-262, 2008.

[15] W. T. Thomson, D. Rankin, D.G. Dorrell "On-Line Current Monitoring to Diagnose Airgap Eccentricity - An Industrial Case History of a Large High-Voltage Three-phase Induction Motor," Proceedings of the IEEE International Conference on Electric Machines and Drives, pp. 10–12, 1997.

[16] A. Barbour and W. T. Thomson, "Finite element study of Rotor Slot Designs with respect to Current Monitoring for Detecting Static Airgap Eccentricity in Squirrel-Cage Induction Motors," Proceedings of IAS '97. Conference Record of the 1997 IEEE Industry Applications Conference Thirty-Second IAS Annual Meeting, Vol. 1, pp. 112–119, 1997.

[17] A. Barbour and W. T. Thomson, "Finite element analysis and on-line current monitoring to diagnose," Proceedings of the IEEE Industry Applications Annual Meeting Conference, Vol. 1, pp. 150–154, 1997.

[18] H. A. Toliyat, M. S. Arefeen, and A. G. Parlos, "A method for dynamic simulation of air-gap eccentricity in induction machines," IEEE Transactions on Industry Applications, Vol. 32,

No. 4, pp. 910–918, 1996.

[19] W. T. Thomson, S. M. Ieee, and A. Barbour, "On-Line Current Monitoring and Application of a Finite Element Method to Predict the Level of Static Airgap Eccentricity in Three-Phase Induction Motors," IEEE Transactions on Energy Conversion, Vol. 13, No. 4, pp. 347–357, 1998.

[20] S. Nandi, X. Li, "Performance Analysis of a 3-Phase Induction Machine with Inclined Static Eccentricity," Proceedings of the international conference on Power Electronic Drives and Energy Systems for Industrial Growth, Vol. 1, pp. 1606–1613, 2005.

[21] K. Yahia, M. Sahraoui, A.J.M. Cardoso, A. Ghoggal," The Use of a Modified Prony's Method to Detect the Airgap Eccentricity Occurance in Induction Motors", IEEE Transaction on Industry Applications, Vol. 52, No. 5 pp. 3869–3877, 2016.

[22] S. Nandi, H.A.Toliyat, "Detection of Rotor Slot and other Eccentricity Related Harmonics in a Three Phase Induction Motor with Different Rotor Cages," Proceedings of the International Conference on Power Electronic Drives and Energy Systems for Industrial Growth, Vol. 1, pp. 135–140, 1998.

[23] J. E. Bangura and N. Demerdash, "Comparison Between Characterization and Diagnosis of Broken Bars End-Ring Connectors and Airgap Eccentricities of Induction Motors in ASD's using a Coupled Finite Element State Space Method," IEEE Transactions on Energy Conversion, Vol. 15, No. 1, pp. 48–56, March 2000.

[24] G. M. Joksimovic, M. D. Durovic, J. Penman, and N. Arthur, "Dynamic Simulation of Dynamic Eccentricity in Induction Machines- Winding Function Approach," IEEE Transactions on Energy Conversion, Vol. 15, No. 2, pp. 143–148, June 2000.

[25] Zhongming Ye, Bin Wu, and A. Sadeghian, "Current Signature Analysis of Induction Motor Mechanical Faults by Wavelet Ppacket Decomposition," IEEE Transactions on Industrial Electronics, Vol. 50, No. 6, pp. 1217–1228, December 2003.

[26] A. M. Knight and S.P.Bertani, "Mechanical fault detection in a medium-sized induction motor using stator current monitoring," IEEE Transactions on Energy Conversion, Vol. 20, No. 4, pp. 753–760, 2005.

[27] B. M. Ebrahimi, J.Fiaz, M.J.Roshtkhari, A.Z.Nejhad,"Static Eccentricity Fault Diagnosis Parmanent Magnet Synchronous Motor Using Time-Stepping Finite Element Method," IEEE Transactions on Magnetics, Vol. 44, No. 11, pp. 4297–4300, 2008.

[28] S. Nandi, R. Bharadwaj, H. A. Toliyat, "Performance Analysis of a Three-Phase Induction Motor under Mixed Eccentricity Condition," IEEE Power Engineering Review, vol. 22, no. 7, p. 49, 2002.

[29] S. Nandi, H. A. Toliyat, "Detection of Rotor Slot and other Eccentricity Related Harmonics in a Three Phase Induction Motor with Different Rotor Cages," IEEE Transactions on Energy Conversion, Vol. 16, No. 3, pp. 253–260, 2001.

[30] W. T. Thomson and A. Barbour, "The on-line Prediction of Airgap Eccentricity Levels in large (MW range) 3-Phase Induction Motors," IEEE International Electric Machines and Drives Conference (IEMDC), pp. 383–385, 1999.

[31] W. T. Thomson, D. Rankin, and D. G. Dorrell, "On-Line Current Monitoring to Diagnose Airgap Eccentricity in Large Three-Phase Induction Motors - Industrial Case Histories Verify the Predictions," IEEE Transactions on Energy Conversion, Vol. 14, No. 4, pp. 1372–1378, 1999.

[32] W. T. Thomson, A. Barbour, and T. Robert, "An Industrial Case Study of On-Line Current Momitoring and Finite Element Analysis to Diagnose Airgap Eccentricity Problems in Large High Voltage 3- Phase Induction Motors," Proceedings of Ninth International Conference on Electrical Machines and Drives, pp. 242–246, 1999.

[33] D. N. Bangura JF, "Improvement of Monitoring and Diagnosis of Broken Bars/End-Ring Connectors and Airgap Eccentricities of Squirrel-Cage Induction Motors in ASDs Using a Time-Stepping Coupled Finite Element-State Space Technique," Electr. Proc. International

IOP Conf. Series: Materials Science and Engineering **331** (2018) 012032 doi:10.1088/1757-899X/331/1/012032

Conference on Machatronics drives, pp. 336–338, 1999.

[34] S. M. A. Cruz and A. J. Marques Cardoso, "Multiple Reference Frames Theory: A New Method for the Diagnosis of Stator Faults in Three-Phase Induction Motors," IEEE Transactions on Energy Conversion, Vol. 20, No. 3, pp. 1340–1346, 2005.

[35] J. F. Bangura, R. J. Povinelli, N. A. O. Demerdash, and R. H. Brown, "Diagnostics of Eccentricities and Bar/End-Ring Connector Breakages in Polyphase Induction Motors Through a Combination of Time-Series Data Mining and Time-Stepping Coupled FE - State-Space Techniques," IEEE Transactions on Industry Applications, vol. 39, no. 4, pp. 1005–1013, 2003.

[36] S. Rajagopalan, W.le Roux, T.G.Habetler, R.G.Harely, "Dynamic Eccentricity and Demagnetized Rotor Magnet Detection in Trapezoidal Flux (Brushless D.C.) Motors Operating Under Different Load Conditions," IEEE Transactions on Power Electronics, Vol. 22, No. 5, pp. 2061–2069, 2007.